P9-CLR-805

COMPARATIVE FARMING SYSTEMS

COMPARATIVE FARMING SYSTEMS

edited by

B. L. TURNER II
Clark University

STEPHEN B. BRUSH
University of California, Davis

THE GUILFORD PRESS
New York London

A Division of Guilford Publications, Inc.
72 Spring Street, New York, N.Y. 10012

Printed in the United States of America

LIBRARY OF CONGRESS CATALOGING IN PUBLICATION DATA

Comparative farming systems.

Bibliography: p.
Includes index.
1. Agricultural systems. I. Turner, B. L. (Billie Lee), 1945– . II. Brush, Stephen, B., 1943– .
S494.5.S95F37 1987 630 86-29554
ISBN 0-89862-780-X

Contributors

Abu Muhammad Shajaat Ali, PhD, Department of Geography, Jahangirnagar University, Savar, Dhaka, Bangladesh.

Jacqueline A. Ashby, PhD, International Fertilizer Development Center and Centro Internacional de Agricultura Tropical, Cali, Colombia.

Stephen Beckerman, PhD, Department of Anthropology, The Pennsylvania State University, University Park, Pennsylvania.

Stephen B. Brush, PhD, International Agricultural Development, University of California, Davis, California.

Gérard Dorel, PhD, Département de Géographie, Université de Paris XII Val de Marne, Creteil, France.

Peter T. Ewell, PhD, Departmento de Ciencias Sociales, Centro Internacional de la Papa (CIP), Lima, Peru.

Nicholas S. Hopkins, PhD, Department of Sociology, Anthropology, and Psychology, American University in Cairo, Cairo, Egypt.

Imréné Karácsony, PhD, Department of Planning and Economic Analysis, Statistical and Economic Survey Center, Hungarian Ministry of Agriculture and Food (Mezögazdasági És Elemezésügyi Minisztérium Statisztikai És Gazdaságelemzö Központ Föosztályvezetö), Budapest, Hungary.

Peter Kunstadter, PhD, Institute for Health Policy Studies, University of California, San Francisco, San Francisco, California.

Murray J. Leaf, PhD, School of Social Science, University of Texas at Dallas, Richardson, Texas.

Deborah Merrill-Sands, PhD, International Services for National Agricultural Research (ISNAR), The Hague, Netherlands.

Douglas Pachico, PhD, Centro Internacional de Agricultura Tropical, Cali, Colombia.

Mary Beth Pudup, PhD, Regional Research Institute, West Virginia University, Morgantown, West Virginia.

Paul Richards, PhD, Department of Anthropology, University College London, London, U.K.

Hans Joachim Späth, PhD, Department of Geography and Institute for Dryland Development, University of Oklahoma, Norman, Oklahoma.

Antal Szathmáry, PhD, Department of Planning and Economic Analysis, Statistical and Economic Survey Center, Hungarian Ministry of Agriculture and Food, Budapest, Hungary.

István Szücs, PhD, Department of Planning and Economic Analysis, Statistical and Economic Survey Center, Hungarian Ministry of Agriculture and Food, Budapest, Hungary.

B. L. Turner II, PhD, Graduate School of Geography, Clark University, Worcester, Massachusetts.

Michael J. Watts, PhD, Department of Geography, University of California, Berkeley, California.

Preface

The origins of this book are several. I (B.L.T.) had been frustrated in attempts to assess empirically certain theories of agricultural change, largely because of the paucity of comparable data for diverse farming systems. Enter Janet Crane. She had surveyed numerous individuals in various disciplines, especially anthropology and geography, who had expressed similar frustrations and the need for a comparative treatment of agriculture from a farming systems perspective. Janet envisioned a reference work on the subject that could also serve as an advanced text. This book would be composed of original works by an interdisciplinary and international set of researchers, headed by two editors representing anthropology and geography. Stephen Brush joined me in this latter role.

The three of us, with some initial assistance from Robert McC. Netting, conceptualized an "ideal" book. It would describe and compare a representative sample of the major types of agricultural systems throughout the world and the forces of change acting on them. This objective would be met through the use of the case-study farming systems/holistic approach. In this way, the diversity of perspectives and interests encompassed within the approach would be demonstrated. Implementation would involve the development of a comparative classification of agricultural systems that focused on the major characteristics used in the farming systems approach to describe the systems and their changes. This classification would be used to identify the types of systems and the critical information, both quantitative and qualitative, to be included. Unique or unusual systems were to be avoided. Selection criteria would also include major regional and environmental representation. Each system would be described in terms of its socioeconomic and environmental contexts, and a consistent set of data would be provided so that comparisons could be made and theories and themes addressed, particularly those dealing with agricultural change. Finally, our ideal book would be composed of original case studies, primarily at the microspatial level (the village or farm). These studies were to be authored by experts from varying disciplines in which the farming systems approach is employed or in which a tradition of holistic approaches to agriculture exists. We would also seek experts in academic and professional positions.

Reality rarely matches ideals, and we were prepared to temper our expectations considerably, particularly for an endeavor of this kind. This temperance was anticipated because of the difficulty of achieving a standardized format and data from authors of diverse backgrounds and interests, and because of the breadth of research that is consumed under the rubric of *farming systems*. Deborah Merrill-Sands (1986, 22: 87–104), a contributor to this volume, has identified six types of farming systems research as applied primarily to the development field. Our book enlarges this number by including farming systems research beyond the development field per se. Hence, as we fully expected, the descriptive, comparative, and explanatory interests of the contributors and their data would vary considerably.

These considerations notwithstanding, we are delighted that so many of the objectives were achieved. Case studies of 12 farming systems were obtained, representing the range of world agricultures in terms of major input–output patterns, technologies, production objectives, cultivars, and physical environments. These systems are located in 11 countries—in the tropics, neotropics, and mid-latitudes, in lowlands and highlands, in xeric and mesic conditions, and in highly to poorly developed economies.

Each system, with one exception, is based on original research and presented as a case study of a small area or group, or farm. The exception is swidden agriculture throughout "greater Amazonia," which is used as a context for information on the Barí system (Chapter 3). Each system is described and, with minor exceptions, the type of information needed to make comparisons and input–output patterns, cropping schedules, and so forth is included. An exception to this format is the lack of references in Chapter 13, which deals with a cooperative farm in Hungary. Here we ask the reader to understand the difficulties of obtaining an original case study of this kind.

We were also able to obtain contributions from the principal social science disciplines in which farming systems research is pursued. The 19 individuals contributing to the book represent the following disciplines: anthropology—7, economics—5, geography—6, and rural sociology—1. Differences also exist among them by subfield and ideological interests, ranging from earth science and energetic analysis to Marxian social relation analysis. Understandably, these differences contribute to variations in chapter styles, foci, and themes of change. This variation, however, was an objective of the book.

The degree to which the volume conforms to our original ideal conceptualization is a result of the heroic efforts of Janet Crane and the authors. Janet was a constant source of assistance, critique, and encouragement. She also provided the occasional nudge the editors needed to complete the project. The authors were overwhelmingly cooperative and responsive to the editors' demands—demands that were exacerbated by the level of chapter consistency sought from such a diverse group. We thank them all.

The number of individuals who contributed to the production of this volume is large. Their efforts are deeply appreciated. A special thanks is extended to William C. Clark and Ferenc Toth of the International Institute of Applied Systems Analysis in Laxenburg, Austria, who introduced us to our Hungarian contributors. Stephen and I are also grateful to the staffs and personnel of the University of California, Davis, and Clark University, for their cooperation and assistance. We are especially indebted to Maureen Hilyard for her diligent and professional assistance in various capacities on this project.

I express my gratitude to my co-editor, Stephen B. Brush, without whom this book would not be possible. Our paths originally crossed in 1971–1972 as students of cultural ecology at the University of Wisconsin–Madison. I admired him then and do more so now. I also acknowledge William M. Denevan, William E. Doolittle, and Gregory Knapp, with whom discussions on farming systems are not only rewarding, but are a major reason for so relishing our get-togethers at the annual meetings of the Association of American Geographers. I am also indebted to the faculty and graduate students of the Graduate School of Geography, Clark University, who have broadened my research and intellectual experiences as they pertain to the subject of this book. Finally, I thank the Simon Guggenheim Foundation for their support of portions of Chapters 1 and 2.

Contents

COMPARATIVE FARMING SYSTEMS

PART I

INTRODUCTION TO FARMING SYSTEMS

Purpose, Classification, and Organization

B. L. TURNER II AND STEPHEN B. BRUSH

Farming systems is a loosely defined, interdisciplinary approach to the study of agriculture that has developed, in part, as a reaction to sectoral and disciplinary approaches. The fundamental premise is that the understanding of agriculture is facilitated by a holistic perspective that integrates the socioeconomic, political, environmental, and technological elements of the system. Emphasis is placed on the micro- or mesospatial scale—the farm, the village, or a small area—as the unit of analysis. The unit and its context are described, although the emphasis of the description varies by research interests. A systems explanation is not usually attempted. Rather, explanations focus on selected facets of the system. Chapter 2 explores the farming systems approach in detail.

The term *farming systems* has different specific meanings in various fields of study, and, depending on their research interests, the various readers of this book will expect certain problem foci, approaches, or modes of analysis. A single volume, however, cannot serve everyone involved in farming systems research, and no such attempt is made here. This book addresses farming systems research in the social sciences, as practiced primarily by anthropologists and geographers, and to a lesser extent by economists and rural sociologists. The farming systems approach is used: (1) to provide case studies that describe a representative range of world agricultural systems; (2) to develop a sufficiently consistent set of data from which comparisons of the systems can be made; and (3) to address some of the forces of change acting on the systems.

Even within these limits, the array of interests and problem foci is considerable, inhibiting comparisons of case-specific data sets. This predicament hinders the broader assessments of various themes, such as agricultural change. Nevertheless, several outstanding comparative studies exist. Ruthenberg (1971) and Norman (1979) have culled from the literature comparative input and output data on farming systems in the tropics, and Bayliss-Smith (1982) has reorganized a small but varied sample of case studies to assess such issues as energetics.

This book follows in the tradition of these works but differs from them in several ways. It attempts to cover a representative array of farming systems,

B. L. Turner II. Graduate School of Geography, Clark University, Worcester, Massachusetts.

Stephen B. Brush. International Agricultural Development, University of California, Davis, California.

unencumbered by geographical or socioeconomic limitations (Figure 1-1). The selected systems range from technologically simple, swidden cultivation among rather isolated groups in Amazonia to computer-based cereal production on a cooperative farm in Hungary. The book also differs from its antecedents in that it is composed primarily of original case studies and includes the perspectives of researchers from four disciplinary backgrounds.

Several geographical zones and types of systems are not represented primarily because of restrictions and/or the paucity of farming systems research for certain regions and countries, especially the Soviet bloc and the People's Republic of China. Most major types of agroecological zones, staple food crops, and technological types are included.

Classifications, Typologies, and Trajectories

Standardized classifications and typologies of agriculture or farming systems have not emerged, despite the recognition that they would be useful (Helburn, 1957; Kostrowicki, 1964; Whittlesey, 1936). Attempts to classify agricultural or farming systems commonly represent a potpourri of formal and informal criteria, and many must be considered typologies because the criteria are not taxonomic, that is, they are not systematically structured to apply to every system. For example, a single classification may refer to *wet-rice* cultivation, *Mediterranean* agriculture, *plantation* agriculture, or *mechanized grain* cultivation, referring respectively to a cultivar and cropping context, a region, an economic structure, and a technology and cultivars.

Four classes of criteria are common to many such schemes: technology, cultivars, multiple factors, and region. Technology and cultivar schemes focus on either one or a set of techniques and procedures or on a cultivar or suite of cultivars. In its broadest use, reference is made to the categories of traditional (paleotechnic) and modern (neotechnic) agriculture, or to horticulture/vegeculture (fruits, tubers, roots, vegetables) and agriculture (grains). More specific technological types include swidden or slash-and-burn cultivation (denoting a cut, burn, and fallow procedure), terrace cultivation, and irrigation agriculture. Subclassifications exist for several technologies (e.g., Denevan & Turner, 1974; Spencer & Hale, 1961).

Multiple-factor classifications are particularly popular. Examples include irrigated paddy cultivation (rice, *Oryza sativa* L., grown with any number of water impoundment and delivery technologies), mechanized grain cultivation, and *milpa* (slash-and-burn cultivation of maize, *Zea mays* L.). Geographers have taken this approach to a macrospatial scale, classifying regional agriculture (e.g., Baker, 1927; Jones, 1928; Spencer & Horvath, 1963; Whittlesey, 1936). Perhaps the best known regional type is Mediterranean agriculture, which refers to the cultivation of wheat (*Triticum* spp.), supplemented by barley (*Hordeum* spp.), and vine and fruit crops in a Mediterranean climate.

Note: Numbers refer to the farming systems as discussed in the text on pages 9–10.

FIG. 1-1. Location of case studies.

Subclassifications, of course, have been identified (Hofmeister, 1971). Other regional types include the Corn Belt and the Wheat Belt of North America.

Most classifications and typologies appear to focus on the differences among agricultural systems, often with such precision as to hinder usefulness for comparative assessments and tests. Interestingly, this observation stands in contrast to the results of a questionnaire on the subject solicited by the International Geographic Union. Those responding to it indicated that while a classification should focus on types (although the meaning of *types* was not defined) and be able to handle dynamic and complex farming systems, it should also be hierarchical (Grigg, 1974:3). The lack of the development of a standard classification reflects in part the diversity of interests in agriculture, both by individual researchers and by fields of study. Among geographers, who are prone to classifications for spatial comparisons, a traditional focus on regional differences, not hierarchical comparability, may be a major reason that this development has been lagging (e.g., Grigg, 1974:1–5; Symons, 1967:195–197; Whittlesey, 1936).

A few hierarchical classifications do exist. One of the better known examples is that produced by Boserup (1965:15–16) in which systems are categorized as forest-fallow, short-fallow, annual cropping, and multicropping types. The particular descriptors used are not crucial to the theme. The categories actually refer to broad classes of frequency of cultivation per unit area (Turner & Doolittle, 1978), such that a move from forest-fallow to multicropping is a move toward increasing cultivation of the land as measured by output per unit area and time. Norman (1979), Turner, Hanham, and Portarero (1977), and others have variants of this scheme for comparative analysis and tests of various relationships. Other criteria that can be standardized are input quantity and quality (per unit area) and input–output relationships, particularly productivity as measured by energic, caloric, or monetary means. Indeed, a multiple classification scheme is possible that uses a number of such criteria.

Here we offer a trial formulation of a classification scheme designed to be applicable to all systems of cultivation. It is based on the broadly defined trajectories of three principal and universal components of farming systems: output intensity, technological type, and production type (Figure 1-2). Regardless of environmental, cultural, or socioeconomic conditions, changes in farming systems typically involve changes in these components. Indeed, the three components are so linked that a change in one of the components generally signifies a change in the other two as well.

Output intensity refers to yield or production per unit area and time. In a general sense, the global history of agriculture is one of increasing output intensification, although instances of systems that may not have changed output intensity for millennia (see Chapter 3) or have actually disintensified can be documented (e.g., Boserup, 1965; Turner, 1983). Output intensity can be measured by monetary value, calories, or weight, and the measure used depends on the type of production and cultivars in question. Here we use weight (kg/ha/

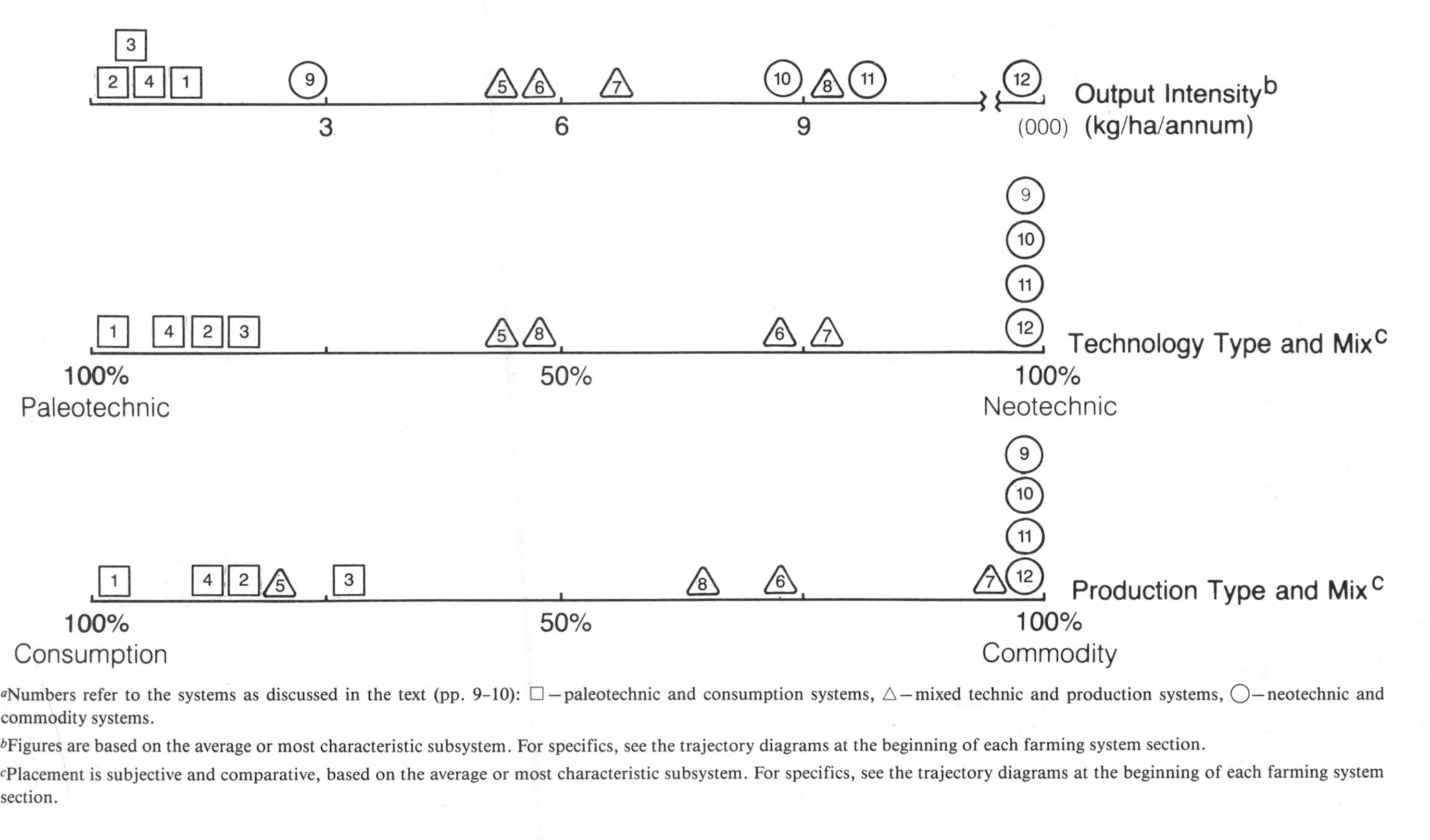

[a]Numbers refer to the systems as discussed in the text (pp. 9–10): □ – paleotechnic and consumption systems, △ – mixed technic and production systems, ○ – neotechnic and commodity systems.

[b]Figures are based on the average or most characteristic subsystem. For specifics, see the trajectory diagrams at the beginning of each farming system section.

[c]Placement is subjective and comparative, based on the average or most characteristic subsystem. For specifics, see the trajectory diagrams at the beginning of each farming system section.

FIG. 1-2. Trajectory of attributes: Paleotechnic and consumption-oriented systems (12 farming systems).[a]

yr)—grain weight for cereals and dry matter weight for root crops (see the Introduction to Part II)—lacking a superior comparative measure. Ideally, some synthesis measure of output intensity is needed that signifies a comparative, global value of production or the material well-being created by the production.

Unless otherwise noted, the estimates of output intensities used in our classification are based on the averages of the system as accounted for by cropping schedules, typical yields, and the amount of land devoted to certain suites of cultivars. If two or more subsystems differ dramatically from one another, both intensities are provided (in the sections devoted to them). The trajectory of output intensities ranges from low values of about 200 kg/ha/yr to high values of about 16,000 kg/ha/yr, with sugar beets in a rotation.

The second trajectory involves the technological component of cultivation, ranging from paleotechnic to neotechnic. This trajectory describes the level and types of inputs of agricultural production. The primary input is labor and its intensity per unit of land. Other inputs are various subsidies to the productive processes, including soil nutrients, irrigation, pest control, mechanization, and genetic material. These technologies are often used as substitutes for labor. The use of greater labor and/or subsidies implies greater management of the productive process and usually greater yields. There is, of course, some dispute about the relative efficiencies (energy input/output ratios) that occur along the trajectory. Paleotechnic systems on the low energy side of the trajectory involve low labor expenditures per unit of land and virtually no external subsidies other than labor and seed. Neotechnic systems may also have low labor inputs, but very high amounts of subsidies. These include chemical fertilizers, pesticides, and hybrid cultivars, in addition to tractors, harvesters, pumps, and so on. The middle ranges of the trajectory involve a mixture of high labor inputs and some subsidies.

In that labor is involved in all agriculture, all farming systems have some mix of paleo- and neotechnic inputs. The mix of the two is important to farming systems because of its significance to output intensity, labor productivity, ecological stability, energetic efficiency, capital costs, and so forth. The placement of any of the farming systems examined here along this trajectory of the classification is determined subjectively. Either pole of the trajectory represents a pure paleotechnic or neotechnic system, and movement away from the poles decreases the ratio of that factor to the other.

The third trajectory involves production type, that is, the intent to produce for direct consumption (producing unit is consumer) or for the market (commodity production). Whereas the first two trajectories deal primarily with material aspects of production, this trajectory deals with the structure of agricultural economies and farm unit decisions within them. The dichotomy between subsistence and market production has been recognized in many different theories of economic change for some time (see Chapter 2). Nonmarket economies, strictly defined, are exceedingly rare today, and in developing economies, most households that produce subsistence also grow for the market. This sometimes involves the same cultivar in the same field and at other times, separate cultivars in separate fields.

Each system's level of involvement in either type of production has been taken directly from the evidence presented, although these assessments are qualitative and may have various degrees of error. For this reason, the precise placement of each system in the trajectory is less significant than is its comparative placement. As in the case of technology, each pole of the trajectory represents a total or 100% commitment to that production type, and movement away from either pole decreases the ratio of that type to the other.

Organization

The 12 farming systems considered in this work represent a full range and mix of the values of the three components trajectories (Figure 1-2). The 12 systems cluster into three groups based on the characteristics of the principal or stable crop(s) production. The text is organized into three parts by these clusters. The first (Figure 1-2, □) involves four farming systems that exhibit relatively low levels of output intensity and a dominance of paleotechnic and consumption production, although some changes have begun to take place in these components. Part II ("Paleotechnic and Consumption-Oriented Systems") includes:

1. Manioc swidden in greater Amazonia
2. *Milpa* or maize swidden in northern Yucatán
3. Dry-rice swidden in northern Thailand
4. Short-fallow rice cultivation in Sierra Leone

The second cluster (Figure 1-2, △) includes four farming systems that exhibit considerable internal diversity in the use of neotechnic and paleotechnic inputs and in the mix of production objectives, either consumption or commodity. These systems also demonstrate considerable variability in output intensities. Part III ("Mixed Technic and Production Systems") involves:

5. Terraced cereal cultivation in Nepal
6. Irrigated cereal–cotton farming in Egypt
7. Irrigated cereal cultivation in Punjab
8. Multicropping, market gardening in Bangladesh

The final four systems (Figure 1-2, ○) are virtually pure neotechnic (all are mechanized) and produce totally for the market. Considerable differences in output intensities exist among them, however, partly because of differences in general land-use intensities and in environmental conditions. Part IV ("Neotechnic and Commodity-Oriented Systems") includes:

9. Rainfed wheat farming on the Great Plains of northeastern Colorado

10. Irrigated rice cultivation in northern California
11. Rainfed mixed/cereal cultivation on a cooperative in southeastern Hungary
12. Rainfed mixed/cereal farming in northern France

The case studies are preceded by a discussion of the farming systems approach as practiced in the social sciences and of the various conceptual themes or theories of agricultural change that are employed in studies using the approach. Each of the three case study sections (Parts II, III, and IV) is introduced by a brief synopsis and comparison of the four systems, including more detailed classification data, and by introductory comments on some of the key themes of that cluster. Beyond this, the case studies stand on their own merit, and the reader can make assessments and comparisons of the data and draw conclusions that are pertinent to individual interests.

References

Baker, O. E. Agricultural regions of North America: Part 20: The South. *Economic Geography*, 1927, 3, 50–86.

Bayless-Smith, T. P. *The ecology of agricultural systems*. Cambridge: Cambridge University Press, 1982.

Boserup, E. *The conditions of agricultural growth*. Chicago: Aldine, 1965.

Denevan, W. M., & Turner, B. L., II. Forms, functions and associations of raised fields in the old world tropics. *Journal of Tropical Geography*, 1974, 39, 24–33.

Grigg, D. B. *The agricultural systems of the world: An evolutionary approach*. Cambridge: Cambridge University Press, 1974.

Helburn, N. The bases for a classification of world agriculture. *The Professional Geographer*, 1957, 9, 2–7.

Hofmeister, B. Types of agriculture with predominant olive-growing in Spain. *Geoforum*, 1971, 5, 15–30.

Jones, C. F. Agricultural regions of South America. *Economic Geography*, 1928, 4, 1–30, 155–187, 259–295.

Kostrowicki, J. Geographical typology of agriculture, principles and methods. *Geographia Polonica*, 1964, 1, 111–146.

Norman, M. J. T. *Annual cropping systems in the tropics: An introduction*. Gainesville: University of Florida Press, 1979.

Ruthenberg, H. *Farming systems in the tropics*. Oxford: Clarendon Press, 1971.

Spencer, J. E., & Hale, G. A. The origin, nature, and distribution of agricultural terracing. *Pacific Viewpoint*, 1961, 2, 1–40.

Spencer, J. E., & Horvath, R. J. How does an agricultural region originate? *Annals of the Association of American Geographers*, 1963, 53, 74–92.

Symons, L. *Agricultural geograpahy*. London: Bell, 1967.

Turner, B. L., II. *Once beneath the forest: Prehistoric terracing in the Rio Bec region of the Maya lowlands*. Boulder: Westview Press, 1983.

Turner, B. L., II, & Doolittle, W. E. The concept and measure of agricultural intensity. *The Professional Geographer*, 1978, 30, 297–301.

Turner, B. L., II, Hanham, R. Q., & Portararo, A. V. Population pressure and agricultural intensity. *Annals of the Association of American Geographers*, 1977, 67, 384–396.

Whittlesey, D. Major agricultural regions of the earth. *Annals of the Association of American Geographers*, 1936, 26, 199–240.

2

The Nature of Farming Systems and Views of Their Change

STEPHEN B. BRUSH AND B. L. TURNER II

Over the last few thousand years, agriculture has radically transformed human life. Humans have developed or carried agriculture into virtually every climatic region of the earth, and in so doing have transformed an originally simple technology, hardly distinguishable from gathering, into a vastly complex and varied one. Cities, civilization, and large human populations are direct results of the emergence of agriculture, and these in turn have greatly altered agriculture itself. There is a long history of studies of agriculture and agrarian societies in both Western and non-Western scholarship. Indeed, few human endeavors have been so studied. Beyond agriculture's role in sustaining human life and in occupying the majority of the world's people, differences between agricultural systems also denote fundamental differences between social and economic systems. Morever, we are challenged to understand agriculture because of familiar problems: hunger, underdevelopment, population growth, and environmental deterioration. Many different approaches have been suggested for understanding agriculture and its changes. Scholarly disciplines and large scientific institutions have evolved to study them. A wide array of public and private organizations operating at levels from villages to the international sphere focus on changing and improving agriculture. Their coexistence suggests both the complexity of agriculture and the inadequacy of any single approach to understand that complexity.

The farming system concept is interdisciplinary and builds on the broad base of social science research that has accumulated largely since 1945. At least three major elements can be identified underlying its emergence.

The first is the development of an interdisciplinary and comparative focus in the several distinct branches of social science with an interest in agriculture: agricultural economics, anthropology, geography, and rural sociology. Examples of single research themes that crosscut these disciplines are the interpretation of peasant economies and agricultural decision making.

Second is the concern for understanding agriculture and its change con-

Stephen B. Brush. International Agricultural Development, University of California, Davis, California.

B. L. Turner II. Graduate School of Geography, Clark University, Worcester, Massachusetts.

ceptually and in the context of social and economic change. Originally this concern emerged under the rubric of farming systems as it evolved in studies of agricultural development in less developed countries. The concept itself, however, has antecedents in studies of so-called primitive or traditional agriculture done in the 1950s and 1960s. By the 1980s this concept had penetrated studies of agriculture in developed contexts. Its importance to development studies has assisted in linking international programs and agencies to social sciences other than agricultural economics.

Third is the use of a holistic or system approach and ecological analysis as a means of analyzing agriculture. Included here are efforts to construct broad comparative frameworks to cover different types of agriculture.

The interaction of the three elements is felt in social science in several ways. There is a large network of social scientists studying agricultural systems, and interdisciplinary working groups have developed around themes such as the nature of traditional and modern agricultural production and the energetic efficiencies of farming systems. Disciplines that did not originally focus on agriculture, such as anthropology, now have strong subdisciplinary components dedicated to understanding it. Social scientists have become regular members of agricultural research and development teams, projects, and institutions. Finally, there is an emerging collaborative field of agricultural ecology or *agroecology*, which links social and biological scientists.

Defining a Farming System

Agriculture can be studied in a number of ways, and even the simplest farm comprises a large number of components and types of relations among them. One sensible response to this complexity is to dissect agriculture into component parts and types of relations in order to study each separately. The organization of most agricultural universities and research wisely follows this course by defining crop type and specific commodity units, by designating programs to study classes and types of biotic and abiotic factors, and by setting social science apart from production science. This type of pedagogic and research organization has worked well in many cases, and it has been copied around the world. This conventional organization of agricultural science has, however, been less successful in underdeveloped regions. Although improved agricultural technology has been generated for these regions, its adoption and use by farmers is very uneven, and technology that is supposedly neutral for size of farm has not been widely adopted on smaller farms. Moreover, there is concern that new agricultural technology may actually have negative consequences, such as higher unemployment and rents, for some sectors of the farm economy (Falcon, 1970).

One way to avoid the creation of unacceptable or inappropriate technology is to adopt a system approach in agricultural research. This approach is interdisciplinary, including social science, and uses information from the small-farm sector to set scientific goals and to evaluate technology. There have been calls in both the production sciences and the social sciences for an integration of different lines of analysis into a system approach that more accurately reflects the complex reality of agriculture (e.g., Edens & Haynes, 1982). The farming system research (FSR) movement in international agricultural research is an attempt from the social science side to build a more integrated approach (Sands, 1986; Simmonds, 1985). Another example is the call for *agroecology* (Altieri, Letourneau, & Davis, 1983). A few examples of success exist in developing integrated teaching and research programs around the notion of farming system, particularly in interdisciplinary programs dealing with such subjects as rural development, resource management, arid lands, and land tenure. Most institutions, however, have maintained their traditional disciplinary and commodity divisions. Given the complexity of agriculture and the sophistication of knowledge about its components, this should not be surprising.

The concept of *farming system* or a *farming system approach* is not new, nor is it the outcome of a single, seminal work that proposed it, nor has it obtained a standard or accepted usage. Rather, the terms have emerged from a large and fragmented literature and have been employed partly for the lack of any better terms. For our purposes, the following definitions are employed:

- A *farming system* is any level of unit(s) engaged in agricultural production as it is wedded in a social, political, economic, and environmental context.
- A *farming system approach* describes the unit(s) in its context and/or explores some characteristics of the unit(s) in terms of all or parts of the context.

A farming system comprises several subsystems. Ruthenberg (1976) describes these as a “hierarchy of systems” and identifies three important subsystems: mechanical, biological, and human. Duckham and Masefield’s (1970) analysis of factors that determine farm location considers at least five subsystems. From the literature on agricultural research in general, three major subsystems can be identified: human, environmental, and genetic. These integrate to form a single farming system, and each is made up of numerous components. Each of these three subsystems has an extensive literature devoted to it.

Social science concentrates on the components of the human subsystem, for instance on the rules that govern resource use (e.g., land tenure), on labor intensity and availability, on human demography, on communication and diffusion of innovation, on the relation between social and economic units, on consumption variables, on decision making, and on links between these features and the environmental subsystems. A number of competing themes have been developed to explain the relations between these components (neoclassical, Marxist, cultural). The environmental subsystem is studied primarily by the earth and agronomic sciences and includes such issues as water, soil, surface geometry, pests, pathogens, and symbiotic organisms. The genetic subsystem, studied by botanists, agronomists, animal scientists, and geneticists, involves genotypes and phenotypes of cultivars and animals, and population dynamics that affect crop and animal evolution.

Each of these subsystems has a largely separate literature and demands a different base of knowledge, theory, and method. Consequently, bridges are difficult to build across all three. Nevertheless, it is increasingly common to encounter their integration in both teaching and research. This is especially true in the study of agricultural ecology and in applied research directed at underdeveloped regions. Their integration has been most advanced in applied agricultural research, such as farming systems research (FSR) in agricultural development agencies, discussed later in this chapter. Basic research has been less successfully integrated, although a major exception may be the study of energy systems and energetics in agriculture by social scientists, agronomists, and population biologists (Odum, 1971). Energy-based system research raises important issues for each of the scientific disciplines dedicated to one of the three subsystems identified here. Energy can be studied in several ways, according to the specific subsystem interest, and it is both an integrating and limiting factor in agriculture. Although social scientists have used energy system analysis to great advantage in modeling specific systems and in comparing them, several caveats have been raised about its ultimate utility. Vayda and McCay (1975), for instance, note that energy is frequently not a significant limiting factor in agriculture. Disagreements exist over the energy usage and energetic efficiencies of different systems (see Chapter 11). Moreover, the relationships examined by many social scientists are not necessarily best illuminated by the use of energetic models. Nevertheless, the integration of farming system analysis has benefited by this mutual concern.

The study of farming systems thus rests on three subsystems and on the effort to understand agriculture through them. This book emphasizes the human subsystem of agriculture as studied primarily by social scientists. Even among these researchers, however, rather varied approaches, questions, and themes are emphasized—as attested by the chapters of this book.

Human Aspects of Farming Systems

Since the neolithic revolution some ten thousand years ago, humans have organized in a large variety of social systems to produce food. The primary concern of these systems has been to mobilize land, labor, and other resources for production. Land and labor are universal resources in farming systems. Additional resources that may be important include water, fertilizer or other subsidies, plant and animal types, and credit. The importance of these resources as reflected in capital inputs and institutional support mechanisms, for example agricultural administration and research, increases as systems become market-oriented and industrialized.

Numerous types of social organization, from kinship groups to governmental departments of nation-states, have been devised to mobilize land and labor for production. Human systems to organize and use land and labor for production may be viewed in many ways. Four different aspects of these systems will be discussed briefly: customs and rules, institutional frameworks, population, and technology.

CUSTOMS AND RULES

All farming societies have *customs* that organize and regulate the use of land and labor for crop production. In developed economies these customs become codified in institutions and laws, and the role of customs as the term is used here becomes less important. At a very basic level, dietary and culinary customs direct which cultivars and animals are favored and kept. Rituals may require that certain crops be grown and processed in prescribed ways, and the agricultural calendar may be ritually organized. Customs may also play a large role in determining how land is distributed and how labor is acquired and rewarded. A general understanding is that this type of ritual organization is one of the salient features distinguishing *traditional* from *modern* or complex societies.

Customs blend into *rules* and/or legal codes, and it is important to understand production systems first as behavioral systems rather than normative ones. This is especially true of traditional societies without written codes or formal judicial systems based on legislation. It is, however, also true of complex societies with modern agriculture and written codes and rules.

One rule that is critical in most agricultural systems is the one that determines *land tenure*: control over and access to land. The simplest agricultural systems, swidden cultivation with impermanent fields and very low population densities, typically have very informal rules governing land control and access. Even in the simplest systems, however, ownership of crops by domestic groups is recognized, although permanent rights over land may not exist. As population density and land-use intensity increase, land tenure rules become more critical, more complex, and more specific. Individual rights and communal rights may be distinguished from one another, and the duration, disposability, and inheritance of rights become important issues.

Land tenure combines two types of rules: control and access. *Control* implies durable rights to use land, expressed as *ownership* in the European tradition. Rules of control are held originally by the social group and delegated to families, households, or individuals. Control means that the holder of the right is permitted to use the land in a manner approved by the group. Even though control may ultimately be vested with the group, many societies recognize that individuals or households have de facto control over land. Control may be loaned, sold, rented, and passed to others, although most societies regulate these passages.

Rules of *access* concern how control is exercised. They are important in translating land control into land use. In comparison to control, access implies temporary rather than durable rights. That is, it refers to usufruct. Having access but not control is a common phenomenon, especially where land pressures are great and intensive agriculture is practiced. Access is generally guaranteed by control, but, in addition, is achieved by the payment of rent or shares. Control may be vested entirely in the social group, which then distributes access. The distinction between control and access is usually indicative of stratification and inequality in the distribution of land. In stratified societies, where control over land is held by a dominant group, access by the majority is obtained by payments of one kind or another to that group. This type of land tenure is found in numerous socioeconomic and historic contexts.

Land tenure rules are evident and important in most farming systems, and they are complemented by rules governing labor and exchange. Although the domestic group is frequently responsible for the labor necessary to produce its subsistence, recruiting labor is a common requirement of all farm systems. Many agricultural tasks exceed the capacity of an individual or household, a circumstance compounded by the cyclic nature of production. Farming systems, therefore, depend on rules for mobilizing labor and for rewarding it. They may also depend on labor to maintain the infrastructure (e.g., roads, irrigation systems).

INSTITUTIONAL FRAMEWORKS

Until the advent of capitalist and socialist states, customs and rules governing land and labor, and production and exchange, were usually embedded in kinship systems. In most agricultural societies, the household, usually a kinship unit, is the primary unit of production and consumption. The nature of the household determines the amount of labor available and necessary for production. Rights to land are commonly held by the household rather than by individuals, and decisions on land use are usually the prerogative of the household rather than more general social institutions, such as the community.

Beyond the family and the household, all agricultural societies have more general social institutions. Many agricultural societies have kin-based institutions beyond the immediate family that control agricultural production. Large kinship groups, such as clans and lineages, may hold rights to land and help organize its use, for instance by setting the agricultural calendar. These may merge into tribal or ethnic groups.

The most common group above the extended kin group in agricultural societies is the *village* or *community*. In many societies these are at least partly based on kinship, as membership in the community and rights to land controlled by the community require kin status. Marriage rules of endogamy and exogamy often define the limits and boundaries of community. The peasant community or village is a common form of social organization whose origins are in antiquity. The norm for these settlements is between 500 and 2,000 people, although modern population growth has raised these figures. Peasant communities are often organized corporately, with definite boundaries and communal control of resources (Wolf, 1966). Boundaries are both territorial and social: The community owns a specific territory and membership is granted according to criteria such as residence or kin status. The community can be internally structured in many ways, for example by kinship, by wealth, or by ritual status. These structures help determine and translate rights of control and access into actual land use. The simultaneous existence of social levels such as household and community imply a hierarchy of control over agricultural production. Communities dominate certain parts of the process, such as rights to use land, and households dominate others, such as labor inputs.

In corporate communities, communal control over the production process is evident in such things as communal ownership and scheduling on the agricultural calendar. Corporate organization is not universal, however, and many peasant communities are open and not tightly organized (Wolf, 1966). This is especially true of peasant communities that are actively involved in commercial agriculture and in frontier areas. Open communities exist without the membership restrictions and the communal control over the productive process found in more closed ones. It is thought that the difference between these two types represents an evolutionary process involving the spread of capitalism, increased population, and increasing regional and national centralization. The difference between the two clearly represents a shift away from communal control toward individual control of production.

Peasant villages, both open and closed, are always connected with larger, regional forms of social organization. These include economic forms, such as market systems, and political forms, such as nation-states. Students of peasant societies believe that this connection between village-based livelihood and wider socioeconomic–political organization is their essence (Wolf, 1966). In most parts of the world and throughout history, social organization above the community level has affected the production process by controlling strategic resources, such as water or transportation, and by extracting surplus from the village.

An important intellectual event in the 1960s and after was the increased awareness that institutions beyond the boundaries of communities or farming regions play a significant role in determining the nature of farming systems. This event can be traced to the rise of theories that stress interconnectedness. Dependency theory (Frank, 1967), world systems theory (Wallerstein, 1974, 1980) and neo-Marxist theory (deJanvry, 1981) have all contributed. Increased interest in connections between different economic sectors has also been spurred by the recognition that change is accelerating with improved communications and transportation technology, population growth, expanding market systems, and rapidly growing urban centers. It is no longer possible to think of folk societies as cut off from national and international economic and political systems.

Institutions that have received particular attention for their role in shaping farming systems are international markets (Orlove, 1977), national political systems (Frank, 1967), class systems (deJanvry, 1981), and land tenure systems (Feder, 1971). The relation that is emphasized in these institutional situations is domination for the purpose of the extraction of surplus (Deer & deJanvry, 1979). In developed economies and neotechnic production, institutions that directly affect farming systems are complex and far-ranging in character. The private sector invests by way of loans to farmers; by vertical integration of production, processing, and distribution; by industrial research on technology; and by corporate farming. The public sector invests by way of such institutions as agricultural universities and experiment stations, agricultural extension services, subsidies for certain technologies (e.g., irrigation, soil conservation), and price supports for crops (Bowers & Cheshire, 1983; Jackson, 1981). Indeed, many modern systems would not be economically viable without the guarantees provided by these institutions (see Chapter 14). In many instances the relationships among farm owners, operators, labor, and the state can be complex, as in the recent lawsuit against the University of California, Davis, for its role in the development of the mechanical tomato harvester (Martin & Olmstead, 1985). Although economists predict that such technologies are necessary to sustain the economic viability of industrialized agriculture, the social consequences are significant (e.g., Buttel & Larson, 1979; Goldschmidt, 1947; Martin, 1983).

POPULATION

The size, density, structure, and change of human populations are major topics in understanding farming systems. In subsistence agriculture (consumption production) the amount of production sought, and hence the land and labor employed, is strongly related to local demographic conditions. In commercial or market agriculture (commodity production) local demographic conditions may not directly affect production goals, but they may play important roles in crop scheduling, selection of cultivars, and so forth. Because of this, the relationships between population change and agricultural change have been crucial topics in studies of farming systems, particularly in the underdeveloped world.

Neo-Malthusians assert that population growth has the ability to outstrip agricultural growth (Grigg, 1982). This means that the physical production capabilities of the system (an individual farm, a village, a region, or a nation) may not be able to keep pace with actual growth in demand. The historical record has been interpreted to both defend and oppose this idea (Grigg, 1980). The general record of expanding agricultural production, however, strongly suggests that a strict interpretation of Malthus is mistaken, at least within known agricultural systems and populations. It should be recognized that Malthusian predictions are difficult to test because it is difficult to separate exogenous influences on agricultural production, such as political disruptions, from the technical capability to meet demand.

Recent expressions of the Malthusian idea are found in several guises. One is the application of the carrying-capacity formula from population biology to agricultural systems (Allan, 1965; Rappaport, 1967). These applications have been generally criticized (Brush, 1975; Street, 1969), although the concept has valuable heuristic uses. Another is the concept of underemployment or decline in real wages as population passes the productive capacity of a farming system (Kao, Anschel, & Eicher, 1964; Wellisz, 1968). This concept has been criticized by economists (Schultz, 1964) and noneconomists (Brush, 1977). A third and extreme expression of the Malthusian idea is triage or lifeboat ethics, in which farming systems, and indeed whole nations, are ranked according to population and agricultural potential (Hardin, 1974). This expression has likewise been questioned (Murdoch, 1980). The challenges to Malthusian formulas all rest on the flexibility of agricultural production, its historic propensity to adjust to higher demands, and the importance of exogenous factors such as the distribution of wealth and industrialization.

The failure of the Malthusian prediction has generated alternative views that relate population growth and agriculture. A widely discussed one is that of Boserup (1965). This theme posits an opposite relation between population and food production from that of the Malthusian formula. Boserup argues that population is actually the limiting factor and that population growth is actually the stimulus to expansion of food-producing potential. Boserup's thesis rests on a number of assumptions: (1) that demand created by population density causes agricultural change; (2) that farmers prefer to use their time in other activities once subsistence has been obtained, rather than increase production of foodstuffs; (3) that traditional farming is flexible and offers numerous alternatives; and (4) that intensification is best measured by the ratio between cultivated and fallowed areas (Grigg, 1982:37–41). The second assumption, often referred to as maximizing leisure, has also been expressed as an assumption of least effort: that farmers seek to minimize the amount of agricultural work necessary to satisfy consumption demands. Boserup's theme has been tested in situations where agriculture is not fully integrated into the market, with generally positive results (e.g., Barlett, 1976; Brown & Podolefsky, 1976; Spooner, 1972; Turner, Hanham, & Portararo, 1977).

Although not as comprehensive as Malthus's or Boserup's views, a third alternative to the relation between population growth and agricultural change is the idea of *involution*. This was perhaps most eloquently stated in Geertz's (1963) comparison of swidden agriculture and irrigated rice in Indonesia. Geertz (1963:80) referred to involution in wet-rice cultivation as "its extraordinary ability to maintain levels of marginal labor productivity by always managing to work one more man in without a serious fall in pre-capita income." The absorption of additional labor was accomplished in many ways, such as moving toward double cropping, careful weeding, and the use of special tools, especially in harvesting and milling. Geertz argues that the process of involution derives from Dutch colonial policies and from the lack of investment in indigenous enterprise. Ultimately, involution was a constraint on Indonesian agricultural development.

The concept of involution seems particularly useful for understanding intensive subsistence agriculture that supports large and dense populations. It has been used especially with reference to intensive cultivation in the humid tropics of Asia. On critical examination, however, the concept of involution and the Javanese case on which it was based are problematic (Collier, 1981; White, 1983). It does not, for instance, include the dependence of off-farm labor, and it does not account for the rapid adoption of new technology by farmers in areas that might be described as involuted. Collier (1981) notes that Indonesian rice farmers have abandoned many labor-intensive methods as population has grown. Nevertheless, involution evokes relevant ideas of the flexibility and malleability of subsistence agriculture and of its capacity to support very dense populations (e.g., Metzner, 1982).

Numerous issues centering on labor as a population component thread through farming systems. A major topic in developing countries has been that of labor bottlenecks (Chapter 6) in agricultural production, and indeed the very system itself, is influenced by insufficient labor at critical times in the crop schedule. This is the same labor that left farms for more lucrative employment elsewhere. In developed contexts, especially in more capital-intensive systems, migrant labor becomes important. Much of California agriculture, for example, relies on cheap Mexican labor, even though in the long run this reliance may be damaging (e.g., Martin, 1983).

Technology

Farming systems should be viewed as goal-driven or teleological systems (Duckham & Masefield, 1970). Technology, be it ideas, behaviors, and/or objects, facilitates production to achieve the system's goals. Indeed, technology can be so dominant in our view of agriculture that it becomes the basis for classifications (see Chapter 1). Goals may be inferred from ethnographic information or by statistical means, or they may simply be assumed according to rational behavioral or optimization models. Most historical and broad evolutionary comparisons of farming systems rely on the assumption of universal goals. Duckham and Masefield (1970), for instance, list six objectives associated with farming systems, stressing the optimization of the use of land and labor, while Boserup (1965) emphasizes the goal of farmers everywhere to minimize labor.

The technology to achieve these goals provides the link between the human, environmental, and genetic subsystems of the farming system. This link must harness necessary resources, especially soil nutrients, moisture and crop germ plasm, for crop production. Technology must also control loss from the system and provide for postharvest storage and processing. The technology for managing soil nutrients includes site selection, clearing, burning, tillage, slope modification, and the application of supplements. Soil nutrients are also managed by the selection of crops and crop mixes and by crop and field rotations. Technology for managing moisture may involve irrigation, but it may also require drainage and slope modification to control runoff. Crop germplasm is managed through the selection and distribution of crops and varieties. Technology aimed at limiting loss includes field and crop rotation, scheduling, selection of precocious or resistant crop varieties, intercropping, fencing, and many soil management practices.

Research among technologically simple agriculturists shows that even the most traditional or primitive systems involve complex knowledge of environmental and crop genetic factors (e.g., Wilken, 1974, 1976). All farming systems are dynamic in that they require constant monitoring of the environment and technology, and adjustment to meet goals. In this sense, all farming systems are equally modern, and innovation and invention are inherent parts of all farming systems.

Although innovation and invention account for much technological change, especially in a historical sense, in farming systems, diffusion or spread is also a major factor. Diffusion of both traits and ideas is important. It has been argued for some time that diffusion processes follow a sigmoid or logistic curve pattern (Rogers, 1962), as, for example, described by Griliches (1957) for the spread of hybrid corn in the United States. Although the shape of this curve is repeated in different situations, its slope and ceiling may vary considerably. Szymanski & Agnew (1981:55) challenge the "unquestioned" use of the curve, noting that several distributions other than logistic can be fitted to diffusion data "with the same or sometimes better goodness-of-fit." Indeed, the curve tells us little about the reasons for adopting a technology or not. Structural characteristics of the system, such as the distribution of wealth, influence greatly diffusion and technological change (Blaut, 1977; Feder, Just, & Zilberman, 1982). Technology transfer is rarely a simple process of the adoption of single traits. Rather, traits are accepted and rejected according to their compatibility with other parts of the system. Technological change must be seen as the product of both internal (evolution) and external (diffusion) factors (e.g., Bayliss-Smith & Wanmali, 1984). Themes of agricultural change will be treated at greater length.

Technological change among farming systems in developed contexts is often taken for granted. That is, farmers and private and public institutions combine to improve constantly the productivity and/or ecological and economic efficiencies of agriculture, such as in the development of hybrid cultivars or major state-sponsored land reclamation projects (e.g., Smith, 1970). Nevertheless, farmers in these highly developed contexts are not necessarily quick to switch to technologies or systems with which they are unfamiliar, despite researchers' evidence that such change would enhance production. The slow move to no-till and organic cultivation in the Great Plains is a case in point (Phillips, Blevins, Thomas, Frye, & Phillips, 1980). As Dillon and Heady

(1960), among others, have shown, even farmers with high levels of information and access to new technologies tend to make choices that affect their "system" on the basis of first-hand experience or tradition. In another vein, decisions to adopt a new technological package may have more to do with short-term economic prospects than with long-term prospects or with conservation-related issues (e.g., Batie & Healy, 1980).

Antecedents in Social Science

The contemporary analysis of farming systems has antecedents in many scientific disciplines. The general antecedents in social science mentioned at the beginning of this chapter are the comparative and international focus in the social science of agriculture, studies of the relation of agricultural change to social and economic change, and holistic or system research on agriculture (e.g., ecology). Instead of going through the intellectual history of four disciplines active in this field, we identify three interdisciplinary foci that crosscut anthropology, geography, agricultural economics, and rural sociology: cultural ecology, peasant economies, and applied farming system research.

Cultural Ecology

The research of cultural ecologists in anthropology and geography has tended to gravitate toward simple or low-energy economic systems (e.g., Rambo & Sajise, 1984). This circumstance is derived from interests in cross-cultural comparison and in cultural evolutionary or landscape modification sequences. As holistic and comparative, both require a data base that only studies of smaller and economically simpler societies could provide.

The works of Wissler (1926) and Kroeber (1939) on the distribution of cultural traits pioneered a path in anthropology toward the study of subsistence systems. Similarly, Sauer (1925) elaborated the study of landscape morphology in geography, focusing on non-Western, nonindustrialized conditions. Kroeber and Sauer were close associates in their formative years at Berkeley, and therefore it is not surprising to find similarities in the subject matter and approaches in their respective forms of anthropology and geography. Their contributions notwithstanding, the modern antecedent of cultural ecology is the product of Steward (1955), whose method of cultural ecology provided a basis for describing and comparing specific subsistence systems.

Steward stressed the importance of the cultural definition of "natural resources" on the one hand and the influence of subsistence activities on other aspects of culture on the other. His notions of adaptation and culture core stressed the human side of human–environmental interaction, but they rapidly pushed anthropologists toward the systematic description and analysis of subsistence systems. Although the Stewardian approach has been challenged for being ecologically weak (e.g., Vayda & Rappaport, 1969), it is still widely practiced by anthropologists (Bennett, 1976; Hardesty, 1977; Orlove, 1980). An important reason for this is its eclecticism and relaxed approach to adaptation (Denevan, 1983). In geography, Brookfield (1962), Denevan and Padoch (in press) and others have applied a strong comparative method of analysis to various subsistence systems.

Cultural ecology in anthropology and geography developed parallel interests and ideas concerning the nature of human adaptation to the environment and change (Grossman, 1977; Mikesell, 1967). Both suffered similar weaknesses, such as their functional emphasis on adaptation with its tendency toward homeostasis. Each tended to emphasize internal dynamics of cultures and communities rather than broader influences such as world trade, although exceptions existed (e.g., Brookfield, 1975). The impact of more formal ecological models (Odum, 1971) and world systems models (Wallerstein, 1974) was similar in both, as each sought greater formalism and connection to socioeconomic and historic research on the same type of societies (e.g. Carlstein, 1982; Ellen, 1982; Gross, Eiten, Flowers, Levi, Ritter, & Werner, 1979; Watts, 1983).

During the 1950s and 1960s a substantive and theoretical base was laid down in cultural ecology for the study of farming systems. Research on simple technological systems, such as swidden cultivation in the tropics (Brookfield, 1962; Conklin, 1957; Geertz, 1963; Spencer, 1966), provided the initial impetus for this. Interest in peasant agriculture did not develop as early, but it soon became equally important (e.g., Clark & Haswell, 1964; Collier, 1981; Gudeman, 1978; Levi & Havinden, 1982; Ortiz, 1973; Ruthenberg, 1976). Although these interests could be subsumed into larger theoretical designs, such as cultural materialism (Harris, 1968), each generated a set of appropriate but more specific questions. For extensive or consumption-based agriculture (e.g, swidden) these questions included demographic and ecological limits and the shifting nature of the agriculture, environmental maintenance or degradation, and the potential for complex social organization and warfare. For more intensive consumption/commodity-producing agriculture (e.g., peasant-based), these questions included the nature of surplus, hierarchical relations within peasant economies, and their relation to larger market systems.

Of the few theoretical concerns that weave these studies together, the overriding ones are energy efficiency and intensification, the increase in labor inputs, use of land, and surplus. The study of energy flows and energy efficiency (energy input/energy output) was borrowed directly from formal ecology (Odum, 1971) for the analysis and comparison of agricultural systems. Rappaport (1967), for instance, described the energy flow in a swidden agricultural system in New Guinea. Nietschmann (1973) and Bergman (1980) demonstrated the temporal inputs for two diverse farming systems that require relatively small levels of labor inputs. Norman (1979) provided an energy analysis of African systems, arguing that the overall trajectory of cropping efficiency was to decrease as output intensity increased, a theme supported by Dalrymple's (1971) study of double cropping systems. Bayliss-Smith (1982) has compared the energy efficiency of different farming systems, concluding, like Pimentel and Pimental (1979), that industrial agriculture is perhaps the least efficient.

Boserup's (1965) thesis will be discussed in more detail in the section on themes of agricultural change. Her ideas on the relation between population, work, productivity, and degree of intensification measured by the frequency of land use challenged the conventional Malthusian formula. This thesis has been addressed in a number of studies by anthropologists, geographers, and agricultural economists (Barlett, 1976; Brown & Podolefsky, 1976; Datoo, 1976; Gleave & White, 1969; Levi, 1976; Netting, 1968; Metzner, 1982; Norman, 1979; Vermeer, 1970).

Other important contributions to the general field of cultural ecology before 1970 include Brookfield's (1962) focus on micro-level comparisons of population–agriculture–environment relationships, Geertz's (1963) comparison of swidden and irrigated rice in Indonesia, Allan's (1965) study of carrying capacity in Africa, and Rappaport's (1967) link of ritual control and agriculture in New Guinea. More recently, the study of hazard and risk has added important contributions (e.g., Watts, 1983).

Peasant Economy

Cultural ecology overlapped with another antecedent to current farming systems research, the study of peasant economies, and this overlap provided a significant link to the large field of economic evolution and to development economics. This tradition of research was dominated initially by the formalist–substantivist debate of the 1950s and 1960s, particularly in anthropology and economics (e.g., Dalton, 1961; Firth, 1967). Substantivists argued that formal economic theory and methods were inadequate to explain nonmonetized economies based on reciprocity and redistribution, where economic transactions were determined by such things as kinship and ritual (e.g., Dalton, 1963). Formalists argued that all people engage in optimizing behavior and thus are subject to neoclassical types of economic analysis (e.g., Schneider, 1974).

The substantivist–formalist debate withered in the 1960s as several new concerns became important. These included the study of household economies, especially under the influence of Chayanov's theory (Chayanov, 1966); an increased interest in economic change and development; the recognition of the importance of demography in change, especially under the influence of Boserup's theory; the impact of ecological research; and the increased recognition of the significance of external economic relations on tribal and peasant economies. The general impact of these factors was to shift the focus of economic anthropology away from exchange and toward production. In agricultural economics, a synthesis developed that merged the special characteristics of behavior in consumption and consumption/commodity production systems (e.g., Fisk, 1964; Scott, 1976) with those that tend to emphasize the economic rationality of traditional farmers (e.g., Mellor, 1966; Schultz, 1964; Wharton, 1969).

The study of agricultural production in technologically simple societies initially merged the interests of cultural ecologists and economists. These interests were to determine a society's definition of natural resources and to describe how these resources are exploited and how they influence behavior and social organization. It was recognized that in many societies, especially peasant ones, the basic unit of production was not a firm but the household. As a theoretical base to research on household economies, scholars (e.g., Sahlins, 1972) turned to the work of the Russian economist Chayanov, who had analyzed Russian peasant behavior in the early 20th century. Chayanov made several contributions to the understanding of this type of economy. First, peasant labor investment must be understood in relation to the ratio between consumers and producers in the household, which changes in relation to the normal domestic cycle. Second, the amount of work done by a household is determined not only by its needs but also by the "drudgery" of the work. Chayanov asserted that a household's labor inputs will increase to meet needs, even if marginal returns are decreasing. Conversely, labor inputs may decrease even under favorable marginal returns if needs are met or if drudgery is too great, an idea inherent in Boserup's (1965) thesis. Chayanov's theories have attracted considerable attention among social scientists studying peasant economies (e.g., Durrenberger, 1984). Grigg (1982:92) notes that Chayanov's theory "remains the only coherent alternative model to that of the capitalist profit-maximizing farmer."

The 1960s and 1970s were decades of renewed interest in demography and its relation to agriculture. This was partly due to the specter of famine in South Asia and along the Sahel belt of Africa. It was also due to the interdisciplinary impact of Boserup (1965, 1981). Rapidly increasing populations helped focus the attention of the social sciences onto production. This focus was also sharpened by the increased awareness of the importance of external economic relations to tribal and peasant societies. This awareness relates to the rise of dependency theory (Frank, 1967) and world systems theory (Wallerstein, 1974), which emphasized the role of trade and economic domination between different societies and economic systems. Both analyzed hierarchical relations between different market centers. Their contribution was in emphasizing the role of political and economic domination in the history and behavior of economics usually described as isolated and self-sufficient. Both examined underdevelopment as a result of domination rather than of endogenous factors. More recently, these theories have been eclipsed by a rise of neo-Marxist or political economy theory (deJanvry, 1981; Deere & deJanvry, 1979; Wolf, 1982; Watts, 1983) that emphasizes exploitation, internal differentiation, and external domination as principal causal factors in determining production.

The Applied Farming Systems Approach

While cultural ecology and peasant economy focused social science disciplines onto basic research questions about agriculture, events in applied science also contributed to the current farming systems theme. In this area of applied social science the most important contributor has been agricultural economics. Rural sociology is a smaller and less visible field, although many rural sociologists have been active in applied work. Although anthropology and geography have applied components, these are smaller than the other two disciplines, and their applied interest in agriculture is relatively recent. The more advanced status of agricultural economics and rural sociology in applied agricultural research is due in part to their historic places in land grant universities, in agricultural experiment stations, and in state and federal agencies in the United States, but more important perhaps is their definition as disciplines.

The contribution of applied social science to the concept of farming systems is especially clear in international agricultural development, where it has led to a much publicized approach called *farming systems research* or FSR (Byerlee, Harrington, & Winkelmann, 1982; Norman, 1980; Sands, 1986; Shaner, Philipps, & Schmehl, 1982; Simmonds, 1985). This approach stresses the need to work upstream from the farm level to the scientist level and emphasizes planning in research and development work in less developed countries. The FSR approach rests on the assumptions that a single farm involves the interaction of myriad elements and that the behavior and decisions of farmers must be analyzed in a framework that does not isolate them from other components or from the general context in which they operate; these assumptions reflect a long-held tradition outside economics. This approach is dedicated to improving agricultural research and to reaching the small-farm sector of underdeveloped regions more effectively. Much FSR research and publication is, unfortunately, not directed at larger theoretical issues in understanding or explaining agricultural change. It does not develop connections to the larger field of science relating to farming systems or to its scientific and scholarly heritage.

The use of the farming system approach, as defined here, to study agriculture in the developed world is rather recent and is dominated largely by economists and agronomists. Of late, rural sociologists, anthropologists, and geographers have begun to explore this line of research, as attested in Part IV of this book. A major contribution has been the application of case studies in context to the theoretically dominated economic literature. Such studies are able to draw on a rich literature and sector-based data developed originally for economic analysis.

Systems Research in Agriculture

The comparative and interdisciplinary analysis of agriculture gained momentum during the same period as did systems theory and ecology. Ecological analysis of agriculture led to significant insights, such as energy efficiencies of particular farming systems (e.g., Bayliss-Smith, 1982; Cox & Atkins, 1979; Leach, 1975; Norman, 1980; Pimentel, Fast, & Berardi, 1983; Pimentel & Hall, 1984; Pimentel & Pimentel, 1979). This coincidence supported the application of systems approaches in social science to the analysis of agricultural production (e.g., Rappaport, 1967). Results of this application are evident in both the descriptive comparison of farming systems (e.g., Duckham & Masefield, 1970; Grigg, 1974; Ruthenberg, 1976) and in the theoretical base of social science (e.g., Bennett, 1976).

After three decades of systems analysis, it is important to recognize two caveats. First, systems must be artificially limited if analysis is to proceed, and second, described systems are, therefore, heuristic and artificial analytical devices rather than natural phenomena. Although the systems approach tends to push the investigator toward descriptive holism, effective limits on the scope of system modeling are imposed by problems of scale. The larger the descriptive model, the weaker it becomes for describing specific behavior. Finer-grained models, on the other hand, may lack the explanatory power of more general ones. It needs to be emphasized that the systems approach is *descriptive* rather than *explanatory*. It helps identify *what* processes exist and *how* sets of interrelated components function together. *Why* systems work the way they do is an explanatory task best performed by theoretical constructs that have traditionally emerged from economics, geography, or anthropology.

Studies of farming systems have involved various levels of investigation, from the household to the international sphere. Nevertheless, a convention of recent farming systems analysis is that the appropriate level is between the household and a subregion or agroclimatic (physiographic) zone, depending on local ecological and socioeconomic conditions. The village or community that falls between these poles is useful for many types of peasant or tribal systems, whereas a farm, cooperative, or economic community may be appropriate for large-holder market systems.

In farming systems research, the system concept is used in its broadest sense—a set of interrelated components and their attributes in which the changes in any component or attribute effect changes in the others. Rarely does it entail general systems theory (e.g., Buckley, 1968; Von Bertanlanfy, 1962) or functional systems explanation (Bergman, 1966; Rudner, 1966). Rather, it is a heuristic concept that draws attention to the descriptions of relations among what may be perceived as disparate elements, and to analysis of how the system works. This usually involves the examination of material flows through the system. Systems are best understood by describing the flow of common entities (e.g., energy, nutrients, money) among components and the changes in one resulting from changes elsewhere. It is important, and somewhat easier, to describe the structure of a system rather than the process of flow that makes it a system. The difficulty of describing flow is increased by the need to limit systems to analytical dimensions and by the tendency to focus on a limited number of components.

Proponents of the systems approach argue that it facilitates study beyond the confines of topical disciplines. Moreover, explication of how a system works tends to lend itself to the study of process and change. Although agriculture appears resilient in most cases, any system, given sufficient time, will pass through various stages of change. Systems approaches are used to analyze this change; explaining it, however, remains the domain of the theory construction associated with traditional disciplines.

Comparative Farming System Studies

Another antecedent in social science to the current framework of farming system analysis is the effort to establish descriptive frameworks for different types of agriculture. The most developed attempt to do this is in agricultural geography, particularly in Europe. Economists, anthropologists, and rural sociologists tend to be less concerned with large-scale and detailed comparison, preferring to work with a more general and evolutionary framework. Some highly influential studies, such as Dumont's (1957), are not rigidly drawn about a central comparative framework. In recent years, several comprehensive comparisons have applied ecological and systems theory (Bayliss-Smith, 1982; Duckham & Masefield, 1970; Ruthenberg, 1976). Others use a loose evolutionary framework (Grigg, 1974).

The utility of systems analysis for comparative agricultural geography is evidenced in Duckham and Masefield's (1970) encyclopedic volume on farming systems of the world and Ruthenberg's (1976) study of farming systems in the tropics. These books cover large topics and provide overviews of specific types of agriculture rather than detailed descriptions of specific cases. Both emphasize that farming is a goal-oriented system defined by human, biological, and abiotic components and boundaries. Duckham and Masefield's (1970) classification of agriculture is done according to two criteria: degree of agricultural intensity and land use. They argue that one of these criteria, intensity, is subjective, arbitrary, and determined by many measures: climate, land use, type and/or rate of production, inputs, size of farm, and stocking rate (Duckham & Masefield, 1970:105). Four levels of intensity are used: very extensive, extensive, semi-intensive, and intensive. Likewise, four types of land use are identified: (1) tree crops; (2) tillage with or without livestock; (3) alternating tillage with grass, bush, or forest; and (4) grassland or grazing of land consistently in indigenous or man-made pasture. This scheme is applied to a worldwide sample of farming systems, the greatest attention being given to temperate and developed regions.

Although Ruthenberg's (1976) scope is somewhat more modest and limited by his focus on tropical agriculture, he too must contend with tremendous diversity. The tropics contain farming systems that are industrialized and commercial as well as technologically simple, subsistence-oriented ones. Ruthenberg (1976:2–3) notes that farms are actually hierarchies of systems ("machine-biological-social"). Five major aspects are necessary to describe a farming system: goals, boundaries, activities and their relations (e.g., inputs and outputs), external relations, and the function between internal and external relations. Ruthenberg's classification of tropical farming systems considers six criteria: type of rotation, intensity of rotation, water supply, cropping pattern and animal activities, implements used for cultivation, and degree of commercialization (Ruthenberg, 1976:13–18). Using this scheme, Ruthenberg describes six major cultivation systems: (1) shifting cultivation systems, (2) fallow systems, (3) ley farming (cultivation/pasture rotation), (4) permanent upland cultivation, (5) systems with arable irrigation farming, and (6) systems of perennial crops.

Regardless of the classification or characterization employed, work in comparative systems is pivotal to the future of agricultural studies in general. This is so because conceptual and theoretical work is enhanced when it can be assessed across space and time, and to achieve this assessment requires frameworks that allow for comparisons on standardized criteria.

Themes of Agricultural Change

There are many ways to compare theories of agricultural change. A few variables are so important that they dominate all theories, and theories can be compared according to how they treat these variables. Included in this list of key variables are population size and density, land availability, agricultural technology, communication, market factors, and economic and political institutions. Farming system change can be defined along two broad axes: technological and structural.

Technological change can be observed in many ways, in terms of the types and amounts used, management practices, productivity, and efficiency. *Intensification* is often used to describe technological change involving greater use of labor or other inputs per unit of land. *Structural* change involves changing social and economic relations in the production process, for instance in land ownership and tenure and in the relation between labor and capital. In a broad view, theories of agricultural change can be grouped into three sets according to the weighting that different variables receive and whether they are viewed as dependent or independent variables. These three sets are (1) technology themes, (2) demand themes, and (3) political economy themes.

Technology Themes

Technology as the principal cause of change in agriculture is central to a number of themes and underlies much of the applied work on agricultural development in the Third World (e.g., Harwood, 1979:3; Ruttan, 1984:108). Antecedents of the theme, however, are attributed to Malthus (1798) and classical economists. Traditionally these themes have assumed the primacy of technology and that its adoption is mainly the result of farming behavior bent on profit maximization or, at least, increased production. More recently, optimization and the role of risk aversion have also been included. Most of the themes are tied to market demand themes of farming behavior, but differ from them by focusing on the creation or invention of technology itself as the necessary precondition for agricultural growth. Two examples of technology themes are the neo-Malthusian theme and the technology diffusion theme.

NEO-MALTHUSIAN THEME

The primacy of technology in the Malthusian theme may be undervalued, especially in comparisons of Malthus's and Boserup's (1965) views. Malthus reasoned that levels of sustenance are directly related to levels of technology. Although this relationship was poorly defined (Grigg, 1980:13), technology is treated as an variable that stimulates changes in agricultural production. Technology holds this position because of the assumption that individual farmers seek to produce toward the maximum possible (e.g., Ghatak & Ingersent, 1984:257). Hence, a new technology that assists in this goal was adopted. While the technology for increased production was envisioned by Malthus to be relatively inelastic in the short run, its long-term trend is unilinear—increasing production.

TECHNOLOGY DIFFUSION THEME

A modern counterpoint to this view, especially as applied in development work, is the technology diffusion theme (e.g., Ruttan, 1984). This theme envisions changes in production and productivity of agriculture in the Third World as the result of the adoption of new technologies. These technologies are adopted because farmers seek to increase output or lower the costs and risks in production. Without the technologies, the growth of agriculture is constrained. In a related argument—induced technological change—technological invention and hence agricultural change are seen as accelerated by the research and development of industry, which constantly attempt to make output more responsive to a new, industrial-made input that has declined in price (Hayami & Ruttan, 1971).

At the micro level the technology diffusion theme is synchronic, dealing with the changes associated with specific technology. At the macro level, however, it involves unilinear change (Browlett, 1980:63–64). Economic (agricultural) change follows a path from traditional (paleotechnic, consumption production) to modern (neotechnic, commodity production) stages of development (e.g., Rostow, 1960).

Critiques of the technology diffusion theme are numerous (Blaut, 1977; Brookfield, 1975, 1978; Browlett, 1980; Mabongunje, 1981). The unilinear view has been questioned for its presumption that Western technology is applicable to other cases and for undesirable consequences of its application such as unemployment and environmental degradation. At the level of individual adopters, the technology diffusion theme has been questioned for its unreasonable and simplistic assumptions, particularly underestimating the constraints to adoption on individual farmers. One response to these critiques has been the emergence of the high-payoff input model, which focuses on institutional and economic changes that reduce constraints on the traditional farmer and allow the opportunities for adoption (Ruttan, 1984:113). Nevertheless, technology remains the key to change.

Demand Themes

These themes examine agricultural change from the perspective of how goals affect the behavior of individual producing units. Demand themes are of two broad types: those that focus on consumption (subsistence) production and those that focus on commodity (market) production. A modified consumption theme or dual consumption–commodity theme may be emerging, although it has yet to be formalized. Boserup is credited with formalizing the consumption demand theme, although antecedents to it exist (see Grigg, 1979; Turner *et al.*, 1977). Commodity demand themes, of course, have a rich history in classical and neoclassical economics. Based on price or market forces, they have expanded to include some nonmarket elements.

CONSUMPTION DEMAND THEMES

Two variants of this theme are Boserup's (1965) concept of agricultural change under population pressure and Chayanov's (1966) theory of intensification as a function of the size and composition of peasant households. Boserup's theme of population pressure or consumption demand has been so thoroughly reviewed, critiqued, and modified that it needs only a brief summary here (see Adams, 1966; Barlett, 1976; Bhatia, 1968; Blitz, 1967; Brookfield, 1972, 1984; Brown & Podolefsky, 1976; Cowgill, 1975; Datoo, 1976, 1978; de Vries, 1972; Doolittle, 1984; Gleave & White, 1969; Grigg, 1979; Jones, 1967; Lagemann, 1977; Levi, 1976; Metzner, 1982; Sheffer, 1971; Spooner, 1972; Robinson & Schutjer, 1984; Turner *et al.*, 1977; Vermeer, 1970). In Boserup's view, agricultural intensity (output or production intensity; see Turner & Doolittle, 1978) is seen as elastic and as changing in relation to consumption or biological demand (population pressure). Output intensity is variable because inputs (labor and technology) are thought to be elastic and because production goals change to meet consumption needs. The primary goal is not maximum production but, rather, some level of output that satisfies needs while incorporating leisure, nutritional quality, and stability over time. Agriculture intensifies or disintensifies, depending on the demands placed on it. An example of this change is the trajectory of cultivation frequency from long fallow to multicropping.

The consumption demand theme assumes that agriculture does not necessarily approach its technological limits and that added inputs of labor, technology, or procedures can increase production. In most cases, however, this increase in production is thought to be obtained at the cost of diminishing returns and decreasing technological efficiency (input/output ratio). These presumed results are fundamental to understanding agricultural change, given the proposed behavior of farmers in consumption production.

Boserup contends that farmers engaged in consumption production do not behave so as to maximize output or profit. Rather, following the substantivist tradition in anthropology, these farmers are seen as seeking "culturally defined output" that is closely related to biological or population demand. The level of output is determined by a least-effort or least-cost mode. As intensity increases, so do effort and costs. Therefore, farmers tend to choose the lowest-intensity system that meets their needs and to increase intensity only under conditions of greater needs.

The behavioral assumptions in Boserup's demand theme were previously articulated by Chayanov (1966) in his studies of peasant farmers in Russia (For discussions of Chayanov's thesis, see Chibnik, 1984; Durrenberger, 1979, 1984; Harrison, 1975, 1977; Hunt, 1979; Lewis, 1981; Littlejohn, 1977; Minge-Kalman, 1977; Patnaik, 1979; Sahlins, 1971; Smith, 1979; Stier, 1982). Chayanov's work, the thrust of which was an attempt to identify a "peasant economy" through an allocation of resources by households, has had a major influence on agricultural change theory of late, particularly among anthropologists (see Chibnik, 1984; Durrenberger, 1984; Hunt, 1979). Chayanov begins with the family farm (no wage labor), production for consumption and market, and elasticity of land and technology (see Hunt, 1979). He concludes that in "peasant economies" a "culturally determined" standard of living (per capita income) prevails, beyond which any increases are considered marginal. This standard, as reflected in return to labor, can be above or below the market wage. Each additional unit of input (labor) to meet production needs is done at the cost of increasing "drudgery" of labor. Therefore, once the minimum production per capita is met, further increases are unlikely. Changes in agricultural production corresponded largely to changes in the size of the production unit, and increased labor per producer changes as the consumer/producer ratio changes.

This view has been assailed primarily by those of socialist and Marxist perspectives. Patnaik (1979) attempts to demonstrate that Chayanov's data were interpreted incorrectly, and Harrison (1975, 1977), among others, finds fault with his assumptions, particularly those dealing with wage labor and capital accumulation (Hunt, 1979; Tannenbaum, 1984). Despite these critiques, elements of Chayanov's theory of peasant behavior have been useful in a number of contemporary studies.

Returning to Boserup's demand theme, least-effort behavior is related to drudgery and to diminishing returns to inputs, particularly labor. Farmers are reluctant to intensify production once the minimum output per production unit is satisfied because of the need for additional inputs and diminishing returns. Moreover, farmers tend to select the lowest input intensity system possible that can fulfill their needs because, according to Boserup, input intensification generally involves lower productivity of labor. Farmers intensify when forced to do so by increasing levels of demand. Conversely, if demand decreases significantly, agriculture disintensifies.

Opposition to this theme has been varied. Early reviews by agricultural economists and historians found it to be too simplistic or not to fit the historical evidence for Europe and Japan (Dovring, 1966; Jones, 1967); later assessments find some utility in the theme (e.g., Ghatak & Ingersent, 1984:256–62). Marxists have rejected the implication, simplistically interpreted, that population growth in poorly developed countries is necessary for development. Indeed, as with Malthusian theory, virtually every assumption and relationship in Boserup's theory has been challenged (Grigg, 1979). Nevertheless, many elements of it have been demonstrated in a variety of studies, and these works combined with the criticisms have led to an informal, modified demand view.

MODIFIED CONSUMPTION THEME

The modified consumption theme has yet to be formalized in a theoretical framework, although it is implicit in the works of many involved in agricultural behavior and decision making. Following Boserup, it asserts that the intensity of agriculture is a direct response to the demands placed on the individual units of production (farming units) and that input and output intensity is usually elastic in relation to demand. However, the sources of demand, the behavioral rationale, and some of the assumptions about various relationships in the system are modified or elaborated from those of the original formulation by Boserup.

Demand is expanded to include biological, social, and market forces. Population remains a principal, if not paramount, variable (Brookfield, 1972, 1984; Chibnik, 1984; Turner *et al.*, 1977), but social variables, such as kinship responsibilities and taxes, are added to the demand ledger. These latter may include extraordinary variables that are specific to a particular place or culture. For example, in parts of the Pacific region a small element of demand includes prestige production; in this case giant yams (*Dioscoria* spp.), for the purpose of display and waste to denote the prosperity of the farmer (e.g., Lea, 1969). In South Asia dowries can have significant effect on demand.

Production strategies to meet these demands are involved with behavior typical of both ideal production types—consumption (or subsistence) and commodity (Hunt, 1979; Turner, 1983). The former emphasis is placed on averting risks (Lipton, 1968; Schulter & Mount, 1976). In contrast to subsistence production, commodity production may involve considerable risk-taking strategies in order to achieve some acceptable or desired return. The key to the "dual farmer" is that the sum agricultural strategy—the farming system—is a product of this mix or blend of purposes (Collinson, 1972; Ortiz, 1973). The resulting system, then, be it in reference to output intensity, crop spacing, or weeding practices, reflects this mix rather than that associated with ideal consumption or commodity production.

The modified theme asserts that as goals and types and of demand change, so does agriculture, but as modified by a series of constraints. These constraints are numerous, and only two are discussed briefly here as examples: environment and technology.

Commodity themes, like Boserup's consumption themes, have paid minimal attention to the influences of the physical environment on agricultural change (Grigg, 1979; Turner *et al.*, 1977). Following Brookfield (1972), the modified consumption model asserts that environment acts as a constraint on agriculture in that it offers various kinds and levels of resistance to production. Increased physical constraints—for example, drought, low soil nutrient levels, or pest problems—require specific inputs to lessen their impacts on production. The results affect output intensity and, in turn, the strategy of cultivation. For instance, one study has found that, given similar levels of demand, high and low constraining environments help create a more intensive system than do medium constrained environments (Turner *et al.*, 1977; see also Dunlap & Martin, 1983). Agricultural intensity is a response to the interaction between demand and the environmental context in which it exists. [It is interesting that rural sociologists investigating the adoption of innovations and agricultural intensity apparently remain in debate over the level of influence that the physical environment has on these issues (Dunlap & Martin, 1983), especially given the rich literature on the subjects by geographers, among others. In rural sociology, see: Ashby (1982), Buttel and Larson (1979), Gartrell and Gartrell (1979), Gilles (1980), Heaton and Brown (1982), and Larson and Buttel (1980).]

A farming system, in this view, balances three factors: output, security, and efficiency. Usually, and particularly within a set of technology, increased output results in diminishing technological efficiency (input/output ratio). Certain technological thresholds are thought to exist, however, which, once adopted, may increase efficiency. Boserup (1965:26) recognized this point but did not elaborate it. Recent arguments provide the elaboration (e.g., Robinson & Shutjer, 1984). Briefly, diminishing returns to sets of technology can be seen as a step curve in which each step raises the level of efficiency and changes the parameters of input–output relationships (see Mellor, 1966). Farmers are seen as reluctant to move to the new set because of the input costs, unless they are forced to do so by demand or changes in such variables as risk.

Unfortunately, field tests to determine the validity of these claims are minimal and hampered by inadequacy of measures with which to compare labor and capital inputs. Converting capital and technology to an energy measure has led to disagreements (Chapter 11), and converting labor to a monetary value is suspect because a production unit's own labor is not necessarily valued as wage rates (Long, 1986). As a result, the types of sets of technology (e.g., irrigation, high-yielding varieties seeds, field raising) that may create the step function are not established.

COMMODITY THEMES

The commodity and land use themes are well-developed products of neoclassical economic analysis of agriculture. Rent, surplus, efficiency, and supply are important facets of this analysis. Farmers are "economic agents" that allocate resources in response to the market so as to maximize profit. Of course, optimum allocation in an economic sense relies on assumptions about individual preferences, backgrounds, situations, and access to information (Dillon & Heady, 1960). Moreover, farmers also have the ability to influence cost and price factors, both directly and indirectly (see Chapter 14). Neoclassical economics has adjusted its models accordingly to account for these "distortions" to the ideal market farm. Nevertheless, the basic theme is that changes in the market create changes in agriculture and, hence, in farming systems. Insufficient historical analysis, however, has limited information on change largely to specific variables such as labor, capital investment, and land tenure.

Market and land-use themes can have a change dimension (Norton, 1984:130–131). One basic theme draws primarily from von Thünen's analysis of economic rent and agriculture, which has been elevated to the status of theory (Chisholm, 1970:20: Norton, 1984:129–135). Simplistically, the "theory of the isolated state" (see Hall, 1966) assumes that farmers are economic agents responding to a central market. The agricultural system that results at any locale is a product of economic rent—the difference in the production (or net value) of the same cultivar on different pieces of land. Given uniformity in all variables, save distance to the market, economic rent will decrease as distance from the market increases because of transportation costs. This phenomenon creates rings of land use around the market in which the next outward ring represents a less intensive form of cultivation than the preceding one.

Elaborations of this theme are numerous as various assumptions have been relaxed to account for more variables (Day & Tinney, 1969; Ewald, 1977; Newbury, 1980:2–23; Sinclair, 1967; Tarrant, 1974:19–68). As commonly employed, von Thünen's theory is static, although some studies have added a temporal dimension (Norton & Conkling, 1974). Several studies have focused on the changing size of the market and the resulting outward expansion of each land-use ring, even at an international scale (Peet, 1979, 1970; Schlebecker, 1960). This expansion results in both quantitative and qualitative changes in farming systems.

COMMODITY DEVELOPMENT THEME

Commodity development themes emerged as an attempt to apply economic concepts to farmers in the Third World engaged either partly or wholly in market production. It was stimulated by research in peasant economies that disputed the view of the "backward peasant." One group attempted to depict such farmers as "penny capitalists" (e.g., Bauer & Yamey, 1959; Behrman, 1966; Dillon & Anderson, 1971; Fisk, 1964; Haswell, 1953; Hopper, 1965; Huang, 1973, 1974; Mellor, 1966; Nakajima, 1969; Schultz, 1964; Southworth & Johnston, 1967; Tax, 1953; Wharton, 1969; Yotopoulos, 1968). A number of studies, particularly among smallholder farmers who were well integrated into the market economy, adopted this position—farmers were poor but efficient in response to the market (e.g., Bauer & Yamey, 1959; Behrman, 1966; Hopper, 1965; Nakajima, 1969; Norman, 1974, 1977; Schultz, 1964; Southworth & Johnston, 1967; Wharton, 1969; Yotopoulos, 1968).

The critics of this theme are many and multidisciplinary, and they have been instrumental in altering the theme, particularly those outside the field of economics (e.g., Chibnik, 1984; Collinson, 1972; Hunt, 1979; Scott, 1976). The economist Lipton (1968) is usually credited with the original challenge to the efficiency concept as applied to farmers faced with considerable risks and uncertainties. With the need to ensure consumption needs, these farmers must be risk-averse, not risk takers as among pure commodity producers (e.g., Roumasset, 1976). A number of studies conclude that these farmers trade off risks against profit and hence maximize utility, not profit (Huang, 1973, 1974; Mellor, 1966; Wolgin, 1975). Schulter and Mount (1974) refer to this kind of behavior as proficient. Of course, this view has not gone unchallenged; some studies conclude that consumption–commodity producers are not risk-averse. The level and duration of market involvement, perhaps the key to the variations in results of these studies, is not sufficiently clear to allow a thorough evaluation.

This brief overview of demand-based themes of agricultural change suggests that the consumption-based and commodity-based behavioral approaches may be merging and that a formal synthesis is forthcoming. The consumption approach has gained significant support among anthropologists, geographers, and some economists as it applies to farmers who are primarily subsistence producers. The commodity approach seems to work well enough in market-dominant situations. A large number of the world's agricultural systems, however, are the product of farmers engaged in various degrees of consumption and commodity production. For the most part, those who work with consumption types have not incorporated well the influence or behavior of market production. Alternatively, many of those working with commodity types seem reluctant to recognize the different goals and objectives of consumption production and/or seem to assume that once some level of entry is made into the market, commodity-based behavior becomes the norm. Some headway has been made in the proposed synthesis. Fisk (1964) and Fisk and Shand (1969) offered pioneering attempts in economics, as has de Vries (1972) from a historical perspective. The next stage involves formal modeling and empirical demonstration of relationships between the "behavioral mixes" and levels of involvement in the two production types (e.g., Grossman, 1985; Gudeman, 1978).

Political Economy Themes

Political economy constitutes a third major class of ideas associated with underdevelopment and change in agriculture. Though not exactly comparable to the technology and demand themes treated earlier, political economy has had a major impact on how we look at conditions in different agricultural societies. Its core is that relations between different economic sectors, economies, and political systems are fundamental in determining local dynamics. Especially important here are unequal relations which allow one sector to extract surplus or some other benefit from another sector. This attention to relations between different economic sectors and classes focuses political economy analysis in a way that is opposed to the technology and demand themes. Whereas those themes lend themselves easily to the analysis of farm-level decisions and to farming systems as an aggregate of these decisions under common conditions, the political economy framework is more useful for larger-scale analysis, one that deemphasizes such factors as farmer rationality, demographic conditions, and the availability of new technology. The power of the political economy theme is its treatment of the disposition of savings and capital, which are, of course, fundamental to technological change. In focusing on unequal development, political economy is prone to ask why certain economic sectors have not changed relative to others and to emphasize economic and political relations as primary determinants. This theme has effectively brought our attention to the importance of the distribution of resources and to exploitation as important factors in shaping decisions and farming systems. Understanding of the process of change is achieved by examining such issues as the penetration of new social relations such as capitalism into the economic system, the accumulation of capital, differentiation among social classes, and the appropriation or extraction of surplus from one class or group.

A very broad base of literature and scholarship on the political economy of agriculture and its change has been made available since 1960 (e.g., Amin, 1976; DeJanvry, 1981; Frank, 1967; Godelier, 1977; Kahn & Llobera, 1981). DeJanvry (1981) and others have traced the origins of this scholarship to the debates on "the agrarian question" in Germany and Russia between 1861 and 1930 (Hussain & Tribe, 1981). This question concerned, in Kautsky's terms, "in what ways is capital taking hold of agriculture," but it can be broadly interpreted as an inquiry about the unequal development of the agricultural sector and especially the peasantry of central Europe. The debate was temporarily eclipsed by World Wars I and II and by the Russian Revolution and Bolshevik reforms, but it was picked up again in the 1960s and 1970s in reference to the peasantry in the Third World. The reemergence of this debate has many sources, important among which are political processes of reformism and revolution in Latin America, Africa, and Asia, and the maturation of a school combining Marxist analysis and social science, especially anthropology, in France and England in the late 1960s. Two major themes stand out in reading this literature with a mind toward farming systems: (1) dependency and domination, and (2) mode of production.

DEPENDENCY AND DOMINATION

In the reemergence of discussions of the agrarian question, perhaps the earliest writings of major influence were those of Frank (1967, 1969) on Latin America (also Foster-Carter, 1976; Leys, 1977; Palma, 1978). Frank laid out the major tenets of the dependency concept that underdevelopment is actually a product of capitalist penetration and domination of rural areas. Capitalist production dominates the world economy in which centers of capitalist wealth and power maintain spheres of influence over peripheries. The centers accumulate wealth by extracting it from the peripheries, creating underdevelopment in the latter. Spheres need not be spatially continuous, and they are transcontinental. Impoverishment occurs because of surplus extraction from marginal or peripheral areas to enrich urban and industrial cores. Through political and economic coercion, the metropolis dominates satellites and creates the conditions of underdevelopment and dependency. Development in satellites occurs so long as it serves the interests of the dominant metropolis, but this often leaves a legacy of political and economic institutions that become identified with underdevelopment. A prime example of this is the large estate system (*hacienda* or *latifundia*) in Latin America.

As noted by Forbes (1984:67), Wallerstein's world systems approach, which is commonly linked to dependency theory, involves a somewhat more detailed explanation. The outcome is similar, however—underdevelopment. This result is the product of the division of labor that accompanies a capitalist, trading economy. The industrialized core, through the division of labor, consolidates wealth at the expense of declining or backward-oriented conditions in the peripheries.

Agriculture presumably changes in agreement with the core–peripheries of these themes. For the peripheries, agriculture changes in at least two ways. The core dictates an extractive system of export crops through the development of plantations and/or depressed rewards to which only large-scale agriculture (landlord–lessee) can adapt. In turn, smallholder farmers cannot compete and eventually lose their ability to operate independently. Production can actually decrease, and the standard of living drops. A related response is for smallholder households to sell their labor at greatly undervalued rates to the nonagricultural core.

DeJanvry (1981:240) summarizes as follows the mechanisms of surplus extraction that play a role in dependent and unequal relations:

> Surplus is extracted via the product market (unfavorable prices due to lack of effective demand for wage foods, cheap-food price policies, monopolistic merchants, and poor infrastructure), the factor market (purchase of means of production at market prices that often include monopolistic margins and reflect industrial protectionism), the labor market (supply of semiproletarian labor at wages below the value of labor power), the land market (payment of rents in labor services, in kind, or in cash), the capital market (payment of interest often at usurious rates), and directly through the payment of taxes (tithes, sales taxes, and retentions). The consequence of this relation of domination is that peasants are incapable of capitalizing, not because they either do not want to . . . or do not generate any surplus . . . , but because the surplus they generate is siphoned out to the benefit of other social classes. . . .

It is important to note that this line of reasoning deals primarily with the lack of change among small farms in underdeveloped countries. It might be argued that other factors that otherwise generate change, such as new technology or demographic change, cannot operate to produce change because of surplus extraction. Dependency theory focuses on exchange relations, or circulation, and it tends to view peasants as passive reactors, concerning itself little with analyzing change that does not involve capitalist penetration. Capitalism as a primary agent of change is a dominant theme throughout the dependency literature, and this is equally apparent in the discussions of modes of production and the articulation between them.

MODE OF PRODUCTION

Marx's primary interest in capitalist and industrial development has proved to be a mixed blessing for political economists interested in noncapitalist and agrarian societies. A relatively loose and somewhat ambiguous theoretical framework has produced many contradictory interpretations of the nature and direction of noncapitalist, agrarian societies, but it has also provided a powerful mechanism for comparing different regions. Going well beyond dependency theory, Marxist analysis has been used effectively to analyze the dynamics of noncapitalist economies as well as their relation with capitalist ones. The *mode of production* concept has been especially useful in this regard. This concept has been elaborated out of Marx's attempt to demonstrate that bourgeois capitalism was not "the perennial core of man's economic society operating at last in accordance with its eternal natural laws" (Lubasz, 1984:457). Marx envisioned a set of classical modes of production—Asiatic, ancient, feudal, and modern bourgeois (capitalist)—that succeeded each other through the dialectical process of resolving contradictions. These are defined by two general types of components: the *forces of production* and the *relations of production*. The former term refers to the actual means of production—nonmechanized versus mechanized agriculture, artisan versus industrial production. The latter refers to the relations between labor and capital and to the manner in which surplus is appropriated. Relations of production have been the dominant focus of the large literature on modes of production.

Although there has been some interest in mode of production, particularly the Asiatic mode of production, as a concept for historical analysis (e.g., Bailey & Llobera, 1981), its most influential use has been on contemporary societies. Roseberry (n.d.), citing Marinez (1981), notes that in Latin America alone, different scholars have described 25 different modes of production. Despite this proliferation, however, the predominant interest is in how to define peasants. An extensive literature exists on this issue (e.g., Ennew, Hirst, & Tribe, 1977; Friedmann, 1980; Kahn & Llobera, 1981). DeJanvry (1981) rejects the notion of a "peasant mode of production," although "simple commodity production" is widely used as a synonym for peasant production (e.g., Friedmann, 1980). Key questions concern differentiation within rural populations and the importance and impact of intercourse with capitalist economies and markets. Issues that are raised include the accumulation of capital among small farmers, the relations between consumption and production in the household, how production is commoditized upon contact and intercourse with capitalist and market economies, and how labor from one sector is valued and used by another.

Throughout this literature, the idea of articulation is crucially important. Simplistically, this articulation occurs where capitalist modes become dominant over precapitalist modes. The former preserves the latter as a source of cheap labor in order to extract surplus (e.g., Laclan, 1971). The results are either underdevelopment or uneven development (Taylor, 1979). In either case, the results are similar to those described earlier for world systems themes (e.g., Watts, 1983; for a review of critiques, see Forbes, 1984:97–100). Whether or not peasants are to be viewed as a unique social category or mode of production, or merely as a class within a larger economic system, political economists generally agree on the significance of capitalism as a major influence on the lives and livelihoods of peasants. Results of this articulation and exploitation include the commoditization of production, differentiation, and export of cheap rural labor and other products to the urban, industrial sector. Increased impoverishment, stagnation, and inequality in the countryside also result.

Essentially, the political economy themes examine the socioeconomic structures in which farming systems exist and farmers must operate. None of them is particularly interested in decision making per se, but focus on how and why structures set the parameters for some type of behavior that is assumed operative for the particular economy in question. Therefore, the questions asked differ from those inherent in the demand-based themes. Moreover, each of these themes involves historical assessments of the relationships among political economies, asserting the negative or uneven outcomes inherent in the conflict among economies.

None of the various themes of agricultural change is intended to explain every facet of a farming system. This fact is obvious in that any system is the product of an array of material variables (tools, cultivars, labor, capital, soil, climate), structural variables (national and international economy, local institutions, social organizations, and responsibilities), and individual behavior, (goals and allocation choices). These elements are in constant states of flux, and farming systems change with them. So-called complete explanations of these changes involve complex descriptions and historical analysis, including analysis of the three elements (material, structure, and behavior). To date, such considerations have been too large for those interested in deriving general explanations of change. As a result, some have focused on structure at the expense of materials variables and individual behavior, and others have done the reverse. Actually, both approaches should complement each other.

References

Adams, J. Review of the conditions of agricultural growth. *Annals of the American Academy of Political Social Sciences*, 1966, 367, 224–225.

Allan, W. *The African husbandman*. London: Oliver and Boyd, 1965.

Altieri, M. A., Letourneau, D. K., & Davis, J. R. Developing sustainable agroecosystems. *Bioscience*, 1983, 33, 45–49.
Amin, S. *Unequal development*. Hassocks: Harvester Press, 1976.
Ashby, J. A. Technology and ecology. Implications for innovation research in peasant agriculture. *Rural Sociology*, 1982, 47, 234–250.
Bailey, A. M., & Llobera, J. R. *The Asiatic mode of production: Science and politics*. London: Routledge and Kegan Paul, 1981.
Barlett, P. F. Labor efficiency and the mechanism of agricultural evolution. *Journal of Anthropological Research*, 1976, 32, 124–140.
Batie, S. S., & Healy, R. (Eds.). *The future of American agriculture as a strategic resource*. Washington, DC: The Conservative Foundation, 1980.
Bauer, P. T., & Yamey, B. S. A case study of responses to price in an underdeveloped country. *Economic Journal*, 1959, 69, 300–305.
Bayliss-Smith, T. P. *The ecology of agricultural systems*. Cambridge: Cambridge University Press, 1982.
Bayliss-Smith, T. P., & Wanmali, S. (Eds.). *Understanding green revolutions: Agrarian change and development planning in South Asia*. Cambridge: Cambridge University Press, 1984.
Behrman, J. R. The price elasticity of the market surplus of a subsistence crop. *Journal of Farm Economics*, 1966, 48, 875–893.
Bennett, J. W. *The ecological transitions: Cultural anthropology and human adaptation*. New York: Pergamon Press, 1976.
Bergman, R. W. *Amazon economics: The simplicity of Shipibo Indian wealth*. Dellplain Latin American Studies (Vol. 6). Syracuse, NY: Syracuse University, Department of Geography, 1980.
Bergmann, G. *Philosophy of science*. Madison: University of Wisconsin Press, 1966.
Bhatia, D. K. Some reflections on the anti-Malthusian thesis of Mrs. Boserup. *Indian Journal of Economics*, 1968, 48, 427–435.
Blaut, J. M. Two views of diffusion. *Annals of the Association of American Geographers*, 1977, 67, 343–349.
Blitz, R. C. Review of *The conditions of agricultural growth*, by E. Boserup. *Journal of Political Economy*, 1967, 75, 212–213.
Boserup, E. *The conditions of agricultural growth*. Chicago: Aldine, 1965.
Boserup, E. *Populations and technological change: A study of long-term trends*. Chicago: University of Chicago Press, 1981.
Bowers, J. K., & Cheshire, P. *Agriculture, the countryside and land use: An economic critique*. New York: Methuen, 1983.
Brookfield, H. C. Local study and comparative method: An example from central New Guinea. *Annals of the Association of American Geographers*, 1962, 52, 242–254.
Brookfield, H. C. Intensification and disintensification in Pacific agriculture: A theoretical approach. *Pacific Viewpoint*, 1972, 13, 30–48.
Brookfield, H. C. *Interdependent development*. London: Methuen, 1975.
Brookfield, H. C. Third world development. *Progress in Human Geography*, 1978, 2, 121–132.
Brookfield, H. C. Intensification revisited. *Pacific Viewpoint*, 1984, 25, 15–44.
Browlett, J. Development, the diffusionist paradigm and geography. *Progress in Human Geography*, 1980, 4, 57–79.
Brown, P., & Podolefsky, A. Population density, agricultural intensity, land tenure, and group size in the New Guinea highlands. *Ethnology*, 1976, 15, 211–238.
Brush, S. B. The concept of carrying capacity for systems of shifting cultivation. *American Anthropologist*, 1975, 77, 799–811.
Brush, S. B. *Mountain, field, and family: The economy and human ecology of an Andean valley*. Philadelphia: University of Pennsylvania Press, 1977.
Buckley, W. *Modern systems research for the behavioral sciences*. Chicago: Aldine, 1968.
Buttel, F. H., & Larson, D. W., III. Farm size, structure, and energy intensity: An ecological analysis of U.S. agriculture. *Rural Sociology*, 1979, 44, 471–488.

Byerlee, D., Harrington, L., & Winkelmann, D. L. Farming systems research: Issues in research strategy and technology design. *American Journal of Agricultural Economics*, 1982, 64, 897–904.

Carlstein, T. *Time, resources, society and ecology: On the capacity for human interaction in space and time*. London: George Allen and Unwin, 1982.

Chayanov, A. V. Peasant farm organization. In D. Thorner, B. Kerblay, & R. E. F. Smith (Eds.), *A. V. Chayanov in the theory of peasant economy*. Homewood, IL: R. D. Irwin, 1966.

Chibnik, M. 1984. A cross-cultural examination of Chayanov's theory. *Current Anthropology*, 1984, 75, 335–340.

Clark, C., & Haswell, M. *The economics of subsistence agriculture*. London: MacMillan, 1964.

Collier, W. Agricultural evolution in Java. In G. Hansen (Ed.), *Agricultural and rural development in Indonesia*. Boulder: Westview Press, 1981.

Collinson, M. *Farm management in peasant agriculture: A handbook for rural development in Africa*. New York: Praeger, 1972.

Conklin, H. C. *Hanunoó agriculture*. FAO Forestry Development Paper No. 12. Rome: Food and Agricultural Organization, 1957.

Cowgill, G. On causes and consequences of ancient and modern population change. *American Anthropologist*, 1975, 77, 505–525.

Cox, G. W., & Atkins, M. D. *Agricultural ecology: An analysis of world food production systems*. San Francisco: W. H. Freeman, 1979.

Dalrymple, D. G. *Survey of multiple cropping in less developed nations*. Washington, DC: U.S. Department of Agriculture (USDA) and U.S. Agency for International Development (USAID), 1971.

Dalton, G. Economic theory and primitive society. *American Anthropologist*, 1961, 63, 1–25.

Dalton, G. Economic surplus once again. *American Anthropologist*, 1963, 65, 389–394.

Datoo, B. A. Relationship between population density and agricultural systems in the Uluguru Mountains, Tanzania. *Journal of Tropical Geography*, 1976, 42, 1–12.

Datoo, B. A. Toward a reformation of Boserup's theory of agricultural change. *Economic Geography*, 1978, 54, 135–144.

Day, R. H., & Tinney, E. H. A dynamic von Thünen model. *Geographical Analysis*, 1969, 1, 137–151.

Deere, C. D., & deJanvry, A. A conceptual framework for the empirical analysis of peasants. *American Journal of Agricultural Economics*, 1979, 61, 602–611.

deJanvry, A. *The agrarian question and reformism in Latin America*. Baltimore: The Johns Hopkins University Press, 1981.

Denevan, W. M. Adaptation, variation, and cultural geography. *Professional Geographer*, 1983, 35, 399–407.

Denevan, W. M., & Padoch, C. (Eds.), *Swidden-fallow agroforestry in the Peruvian Amazon*. Advances in Economic Botany. New York: New York Botanical Garden, in press.

de Vries, J. Labor/leisure trade-off: A review of Ester Boserup, *Conditions of agricultural growth*. *Peasant Studies Newsletter*, 1972, 1, 45–50, 62.

Dillon, J. L., & Anderson, J. R. Allocative efficiency, traditional agriculture, and risk. *American Journal of Agricultural Economics*, 1971, 53, 26–32.

Dillon, J. L., & Heady, E. O. Theories of choice in relation to farmer decisions. *Iowa Agricultural and Home Economics Experimental Station Research Bulletin*, Ames, 1960, 485.

Doolittle, W. E. Agricultural change as an incremental process. *Annals of the Association of American Geographers*, 1984, 7, 124–137.

Dovring, F. Review of *The conditions of agricultural growth* by E. Boserup. *Journal of Economic History*, 1966, 26, 380–381.

Duckham, A. N., & Masefield, G. B. *Farming systems of the world*. London: Chatto and Windus, 1970.

Dumont, R. *Types of rural economy: Studies in world agriculture*. London: Methuen, 1957.

Dunlap, R. E., & Martin, K. E. Bringing environment into the study of agriculture: Observations and suggestions regarding the sociology of agriculture. *Rural Sociology*, 1983, 48, 201–218.

Durrenberger, E. P. Rice production in a Lisu village. *Journal of Southeast Asian Studies*, 1979, 10, 139–145.

Durrenberger, E. P. Chayanov's economic analysis in anthropology. *Journal of Anthropological Research*, 1980, 36, 133–148.

Durrenberger, E. P. (Ed.). *Chayanov, peasants, and economic anthropology*. New York: Academic Press, 1984.

Edens, T. C., & Haynes, D. L. Closed system agriculture: Resource constraints, management options, and design alternatives. *Annual Review of Phytopathology*, 1982, 20, 363–395.

Ellen, R. *Environment, subsistence and system: The ecology of small-scale social formations*. Cambridge: Cambridge University Press, 1982.

Ennew, J., Hirst, P., & Tribe, K. "Peasantry" as an economic category. *Journal of Peasant Studies*, 1977, 4, 295–322.

Ewald, V. The von Thünen principle and agricultural functions in colonial Mexico. *Journal of Historical Geography*, 1977, 3, 123–134.

Falcon, W. P. The green revolution: Second generation problems. *American Journal of Agricultural Economics*, 1970, 52, 698–710.

Feder, E. *The rape of the peasantry: Latin America's landholding system*. Garden City, NY: Anchor Books, 1971.

Feder, G., Just, R. E., & Zilberman, D. Adoption of agricultural innovations in development countries: A survey. World Bank Staff Working Papers No. 542. Washington, D.C.: World Bank, 1982.

Firth, R. (Ed.). *Themes in economic anthropology*. Association of Social Anthropologists, Monograph No. 6. London: Tavistock, 1967.

Fisk, E. H. Planning in a primitive economy: From pure subsistence to the production of a market surplus. *Economic Record*, 1964, 40, 156–174.

Fisk, E. H., & Shand, R. T. The early stages of development of a primitive economy: The evolution from subsistence to trade and specialization. In C. R. Wharton, Jr., (Ed.), *Subsistence agriculture and economic development*. Chicago: Aldine, 1969.

Forbes, D. K. *The geography of underdevelopment: A critical survey*. Baltimore: The Johns Hopkins University Press, 1984.

Foster-Carter, A. From Rostow to Gunder Frank: Conflicting paradigms in the analysis of underdevelopment. *World Development*, 1976, 4, 167–180.

Frank, A. G. *Capitalism and underdevelopment in Latin America*. London: Pelican, 1967.

Frank, A. G. *Latin America: Underdevelopment or evolution*. New York: Monthly Review Press, 1969.

Frank, A. G. *Dependent accumulation and underdevelopment*. London: Macmillan, 1978.

Friedmann, H. Household production and the national economy: Concepts for the analysis of agrarian formations. *Journal of Peasant Studies*, 1980, 7, 158–184.

Gartrell, J. W., & Gartrell, C. D. Status, knowledge, and innovation. *Rural Sociology*, 1979, 44, 73–94.

Geertz, C. *Agricultural involution: The processes of ecological change in Indonesia*. Berkeley: University of California Press, 1963.

Ghatak, S., & Ingersent, K. *Agriculture and economic development*. Baltimore: The Johns Hopkins University Press, 1984.

Gilles, J. Farm size, farm structure, energy, and climate: An alternative ecological analysis of United States agriculture. *Rural Sociology*, 1980, 45, 332–339.

Gleave, M. B., & White, H. P. Population density and agricultural systems in West Africa. In M. F. Thomas & G. W. Whittington (Eds.), *Environment and land use*. London: Methuen, 1969.

Godelier, M. *Perspectives in Marxist anthropology*. Cambridge: Cambridge University Press, 1977.

Goldschmidt, W. R. *As you sow*. New York: Harcourt, Brace, 1947.

Grigg, D. B. *The agricultural systems of the world*. Cambridge: Cambridge University Press, 1974.

Grigg, D. B. Ester Boserup's theory of agrarian change: A critical review. *Progress in Human Geography*, 1979, 3, 64–82.

Grigg, D. B. *Population growth and agrarian change: An historical perspective*. Cambridge: Cambridge University Press, 1980.

Grigg, D. B. *The dynamics of agricultural change: The historical experience*. New York: St. Martin's Press, 1982.

Griliches, Z. Hybrid corn: An exploration in the economics of technological change. *Econometrica*, 1957, 25, 501–502.

Gross, D. R., Eiten, G., Flowers, N. M., Levi, F. M., Ritter, L., & Werner, D. W. Ecology and acculturation among native peoples of central Brazil. *Science*, 1979, 206, 1043–1050.

Grossman, L. S. Man–environment relationships in anthropology and geography. *Annals of the Association of American Geographers*, 1977, 67, 126–144.

Grossman, L. S. *Peasants, subsistence, ecology, and development in the highlands of Papua New Guinea*. Princeton: Princeton University Press, 1985.

Gudeman, S. *The demise of a rural economy: From subsistence to capitalism in a Latin American village*. London: Routledge and Kegan Paul, 1978.

Hall, P. G. (Ed.). *Von Thünen's isolated state*. Oxford: Pergamon Press, 1966.

Hardesty, D. L. *Ecological anthropology*. New York: Wiley, 1977.

Hardin, G. Living on a lifeboat. *Bioscience*, 1974, 24, 561–568.

Harris, M. *The rise of anthropological theory: A history of theories of culture*. New York: Thomas Y. Crowell, 1968.

Harrison, M. Chayanov and the economics of Russian peasantry. *Journal of Peasant Studies*, 1975, 2, 389–417.

Harrison, M. The peasant mode of production in the work of A. V. Chayanov. *Journal of Peasant Studies*, 1977, 4, 323–336.

Harwood, R. R. *Small farm development: Understanding and improving farming systems in the humid tropics*. Boulder: Westview Press, 1979.

Haswell, M. R. Economics of agriculture in a savannah village. London: The Colonial Office, Her Majesty's Stationery Office, 1953.

Hayami, Y., & Ruttan, V. W. *Agricultural development: An international perspective*. Baltimore: The Johns Hopkins University Press, 1971.

Heaton, T. B., & Brown, D. L. Farm structure and energy intensity: Another look. *Rural Sociology*, 1982, 47, 17–31.

Hopper, W. D. Allocative efficiency in a traditional Indian agriculture. *Journal of Farm Economics*, 1965, 47, 611–624.

Huang, Y. Risk, entrepreneurship, and tenancy. *Journal of Political Economy*, 1973, 81, 1241–1244.

Huang, Y. The behavior of indigenous and non-indigenous farmers: A case study. *Journal of Development Studies*, 1974, 10, 175–187.

Hunt, D. Chayanov's model of peasant household resource allocation. *Journal of Peasant Studies*, 1979, 6, 247–285.

Hussain, A., & Tribe, K. *Marxism and the agrarian question* (2 vol.). London: MacMillan, 1981.

Jackson, R. H. *Land use in America*. New York: V. H. Winston and Sons/John Wiley and Sons, 1981.

Jones, W. O. Reviews of *The conditions of agricultural growth* by E. Boserup. *American Economic Review*, 1967, 57, 679–680.

Kahn, J., & Llobera, J. *The anthropology of pre-capitalist societies*. London: MacMillan, 1981.

Kao, C., Anschel, K., & Eicher, C. Disguised unemployment in agriculture: A survey. In C. Eicher & L. Witt (Eds.), *Agriculture in economic development*. New York: McGraw-Hill, 1964.

Kroeber, A. *Cultural and natural areas of native North America*. Berkeley: University of California Press, 1939.

Laclan, E. Feudalism and capitalism in Latin America. *New Left Review*, 1971, 67, 19–38.

Lagemann, J. *Traditional African farming systems in eastern Nigeria*. Munich: Weltforum-Verlag, 1977.
Larson, O. W., III, & Buttel, F. H. Farm size, farm structure, climate, and energy: A reconsideration. *Rural Sociology*, 1980, 45, 340–348.
Lea, D. Some non-nutritive functions of food in New Guinea. In F. Gale & G. H. Lawton (Eds.), *Settlement and encounter*. Melbourne: Oxford University Press, 1969.
Leach, G. *Energy and food production*. London: International Institute for Environment and Development, 1975.
Levi, J. F. S. Population pressure and agricultural change in the land-intensive economy. *Journal of Development Studies*, 1976, 13, 61–78.
Levi, J., & Havinden, M. *Economics of African agriculture*. Harlow, UK: Longman, 1982.
Lewis, J. V. D. Domestic labor intensity and the incorporation of Malayan peasant farmers into localized descent groups. *American Ethnologist*, 1981, 8, 53–73.
Leys, C. Underdevelopment and dependency: Critical notes. *Journal of Contemporary Asia*, 1977, 7, 92–107.
Lipton, M. The theory of the optimizing peasant. *Journal of Development Studies*, 1968, 4, 327–351.
Littlejohn, G. Peasant economy and society. In B. Hindess (Ed.), *Sociological theories of the economy*. London: MacMillan, 1977.
Long, N. *Family and work in rural societies: Perspectives on non-wage labour*. New York: Methuen, 1986.
Lubasz, H. Marx's concept of the Asiatic mode of production. *Economy and society*, 1984, 13, 456–483.
Mabogunje, A. L. *The development process: A spatial perspective*. New York: Holmes and Meier, 1981.
Malthus, T. R. *Essay on the principle of population as it affects the further improvement of society*. London: Royal Economic Society, 1798.
Marinez, P. Acerca de los modos de producción precapitalistos en America Latina. *Estudios sociales Centroamericanos*, 1981, 10, 121–140.
Martin, P. L. Labor-intensive agriculture. *Scientific American*, 1983, 249, 54–59.
Martin, P. L., & Olmstead, A. L. The agricultural mechanization controversy. *Science*, 1985, 227, 601–606.
Mellor, John W. *The economics of agricultural development*. Ithaca: Cornell University Press, 1966.
Metzner, J. K. *Agriculture and population pressure in Sikka, Isle de Flores*. Development Studies Center Monograph No. 28. Canberra: Australian National University Press, 1982.
Mikesell, M. W. Geographical perspectives in anthropology. *Annals of the Association of American Geographers*, 1967, 57, 617–634.
Minge-Kalman, W. On the theory and measurement of domestic labor intensity. *American Ethnologist*, 1977, 4, 273–284.
Murdoch, W. W. *The poverty of nations: The political economy of hunger and population*. Baltimore: The Johns Hopkins University Press, 1980.
Nakajima, C. Subsistence and commercial farms: Some theoretical models of subjective equilibrium. In C. R. Wharton, Jr. (Ed.), *Subsistence agriculture and economic development*. Chicago: Aldine, 1969.
Netting, R. McC. *Hill farmers in Nigeria: Cultural ecology of the Kofyar of the Jos Plateau*. Seattle: University of Washington Press, 1968.
Newbury, R. A. *Geography of agriculture*. Plymouth, England: MacDonald and Evans, 1980.
Nietschmann, B. Q. *Between land and water: The subsistence ecology of the Miskito Indians, eastern Nicaragua*. New York: Seminar Press, 1973.
Norman, D. W. An economic study of three villages in Zaria province. Part 1: Land and labour relationships. *Samaru Miscellaneous Paper 19*. Institute for Agricultural Research, Samaru, Ahnadu Bello University, 1974.
Norman, D. W. Economic rationality of traditional Hausa dryland farmers in the north of Niger-

ia. In R. D. Stevens (Ed.), *Tradition and dynamics in small farm agricultural economic studies in Asia, Africa, and Latin America*. Ames: Iowa State University Press, 1977.

Norman, D. W. The farming systems approach: Relevancy for the small farmer. MSU Rural Development Paper No. 5. East Lansing: Michigan State University, 1980.

Norman, M. J. T. *Annual cropping systems in the tropics: An introduction*. Gainesville: University of Florida Press, 1979.

Norton, W. *Historical analysis in geography*. New York: Longman, 1984.

Norton, W. & Conkling, E. C. Land-use theory and the pioneering economy. *Geografiska Annaler*, 1974, 568, 44–56.

Odum, H. T. *Environment, power, and society*. New York: Wiley, 1971.

Orlove, B. S. *Alpacas, sheep and men: The wool export economy and regional society in southern Peru*. New York: Academic Press, 1977.

Orlove, B. S. Ecological anthropology. *Annual Review of Anthropology*, 1980, 9, 235–273.

Ortiz, S. R. *Uncertainties in peasant farming: A Colombian case*. London School of Economics Monographs in Social Anthropology No. 46. London: University of London, Athlone Press, 1973.

Palma, G. Dependency: A formal theory of underdevelopment or a methodology for the analysis of concrete situations of underdevelopment? *World Development*, 1978, 6, 881–924.

Patnaik, U. Neo-populism and Marxism: The Chayanovian view of the agrarian question and its fundamental fallacy. *Journal of Peasant Studies*, 1979, 6, 375–420.

Peet, J. R. The spatial expansion of commercial agriculture in the nineteenth century: A von Thünen interpretation. *Economic Geography*, 1969, 45, 283–301.

Peet, J. R. Von Thünen theory and the dynamics of agricultural expansion. *Explorations in Economic History*, 1970, 8, 181–201.

Phillips, R. E., Blevins, R. L., Thomas, G. W., Frye, W. W., & Phillips, S. H. No-tillage agriculture. *Science*, 1980, 208, 1108.

Pimentel, D., Fast, S., & Berardi, G. Energy efficiency of farming systems: Agriculture. *Ecosystems and environment*, 1983, 9, 359–372.

Pimentel, D., & Hall, C. W. (Eds.). *Food and energy resources*. New York: Academic Press, 1984.

Pimentel, D., & Pimentel, M. *Food, energy and society*. London: Arnold, 1979.

Rambo, A. T., & Sajise, S. E. (Eds.). *An introduction to human ecology research on agricultural systems*. Los Baños, Philippines: University of Philippines of Los Baños, 1984.

Rappaport, R. A. *Pigs for the ancestors: Ritual in the ecology of a New Guinea people*. New Haven: Yale University Press, 1967.

Robinson, W., & Schutjer, W. Agricultural development and demographic change: A generalization of the Boserup model. *Economic Development and Cultural Change*, 1984, 32, 355–366.

Rogers, E. *Diffusion of innovations*. New York: Free Press, 1962.

Roseberry, W. Anthropology, history, and modes of production. In B. Orlove, K. Yambert, & T. Love (Eds.), *Dependency and its successors: An anthropological view of political economy in Latin America*. Manuscript.

Rostow, W. W. *The stages of economic growth: A non-communist manifesto*. Cambridge: Cambridge University Press, 1960.

Roumasset, J. *Rice and risk*. Amsterdam: North-Holland Press, 1976.

Rudner, R. S. *Philosophy of social science*. Englewood Cliffs, NJ: Prentice-Hall, 1966.

Ruthenberg, H. *Farming systems in the tropics*. Oxford: Clarendon Press, 1976.

Ruttan, V. W. Induced innovations and agricultural development. In G. K. Douglas (Ed.), *Agricultural sustainability in changing world order*. Boulder: Westview Press, 1984.

Sahlins, M. D. The intensity of domestic production in primitive societies: Social inflections of the Chayanovian slope. In G. Dalton (Ed.), *Studies in economic anthropology*. Washington, DC: American Anthropological Society, 1971.

Sahlins, M. D. *Stone age economics*. Chicago: Aldine-Atherton, 1972.

Sands, D. M. Farming systems research: Clarification of terms and concepts. *Explanation in Agriculture*, 1986, 22, 87–104.

Sauer, C. O. *The morphology of landscape*. University of California Publications in Geography, 1925, 2, 19–54.
Schlebecker, J. T. The world metropolis and the history of American agriculture. *Journal of Economic History*, 1960, 20, 187–208.
Schluter, M. G. G., & Mount, T. D. *Management objectives of the peasant farmer: An analysis of risk aversion in the choice of cropping pattern, Sura District, India*. Occasional Paper No. 78. Cornell University, Department of Agricultural Economics, 1974.
Schluter, M. G. G., & Mount, T. D. Some management objectives of the peasant farmer: An analysis of risk aversion in the choice of cropping patterns. *Journal of Development Studies*, 1976, 12, 246–267.
Schneider, H. K. *Economic man: The anthropology of economics*. New York: Free Press, 1974.
Schultz, T. W. *Transforming traditional agriculture*. New Haven: Yale University Press, 1964.
Scott, J. D. *The moral economy of the peasant: Rebellion and subsistence in Southeast Asia*. New Haven: Yale University Press, 1976.
Shaner, W. W., Philipps, P. F. & Schmehl, W. R. *Farming systems research and development: A guideline for developing countries*. Boulder, CO: Westview Press, 1982.
Sheffer, C. Review of Boserup: *The conditions of agricultural growth*. *American Antiquity*, 1971, 36, 377–389.
Simmonds, N. W. *The state of the art in farming systems research*. Technical Paper No. 43. Washington, DC: World Bank, 1985.
Sinclair, R. J. Von Thünen and urban sprawl. *Annals of the Association of American Geographers*, 1967, 57, 72–87.
Smith, A. E. Chayanov, Sahlins, and the labor-consumer balance. *Journal of Anthropological Research*, 1979, 35, 477–480.
Smith, H. Land reclamation in the former Zuyder Zee in the Netherlands. *Geoforum*, 1970, 4, 37–44.
Southworth, H. M., & Johnston, B. F. (Eds.). *Agricultural development and economic growth*. Ithaca: Cornell University Press, 1967.
Spencer, J. E. *Shifting cultivation in Southeast Asia*. University of California Publications in Geography No. 10. Berkeley: University of California, 1966.
Spooner, B. (Ed.). *Population growth: Anthropological implications*. Cambridge, MA: MIT Press, 1972.
Steward, J. H. *Theory of culture change: The methodology of multilinear evolution*. Urbana: The University of Illinois Press, 1955.
Stier, F. Domestic economy: Land, labor, and wealth in a San Blas community. *American Ethnologist*, 1982, 9, 519–537.
Street, J. An evaluation of the concept of carrying capacity. *The Professional Geographer*, 1969, 21, 104–107.
Szymanski, R., & Agnew, J. A. *Order and skepticism: Human geography and the dialectic of science*. Washington, DC: Association of American Geographers, 1981.
Tannenbaum, N. Chayanov and economic anthropology. In E. P. Durrenberger, (Ed.), *Chayanov, peasants, and economic anthropology*. New York: Academic Press, 1984.
Tarrant, J. R. *Agricultural geography*. Newton Abbot, England: David and Charles, 1974.
Tax, S., *Penny capitalism*. Smithsonian Inst., Institute of Social Anthropology. Publication No. 16. Washington, DC: U.S. Government Printing Office, 1953.
Taylor, J. *From modernization to modes of production*. London: MacMillan, 1979.
Turner, B. L., II. Micro-sale modeling of behavior in natural resource use systems: A trial formulation. In E. Berry & B. Thomas (Eds.), *Natural Resource Management Workshop: Collected papers*. International Development Program, Clark University and Institute for Development Anthropology, Binghamton, NY, 1983.
Turner, B. L., II, & Doolittle, W. E. The concept and measure of agricultural intensity. *The Professional Geographer*, 1978, 30, 297–301.
Turner, B. L., II, Hanham, R. Q., & Portararo, A. V. Population pressure and agricultural intensity. *Annals of the Association of American Geographers*, 1977, 67, 384–396.

Vayda, A. P., & McCay, B. New directions in ecology and ecological anthropology. *Annual Review of Anthropology*, 1975, 4, 293–306.

Vayda, A. P., & Rappaport, R. A. Ecology, cultural and noncultural. In J. A. Clifton (Ed.), *Introduction to cultural anthropology*. Boston: Houghton Mifflin, 1969.

Vermeer, D. E. Population pressure and crop rotational changes among the Tiv of Nigeria. *Annals of the Association of American Geographers*, 1970, 60, 199–314.

Von Bertalanffy, L. General systems theory: A critical review. *General Systems*, 1962, 7, 1–20.

Wallerstein, I. *The modern world-system: Capitalist agriculture and the origins of the European world-economy in the sixteenth century*. New York: Academic Press, 1974.

Wallerstein, I. *The capitalist world-economy*. Cambridge: Cambridge University Press, 1980.

Watts, M. *Silent violence: Food famine and peasantry in northern Nigeria*. Berkeley: University of California Press, 1983.

Wellisz, S. Dual economies, disguised unemployment and the unlimited supply of labor. *Economica*, 1968, n.s., 35, 22–51.

Wharton, C. R., Jr. (Ed.). *Subsistence agriculture and economic development*. Chicago: Aldine, 1969.

White, B. "Agricultural involution" and its critics: Twenty years after. *Bulletin of Concerned Asian Scholars*, 1983, 15, 18–31.

Wilken, G. C. Some aspects of resource management by traditional farmers. In H. H. Biggs & R. L. Tinnermeier (Eds.), *Small farm agricultural development*. Fort Collins: Colorado State University, 1974.

Wilken, G. C. Management of productive space in traditional farming. *Actos du XLII[e] Congres International des Americanistes*, 1976, 2, 407–419.

Wissler, C. *The relation of nature to man in aboriginal America*. New York: Oxford University Press, 1926.

Wolf, E. R. *Peasants*. Englewood Cliffs, NJ: Prentice-Hall, 1966.

Wolf, E. *Europe and the people without history*. Berkeley: University of California Press, 1982.

Wolgin, J. M. Resource allocation and risk: A case study of smallholder agriculture in Kenya. *American Journal of Agricultural Economics*, 1975, 57, 622–630.

Yotopoulos, P. A. On the efficiency of resource utilization in subsistence agriculture. *Food Research Institute Studies*, 1968, 8, 125–135.

PART II

PALEOTECHNIC AND CONSUMPTION-ORIENTED SYSTEMS

Introduction to Part II

The four farming systems represented in this section have emerged, or are in the process of so doing, from so-called traditional ones characterized by consumption (subsistence) production and a paleotechnic (non-fossil-fuel) base. Each system remains largely embedded in a swidden or slash-and-burn mode in which total inputs and outputs are low. Not surprisingly, each system is associated with relatively low population densities and, until recently, with the margins or peripheries of the spatial domain of market economies. Each system has been penetrated by the market to different degrees and for different lengths of time.

The attributes of these systems display tendencies toward the left side of our trajectory (Figure II-1). Output intensities are low, and vary according to cultivars and environment. Technologies vary primarily by the degree of commodity production involved, as has been noted in detail for agriculture in tropical Africa (Pingali, Bigot, & Binswanger, 1986). Even within a single system, two technologies may exist—one for subsistence production and one for market production.

The first case study is a synthesis of swidden farming systems as practiced in the wet tropical lowlands of South America, but focusing on the Barí of the Lake Maracaibo Basin of Venezuela and Colombia. These systems are ancient, and many have remained relatively unchanged for centuries, practiced by small groups in rather isolated and extremely land-extensive circumstances. The Barí, for example, have population densities below 1 person/km^2. Modern agricultural technology is limited primarily to steel cutting tools. The Barí cultivate predominantly for themselves, although some marginal market activity has begun.

Among the Barí, cropping frequencies (ratio of cultivation to fallow) range from 1:5 to 1:7 based on a cycle of 10 years of cultivation and 50–70 years of fallow. The years of continuous (annual) cultivation are lengthy compared to swidden cultivators in general, demonstrating the capacity of manioc (also cassava or *Manihot* spp.) to produce acceptable yields with minimal land improvements (inputs). The Barí typically obtain 18 tons per hectare per year (t/ha/yr) of this root crop. Adjusting for the 6- to 8-year agricultural cycle, the output intensity by raw field weight ranges from 2,250 to 3,000 kilograms per hectare (kg/ha). As with most root crops, however, manioc tubers contain more water than edible dry matter, in this case 60% or more water (Norman, Pearson, & Searle, 1984:229). Taking this into account, output intensity drops to about 900–1,200 kg/ha for the Barí.

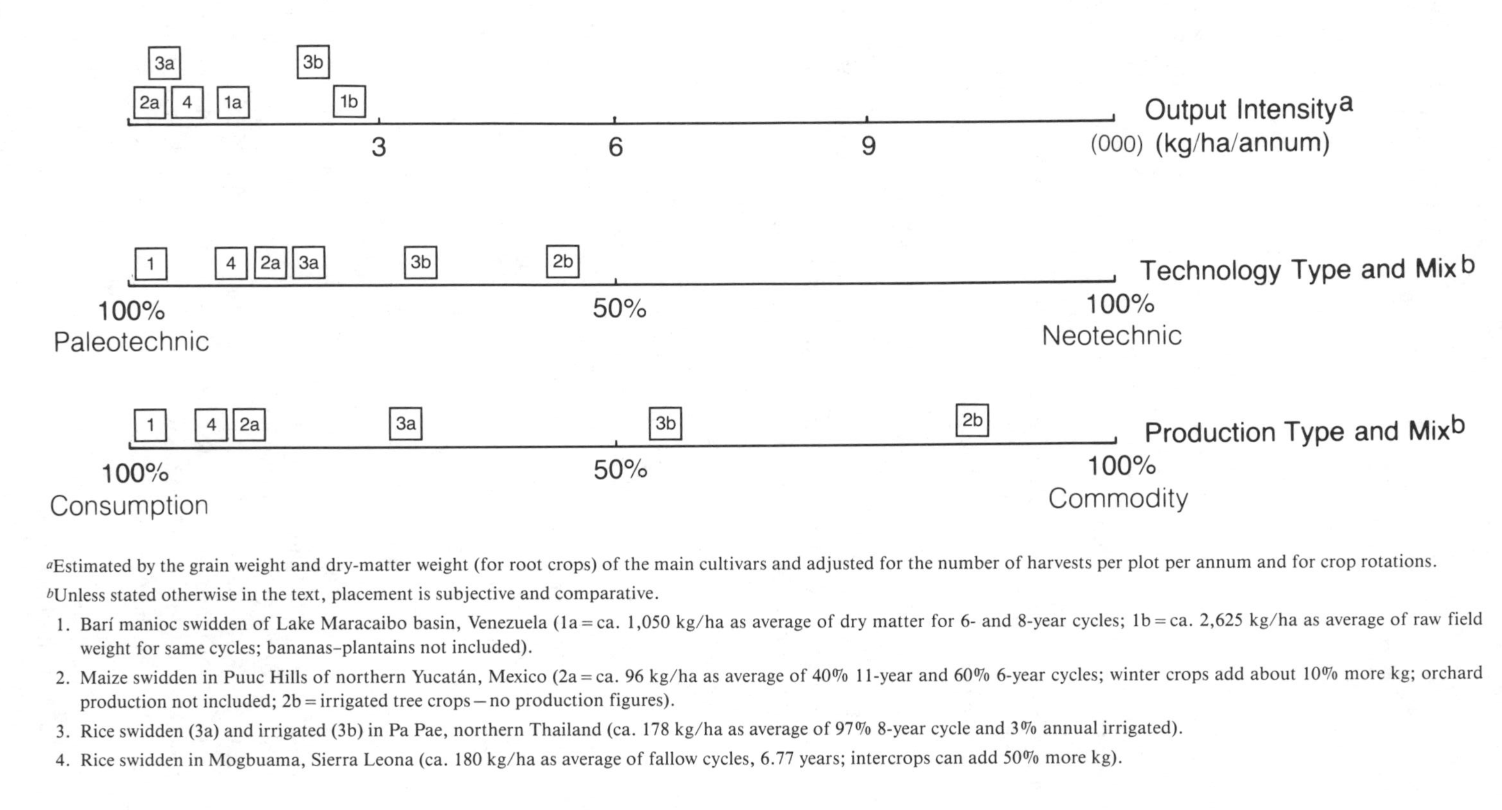

[a]Estimated by the grain weight and dry-matter weight (for root crops) of the main cultivars and adjusted for the number of harvests per plot per annum and for crop rotations.

[b]Unless stated otherwise in the text, placement is subjective and comparative.

1. Barí manioc swidden of Lake Maracaibo basin, Venezuela (1a = ca. 1,050 kg/ha as average of dry matter for 6- and 8-year cycles; 1b = ca. 2,625 kg/ha as average of raw field weight for same cycles; bananas–plantains not included).
2. Maize swidden in Puuc Hills of northern Yucatán, Mexico (2a = ca. 96 kg/ha as average of 40% 11-year and 60% 6-year cycles; winter crops add about 10% more kg; orchard production not included; 2b = irrigated tree crops – no production figures).
3. Rice swidden (3a) and irrigated (3b) in Pa Pae, northern Thailand (ca. 178 kg/ha as average of 97% 8-year cycle and 3% annual irrigated).
4. Rice swidden in Mogbuama, Sierra Leona (ca. 180 kg/ha as average of fallow cycles, 6.77 years; intercrops can add 50% more kg).

FIG. II-1. Trajectory of attributes: Paleotechnic and consumption-oriented systems.

The second case study involves the recent changes in and additions to the *milpa* system of the Maya of the Puuc Hills of northern Yucatán, where population densities range from less than 10–25 people/km^2. The colonial history of the system is one of a moderately extensive maize (*Zea mays* L.) swidden, culminating in current 3.5 (2 : 5) and 6-year (1 : 5) cycles. Based on an average yield of maize (grain) of 700 kg/ha, the system's staple output is about 168 kg/ha/cycle. Both the yield and output intensity figures are low, even for the tropics. The total output figure would be higher, of course, if the other cultivars were considered (e.g., legumes, cucurbits). The system has been enveloped by a market economy, resulting in a diminution of staple food production. Farmers have increasingly emphasized the production of honey and fruit for the market, the latter with the aid of small-scale irrigation and other neotechnic inputs. Production of maize, beans, and squash remains essential, however. Interestingly, the recent emphasis on small orchards may be a resurrection of an ancient Maya practice (Turner & Miksicek, 1983).

The Lua' of Pa Pae, Thailand, are the third case study. With a current population density of about 12.6 people/km^2, the Pa Pae villagers have long practiced an extensive, paleotechnic system of swidden cultivation based on upland or dry rice (*Oryza sativa* L.) for subsistence. A typical cultivation cycle was 26 years (1 : 25), and with typical yields (grain) of 1,000 kg/ha, the resulting output intensity per cycle was 38.5 kg/ha/cycle. Both population growth and government inducements have changed this traditional system. Fallow cycles have been reduced in the swiddens and currently average about 8 years (1 : 7). With no major input changes, output has dropped precipitously to about 125 kg/ha/cycle on 688 ha. This result has made the traditional alternative, irrigated rice, economically feasible and acceptable. The shift to irrigation has led to major increases in output intensity, to 2,200 kg/ha. This transformation, however, involves substantial labor and capital investments that have acted as barriers or constraints to adoption, and currently irrigated rice is restricted to only 18 ha. This result lends some support to the *threshold concept*, which asserts that the high levels of inputs necessary to initiate certain technological transformations, in this case irrigation, may act as a barrier to change, but that once the transformation takes place, cultivation efficiency may increase (e.g., Robinson & Schutjer, 1984). Considering the hectarage devoted to both systems, the village averages about 178 kg/ha.

The fourth case study involves West African rice cultivation as practiced in the village of Mogbuana in central Sierra Leone. With a population density approaching 17 people/km^2, extensive fallow systems have been the tradition. Recently, however, the longer, forest-fallow system has almost disappeared, perhaps because of local labor shortages, which inhibit forest clearance. Herein is a case of "intensification" not created so much by changes in demand but by labor bottlenecks—insufficient labor at critical times. Current cycles average about 6.77 years (2 : 8–15), with average rice yields (grain) of about 1,225 kg/ha, or 181 kg/ha/cycle. Production remains primarily for consumption, and paleotechnic inputs are mainly used.

References

Norman, M. J. T., Pearson, E. J., & Searle, P. G. E. *The ecology of tropical food crops*. Cambridge: Cambridge University Press, 1984.

Pingali, P. L., Bigot, Y., & Binswanger, H. P. *Agricultural mechanization and the evolution of farming systems in sub-Saharan Africa*. Discussion Paper, Report No. ARU 40. Research Unit, Agricultural and Rural Development Department, World Bank, Washington, DC, 1986 (revised).

Robinson, W., & Schutjer, W. Agricultural development and demographic change: A generalization of the Boserup model. *Economic Development and Cultural Change*, 1984, 32, 356–366.

Turner, B. L., II, & Miksicek, C. Economic plant species associated with prehistoric agriculture in the Maya lowlands. *Economic Botany*, 1983, 38, 179–193.

3

Swidden in Amazonia and the Amazon Rim

STEPHEN BECKERMAN

Swidden (slash-and-burn, shifting, transient, itinerant, *tumba y quema*, *tumba y roça*, *tala y quema*, extensive) agriculture is the most widespread aboriginal farming system in the tropical forests of the New World. It is also widely practiced in the tropical Old World (Manshard, 1974:52–66), and was the major route to the clearing of the forests of Europe and North America in historic times (Sigaut, 1975). As practiced today in greater Amazonia, its essential features are the felling and burning of a patch of forest, which is allowed to regrow in trees after a small number of years under crop. Swidden shades imperceptibly into other forms of agriculture, which differ from it in the length of the fallow period (the major "development" of swidden is to shorten the fallow period until trees can no longer return [Boserup, 1965; Greenland, 1974]), the length of the cropping period, and the question of the burn. This chapter excludes market-oriented "bush fallow" systems in which trees do not come back, because of the recency of such systems in Amazonia, but includes subsistence bush fallow; excludes lengthy cropping cycles on improved lands, except for passing mention, because such systems are largely extinct (Denevan & Schwerin, 1978); and includes two examples of "slash-and-mulch" agriculture in which for one reason or another the forest is not burned after cutting, but is simply allowed to rot.

This introduction presents a brief description of a typical Amazonian aboriginal swidden farming system, and a summary sketch of the Barí of the southwestern Maracaibo Basin (Colombia–Venezuela), who are used to exemplify the aspects of the system that are discussed in the body of the chapter. Following the introduction, the chapter is organized under six additional headings: (1) natural environment, (2) human modifications, (3) labor costs, (4) productivity, (5) options and opportunities of swidden, and (6) directions of change.

It should be emphasized that this review is in no sense a complete discussion of Amazonian swidden. A generation ago Conklin (1961) published a guide to the study of shifting cultivation that included a massive outline of questions to be asked. To the best of my knowledge, no one has yet answered all these questions for even a single society, although Bergman's (1980) study of

Stephen Beckerman. Department of Anthropology, Pennsylvania State University, University Park, Pennsylvania.

the Shipibo comes close. This chapter concentrates on the data that have become available over the last decade or so, which reflect current trends and interests in human ecology.

General Characteristics. A typical Amazonian swidden farming system is practiced by indigenous peoples today living at a population density of considerably less than 1/km^2, and, in traditional times, inhabiting a communal dwelling. A plot in the nearby forest is selected by a man or men, who then clear it in two stages, removing first the undergrowth ("to brush out" is the inelegant English equivalent of the Spanish verb *socolar*, "to clear the brush from under a forest") and later the trees. After a drying period, the fallen vegetation is burned, a step that accomplishes a number of goals: The wood is removed from the way of planting; the weeds that have begun to sprout are depressed; and the resulting ash fertilizes the acid and nutrient-poor soil of the tropical forest.

The field is then planted, by men and/or women, to manioc and/or bananas and plantains, often with a secondary crop of maize. Minor crops may be various but seldom occupy more than a small fraction of field space. More often than not, the field is largely planted in such a way that crops occur in single-species stands, contrary to the pioneering observations of D. Harris (1971). Weeding is highly variable in intensity and in the gender of the major participant in this stage of labor. Harvesting is generally the province of women, and the major processing of garden produce is universally so. Harvesting goes on bit by bit, with only enough food for a few days being taken at one time; replanting occurs simultaneously.

After 3 or 4 years weeding stops and harvesting diminishes considerably, concurrently with a decline in productivity. Replanting of staples also ceases, although long-lived tree crops may be set out at this stage, if not already present. Some harvesting may continue for years after, as many swidden cultivars are strong competitors. Eventually, a secondary forest reappears and shades out surviving cultivar and bushy weeds alike, permitting only the normal forest understory. At a highly variable time in the future—something on the order of one or two human generations may be as close as one can come to a median figure—the forest is appropriate for reclearing. Many peoples, however, prefer to seek climax forest.

Specific Case. The Barí are one of a number of South American Indian societies who follow a swidden system. They are tropical rain forest people, speaking a Chibchan language, who live in the southwestern corner of the Maracaibo Basin (Figure 3-1). Although sharing the general South American tropical forest adaptation of swidden agriculture, fishing, hunting, and communal dwellings, they are different enough from many of the indigenous peoples of the Amazon Basin proper that they provide a check as to what is intrinsic to swidden agriculture of the wet lowlands and what is merely often associated with it.

For instance, in contrast to swiddeners in the Amazon Basin proper, the

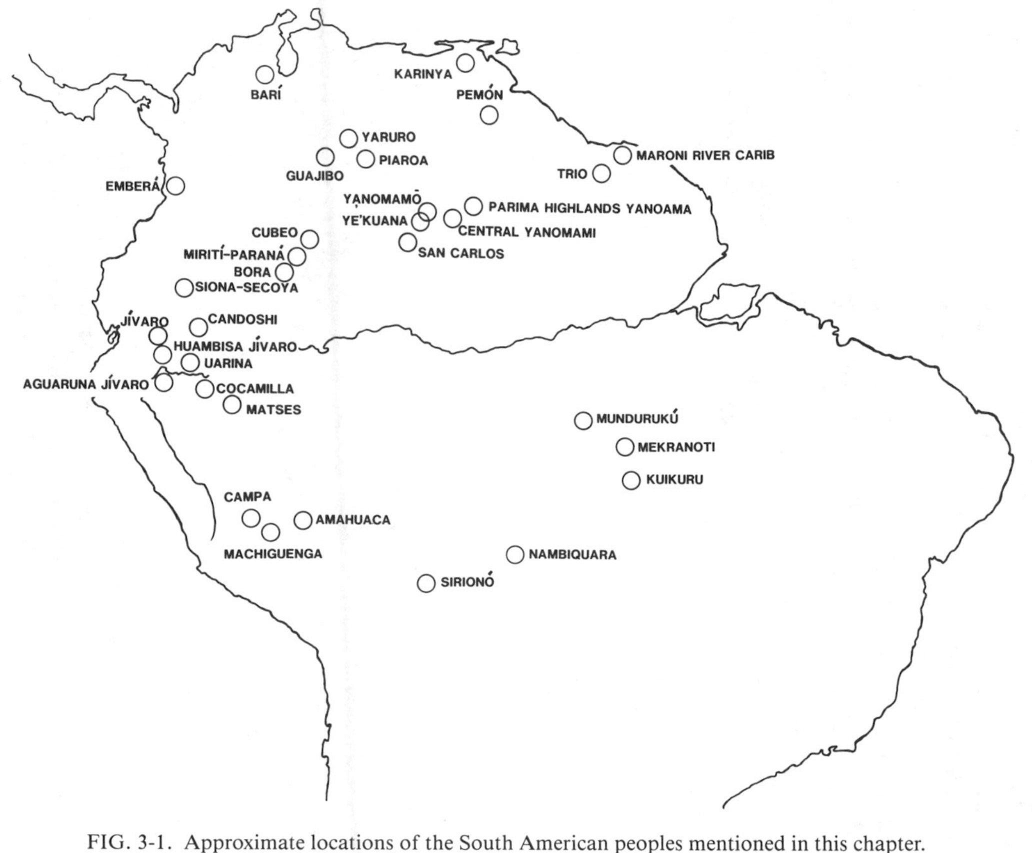

FIG. 3-1. Approximate locations of the South American peoples mentioned in this chapter.

Barí are semisedentary; that is, each local group of around 50 people has (in the ethnographic present) several communal dwellings, and the group moves among them over the course of the year. This pattern means that the fields associated with each house are used for less than the full year, a fact that helps in untangling the factors resulting in field abandonment. In another vein, the Barí are typical of Amazonian swiddeners in that their staple crop is manioc, and their major secondary crop is bananas. The fact that these cultivars dominate the subsistence of these people, linguistically unrelated to and geographically distant from Amazon valley groups, speaks strongly to the advantages of using these cultivars in the tropical rain forest.

As is unexceptional but not universal among Amazonian swiddeners, the Barí use their fields almost entirely to produce carbohydrates, which provide the great majority of the calories in their diet (Table 3-1). Protein comes mostly from fishing and secondarily from hunting (Beckerman, 1983c). Agricultural returns are steady around the year, whereas fishing returns vary dramatically with monthly rainfall, although hunting returns do not. Gathering is relatively unimportant, averaged over the year, but may be significant as a secondary source of fat and protein in poor fishing months (Beckerman, 1977).

The Natural Environment

Land and Water

The areas of greater Amazonia in question have 1–2% of their surface occupied by bodies of water (Geisler, Knoppel, & Sioli, 1973:147; Beckerman, 1979:535). These ratios are for large areas, and the percentage of water surface in the territory of a single community may vary considerably. In some regions, particularly those with marked seasonality, the rainy season may involve large-scale inundation.

Temperature

All the regions of greater Amazonia share the same basic climatic pattern: They are hot and wet. Temperature regimes in particular are remarkably similar. Diurnal variation is much wider than annual, with mean monthly temperature remaining almost constant around the year. In the area of the southwestern Maracaibo Basin inhabited by the Barí, the annual mean temperature is 27.5°C, with a range only from the January mean of 26.4° to the August mean of 28.3° (Beckerman, 1975:249–251). The pattern of daily cycle is unexceptional for the region. The Barí values are a degree or two too high to be typical of Amazonia, but the relationship of daily to yearly fluctuation is universal in the area.

TABLE 3-1. Cultivars, diet, and field size.

Language family	People	Major cultivar	Contribution to diet (%)	Mean field size (ha)	Range (ha)
Macro-Chibchan	Barí	manioc (sweet)	90	0.4	0.30–2.0
	Emberá	plantain		0.49	0.09–1.28
	Parima Highlands Yanoama	plantain		2.6	1.40–4.6
	Central Yanomami	plantain	55	1.3	0.20–2.5
Jivaroan	Jívaro	manioc (sweet)	65	0.6	0.40–0.7
	Huambisa Jivaro	manioc (sweet)	60	0.5	
	Aguaruna Jívaro	manioc (sweet)	60	0.25	
	Yaruro	manioc (bitter)	50	0.14	0.01–0.6
	Candoshi	manioc		0.5	
Macro-Caribe	Kuikuru	manioc (bitter)	80	0.65	0.40–1.1
	Maroni River Carib	manioc (bitter)		0.5	0.30–0.7
	Ye'kuana	manioc (bitter)	70	1.6	
	Bora	manioc			0.25–1.0
Macro-Tukanoan	Cubeo	manioc (bitter)		0.4	
	Siona-Secoya	manioc	45	0.6	0.30–2.0
	Uarina	plantain		0.5	0.25–1.0
Arawakan	Machiguenga	manioc (sweet)	65	0.49	0.05–1.12
	Campa	manioc (sweet)	75	0.7	0.50–1.0
Macro-Pano	Shipibo	plantain	70		0.02–10.0
	Amahuaca	manioc, maize	50	0.6	
Tupí	Cocamilla	manioc		1.0	

Note: The peoples listed in this table are grouped, as are those in all the following tables, according to the recent classification of South American Indian languages by Pottier (1983); it should be borne in mind that aspects of this classification are controversial. Data sources for the figures are: Barí (Beckerman, 1975); Emberá (Isacsson, 1975); Parima Highlands Yanoama (Smole, 1976); Central Yanomami (Lizot, 1971); Jívaro (Harner, 1973); Huambisa Jívaro and Aguaruna Jívaro (Boster, 1980; personal communications); Yaruro (Leeds, 1961); Candoshi (Stocks, 1983); Kuikuru (Carneiro, 1983); Maroni River Carib (Kloos, 1971); Yekuana (Hames, 1983); Bora (Denevan, Treacy, Alcorn, Padoch, Denslow, & Flores Paitan, 1984); Cubeo (Goldman, 1948); Siona-Secoya (Vickers, 1983a); Machiguenga (Johnson, 1983); Campa (Denevan, 1971); Shipibo (Bergman, 1980); Amahuaca (Carneiro, 1964); Cocamilla (Stocks, 1983).

The stable, high mean temperatures of Amazonia promote rapid decomposition of soil organic material, and remove any need for overwintering strategies by the biota. Seasonal changes in heat budget have essentially no effect on soils or vegetation.

Precipitation

Rainfall is considerably more problematic in Amazonia than temperature. Issues relevant to farming are seasonality and reliability (the latter usually unmentioned in the literature), as well as mean annual total. The Barí area receives an average yearly rainfall of 3,000–3,500 mm, the figure varying among weather stations no more than a few dozen kilometers distant from one another (Beckerman, 1975:240–241). At all stations there is a stable monthly pattern (Figure 3-2). As even February, the driest month, has 75 mm of rain, there is no true dry season in this area. (The rule of thumb is that as long as rainfall [mm] is more than twice the mean temperature [°C] for the month, water is not a limiting factor [Walter, 1973:21].)

This absence of a true dry season, though widespread in Amazonia, is not characteristic of it as a whole (Figure 3-3). Wherever the monthly rainfall curve dips near or below roughly the 50-mm line, plants may experience a water deficit at that time unless local conditions, such as the aquifer of a perennial stream, alleviate the stress. Such a relatively dry season has a pronounced effect on fruiting and flowering schedules of plants and on the activities of the animals that eat them (Smith, 1982:38–42), and thus the yearly distribution of rainfall has a direct effect on the seasonal abundance of wild as well as cultivated foods. Insofar as agriculture itself is concerned, the best drying of the slash for a maximum burn is accomplished by letting it sun during the dry season. Manioc, the most common crop in Amazonian swidden, gains little root weight during dry periods and may actually lose weight (Mayobre, San Jose, Orihuela, & Acosta, 1982a, 1982b).

In a region such as Amazonia, where nutrient reserves in the soil are ordinarily quite low, the fertilizing effect of rainfall can be expected to be important. The nutrient content of rainwater, however, is related to the nutrient content of the soil of the area where it falls (Ericksson, 1955). Thus the sets of rainfall nutrient figures provided by Ungemach (1970) and Stark (1972), both of which come from near Manaus, Brazil, on the fringes of an area of impoverished soil and blackwater rivers, should be taken as near minimal for Amazonia, not as typical. These investigators found that rain delivered 6.0–10.0 kg/ha of nitrogen per year, along with 0.2–0.3 kg/ha of phosphorous, 2.0–3.0 kg/ha of magnesium, and 0.8–3.7 kg/ha of calcium. Potassium, of special interest in the cultivation of manioc, was not determined. Salati and Vose (1984:132) estimate an "average" rainfall contribution to the soil of 0.8 kg/ha of potassium per year in Amazonia.

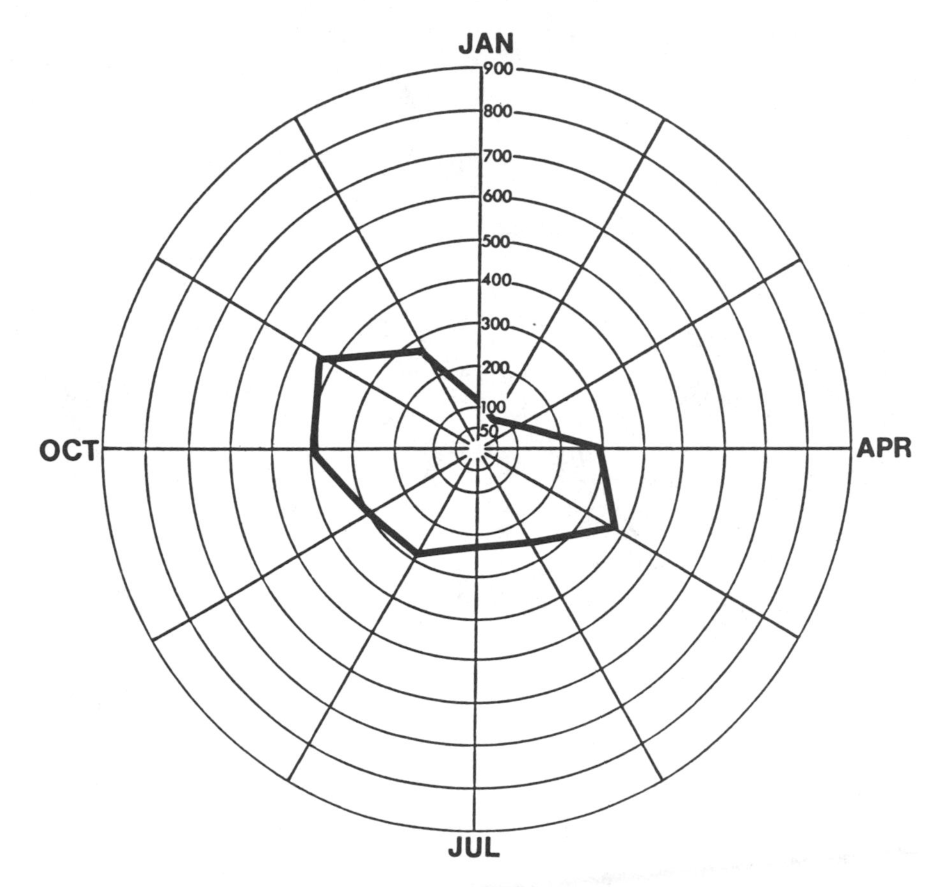

FIG. 3-2. Rainfall rose for annual precipitation regime in Barí area (monthly rainfall figures in mm).

Soils

In comparison to the unimproved agricultural soils in temperate regions, soils throughout Amazonia tend to be poorer and more vulnerable to degradation. Most Amazonian soils are old and deep, and parent rock is located far below the surface, impeding the supply of weathered, nonorganic nutrients to the surface horizons. Soil nutrients, therefore, are derived primarily from surface organic matter, which concentrates in the shallow topsoil (A and B horizons) (Young, 1976:101), typically in high quantities (Salati & Vose, 1984:133; Young, 1976:42). Constant high temperatures and abundant moisture promote rapid decomposition of organic matter by microbal action, while rainfall leaches nonorganic nutrients to depths that are not root accessible.

Additionally, much of the soil of the Amazon Basin proper is underlain by

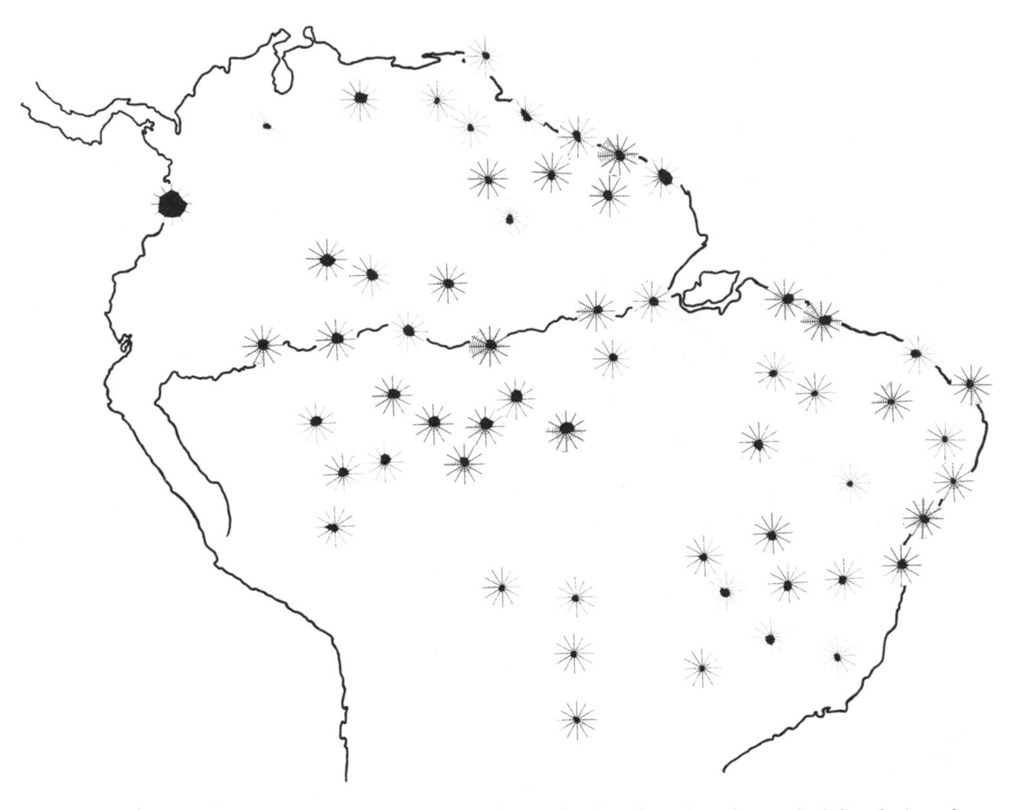

FIG. 3-3. Rainfall rose for annual precipitation regimes of selected stations in tropical South America, scale for monthly rainfall figures as in Figure 3-2. Redrawn from data assembled in Walter and Lieth (1967). A notable feature is the location of the rainy season in the (Northern Hemisphere) summer months north of the equator, and in the (Southern Hemisphere) summer months south of the equator.

parent materials that themselves are poor in nutrients (Schubart & Salati, 1982:212–215). Cochrane and Sanchez (1982:164) conclude that

> the vast majority of the Amazon soils are acid and with low native fertility in their undisturbed state . . . only about 8 percent of the region has high base status soils with relevantly high native fertility. The main chemical soil constraints in the region are soil acidity, P deficiency, low effective cation exchange capacity, and widespread deficiency of the following nutrients: N, K, S, Ca, Mg, B, Cu, Zn, and occasionally others. . . .

The major exceptions are the alluvial soils formed from Andean sediments (*varzeas*); the soils (*terras roxas*) formed from mostly extrusive basic rocks and their decomposition products; and the black soils (*terras pretas do indio*), which are partly anthropogenic in origin (Cochrane & Sanchez, 1982:155; Falesi, 1974:227). These agriculturally desirable soils are predictably distributed in strips along the rivers that drain the geologically young Andes chain (the *varzeas*), and in pockets throughout the basin and elsewhere (the *terras roxas* and *terras pretas*). Cochrane and Sanchez (1982:154) caution that even in the case of the *varzea*, the variability in source of sediments is such that "it cannot be generalized that the alluvial soils of the Amazon are always of high native fertility."

Most surveys indicate that in the Amazon basin proper, the proportion of soils desirable for commercial agriculture is between 5 and 10%, with current estimates tending toward the lower figure. Regional variability is nonetheless important. Certain lowland tropical forest regions, such as areas around Altamira, Porto Velho, and Rio Branco in Brazil, and locations in eastern Ecuador, have relatively high proportions of fertile nonalluvial soils, largely *terras roxas* (Cochrane & Sanchez, 1982:155). Even in these areas, however, the superior soils are irregularly distributed and discernible by gross inspection only to someone familiar with the region and the indicator vegetation for local soil variability (Moran, 1981:40, 109–112). The *terras pretas do indio* (black soils) are still imperfectly known (Sombroek, 1966:175–176, 252–256; Falesi, 1974:227; Sternberg, 1975:32–33; Smith, 1980). These patches tend to be larger on or near floodplains than on *terra firme* (upland forest) away from the rivers, but they do not appear to be tied to rivers or any particular parent material. They are very likely the remains of former aboriginal settlements (Smith, 1980).

There are also areas of greater Amazonia, particularly those drained by Rio Negro and its tributaries, floored by soils of remarkable infertility—hardly more than bleached sand (Herrera, 1979). No agriculture is practiced on these Spodosols, and even the Oxisols of the area give small yields and recover slowly, if at all (Uhl, 1980).

The agricultural distinctions between the rich *varzea* and the impoverished *terra firme* has been forcefully formulated by Lathrap (1968:27) and Meggers (1971:14–32), among others. Recent work (e.g., Cochrane & Sanchez, 1982), however, directs attention to variability in agricultural potential within each of these broad classes, pointing out that some *terra firme* soils are better than some *varzea* soils. The fact that these richer *terra firme* soils are not easily located or identified by temperate-trained scholars (who may also err in judging some alluvial soils too favorably) tells us nothing about the ability of native Amazonians to find or utilize them. The emerging picture is one of mosaics nested within mosaics, with significant variability in fertility at each level of scale. The enduring conclusion appears to be that, at all scales except the smallest, the proportion of agriculturally desirable soils is small, usually around 5%. The typical problems for all but the best soils are acidity, aluminum toxicity, low levels of available potassium and phosphorus, and a low effective cation exchange capacity (Cochrane & Sanchez, 1982:168).

Flora

The reason that the Amazonian forests are as luxurious as they are, despite the relative poverty of the soils on which they grow, is that the great majority of the nutrients in the ecosystem are tied up in the living matter itself. Nitrogen and phosphorus are the only nutrients found primarily in the soil (Salati & Vose, 1984:131). Forest litter rots on the forest floor, and the nutrients from it are recycled, largely through microrrhyzae attached to the very shallow roots, without ever becoming part of the soil proper (Went & Stark, 1968). This process, along with a year-round growing season, an extraordinary species diversity, and a generally closed canopy which provides a shaded and mostly open ground surface, are among the major features that characterize mature tropical rain forests.

On walking through the forests inhabited by the Barí, one is first struck by the size of the trees, some 40–50 meters tall. Trunks are relatively slender and do not branch below the nearly closed canopy. A profusion of vines ties the trees together. An inevitable observation is the sameness of the leaves. With the exception of the palms and a few understory and secondary-growth species, they all appear to be shaped like laurel leaves. Occasional enormous plank-buttressed trunks make one despair of ever clearing a substantial area of ground with hand tools. The few flowers are inconspicuous but may appear on nearly any part of a tree. The ground is covered with a soft, corky layer of litter in most places, and a few hours observation make it clear that much of the animal life is high above one's head, where the canopy leaves and fruit provide food for insects, birds, squirrels, sloths, monkeys, and their predators.

Estimates of the biomass of the tropical forests range from those for the stunted blackwater river forests around San Carlos de Rio Negro (Uhl & Jordan, in press) through those from the *terra firme* forests around Manaus (Klinge & Rodrigues, 1974:343) to those from the large forests on the Jarí River (Pires, 1978:615). The figures are presented in different forms and must be transformed to make them comparable. Uhl and Jordan (in press) found an above-ground living biomass of 260 t/ha and a belowground living biomass of

50 t/ha, both expressed as dry weight. Klinge and Rodrigues (1974) found an above-ground living biomass of 687 t/ha and a below-ground living biomass of 255 t/ha, both expressed as *wet* weight. The tropical forester's rule of thumb that wet weight of living biomass is roughly twice dry weight allows the conversion of Klinge and Rodrigues's figures to 343 t/ha above ground and 128 t/ha below ground, dry weight. Pires (1978) gives a figure for the above-ground *volume* of the Jarí forests of 584 m^3/ha. If this estimate is projected to weight in the proportions Klinge and Rodrigues found around Manaus, it yields an above-ground living biomass of 875 t/ha wet weight, or 438 t/ha dry. From the forests around Manaus there also comes the figure for *dead* biomass of 285 t/ha, wet weight (Klinge & Rodrigues, 1974:348).

Productivity is high in the Amazonian rain forests, though lower than the world mean for all tropical forests, which is about 20 t/ha/year (Lieth, 1975:205). Pires (1978:615) estimates the Amazonian forests to have a net primary productivity of about 10 t/ha/year. In climax forest this productivity is balanced by a yearly litter fall of approximately the same magnitude. Klinge and Rodrigues (1968a:293) found yearly litter fall of 6.7–7.9 t/ha dry weight in a forest near Manaus, not counting "bigger branches and stems which contribute a large portion to total dead plant material returned yearly to the forest floor." Of the litter fraction that was measured, leaves accounted for 70%. This balance is important because leaves contain the highest proportion of nutrients of any of the litter fractions (Klinge & Rodrigues, 1968b:306). Of all measured litter, the nutrient return to the forest floor was, on a yearly basis, 76.0–96.5 kg/ha of nitrogen, 2.0–2.3 kg/ha of phosphorus, 11.9–13.4 kg/ha of potassium, 11.4–13.8 kg/ha magnesium, and 17.9–18.6 kg/ha of calcium (Klinge & Rodrigues, 1968a:297, 1968b:304). These figures can be compared with the nutrient contribution of rainfall. The leaves, the most easily burned fraction of the forest, contain the highest concentration of nutrients.

The huge, productive rain forests, with their enormous species diversity, contain numerous plants of use to man. The Barí, in common with the other peoples of South American tropical forests, exploit many species of trees—palms in particular—for their fruits, leaves, and wood. The seasonality of rainfall strongly affects the availability of wild fruits.

Fauna

In addition to an exceedingly rich flora, the tropical forest also contains a diverse fauna. For most of the aboriginal peoples of greater Amazonia, fish are more important than birds and mammals, supplying the plurality if not the majority of the protein in the diet (Gross, 1975). Animals of the forest proper remain important, however, during periods of poor fishing. For swiddening peoples who live away from major rivers, game is practically the only source of meat.

The Barí obtain about three-quarters of their meat from fish and the rest from game. As with many other greater Amazonian swiddeners, many of the animals they hunt are also garden pests. Such is the case with both species of peccary (*Tayassu pecari*, the white-lipped, and *T. tajacu*, the collared peccary); with at least two of the three major rodents (*Dasyprocta punctata*, the agouti, and *Agouti paca*, the paca); with the brocket deer, *Mazama americana*; and with several species of parrot. Of their major terrestrial game animals, only the tapir (*Tapirus bairdii*) does not raid cultivated fields at least occasionally. The Barí do not grow maize. Swiddeners who do are also subject to garden raiding from the coati (*Nasua narica*). Thus the forest around a swidden plot serves as a reserve of meat which is actually drawn to the agricultural modification; abandoned fields in particular, with their diminished frequency of frightening human presence, are "game magnets" (Beckerman, 1984; Irvine, 1981; Linares, 1976). In the sense of providing game as well as nutrients for the crops, the forest is in continual interaction with the swidden field.

The Human Modifications

Crops

Only the three most common Amazonian staple crops are considered in this section, for reasons of space. Treatments of many others can be found in Purseglove (1968, 1972) and Alvim and Kozlowski (1977); and virtually all are meticulously examined in Patino's (1963–1974) monumental *Plantas cultivada y animales domésticos en América equinoccial.*

The most common South American swidden crop is manioc (*Manihot esculenta*), a very old New World domesticate of the Euphorbiaceae which produces large starchy roots (also known as cassava). It has been customary to divide manioc into two agricultural types, bitter and sweet, depending on the amount of hydrocyanic acid present in the flesh of the root; however, with the realization that "prussic acid content varies continuously from innocuous to lethal concentrations depending on soil and climatic conditions as well as genetic factors [Schwerin, 1970]" (Boster, 1980:23), current practice (Rogers & Fleming, 1973) is to ignore this chemical factor in botanical classification, although the distinction is extremely important in the processing of the food by its cultivators.

The Barí have only sweet manioc, and only two traditional varieties of it.[1] These cultivars are ready for harvesting at something over a year from planting (bits of stem are used as propagules) and give an average of at least 5 kg of roots (fresh weight) per planting, this figure comprising about half the weight of the whole plant (Beckerman, 1983a:97). Unharvested plants can be left in the ground for well over 2 years, a length of time unusual even for manioc, which is noted for its ability to store itself in the ground. The Barí speak with some disdain of "homesteader manioc," which begins to rot less than a year from planting.

The dietary value of manioc is in its starch. Typical figures (there is not much variation) for raw peeled roots are about 62% water, 35% carbohydrates (almost all starch), 1% fiber, 1% ash; 0.8% protein, and 0.2% fats; calories run around 150/100 gm (Montaldo, 1973:21, 25).

Manioc gives one of the highest yields of calories per unit land per unit time of any cultivar, but perhaps its greatest advantage for tropical forest swidden agriculture is not its productivity but its tolerance. It will grow in soils that are poor in nutrients and of such acidity and high levels of available aluminum ions that most other crops would find them toxic. It does, however, tend to be potassium demanding.

For the swiddeners emphasized in this chapter, only bananas and plantains (various spp. of *Musa*, family Musaceae), are of an importance comparable to that of manioc. The New World *Musa* cultivars arrived with the Spanish and Portuguese conquerors, who brought them from Africa. They spread rapidly across South America, far in advance of the front of European colonization, and arc now completely integrated in traditional aboriginal swidden systems, with many native varieties.

The Barí recognize at least four varieties of banana and one of plantain. At least one of their banana cultivars produces fruit in 9–10 months, with bunches of fruit averaging 6 kg fresh weight (Beckerman, 1983a:98) Plantain bunches are, of course, much heavier. Bananas and plantains, like manioc, are propagated vegetatively. The underground buds from the tuberous rhizome under a mature plant are cut away from the parent plant and used as propagules. Bananas are broadly similar to manioc in their dietary value, the following rounded figures being typical: 75% water, 22% carbohydrates (mostly sugar rather than starch), 1.2% protein, 1% ash, 1% fiber, 0.8% fat; calories are about 90/100 g.

Bananas and plantains share many of the diagnostic characteristics of root crops. They are propagated by vegetative parts, not seeds; the edible part is high in carbohydrate and low in protein. They are not seasonal, but are ready a given amount of time from the moment of planting. An important difference, however, is that the *Musa* spp. cannot be stored on the living plant. Parrots, in particular, are fond of the ripe fruits and will destroy them even before they begin to rot and fall to the ground.

Bananas and plantains are considerably more demanding of soil conditions than is manioc. In particular, they prefer soft, deep, well-watered alluvial soils, and do not do well in hard or shallow soil. Aside from their rather high demands for soil texture, there is some evidence that bananas are particularly demanding of potassium (Tai, 1977:453).

Maize is an important secondary crop for many of the farmers to be discussed, although only among the Amahuaca does it apparently take first place in the diet. *Zea mays* is an ancient New World seed crop of the Graminae. There are many varieties and enough variation in their composition that a discussion of the nutritive value of maize without knowledge of the variety under consideration is not very informative. In very broad terms, the kernels tend to have about 70% carbohydrates and 10% protein. This protein is deficient in lysine and tryptophan, two of the essential amino acids. Maize is demanding of soil nutrients, particularly nitrogen, but also phosphorous and potassium.

Field Architecture

The distribution of crops within a swidden field has been an issue for some years (cf. Geertz, 1963; D. Harris, 1971). Fields may be monocultured (dedicated to a single cultivar); zoned in various ways; or polycultured (intercropped) in numerous spatial arrangements using varying numbers of cultivars, the latter belonging to several species or being varieties of the same species (cf. Hames, 1983; Kass, 1978; Sanchez, 1976; Stelly, 1976; Wiley, 1979).

The Barí, with minor exceptions, practice whole-field monoculture and zoned (in concentric rings) monoculture (Beckerman, 1983a). The fields that surround their longhouses, where these houses are surrounded by alluvial soils, typically have an outer ring of bananas and plantains, a wide middle ring of manioc, and two or more inner rings of sugarcane–cotton–peppers, *Tephrosia* sp. (a barbasco), and/or squash–yams–sweet potatoes. The fields located away from the houses are either monocropped in manioc or bananas, or composed of a ring of bananas around a core of manioc.

A number of Amazonian swiddeners have mainly monocropped fields. Besides the Barí they include the Kuikuru (Carneiro, 1983) and the Piaroa (Eden, 1974), both manioc cultivators, and the Emberá (Isacsson, 1975) and Shipibo (Bergman, 1980), who subsist on plantains and bananas.

The planting of different crops in concentric rings is reported from widely distant areas of greater Amazonia (Beckerman, 1983b, 1984). This pattern is found among the Barí (Beckerman, 1983a), the Candoshi (Stocks, 1983), the Mekranoti (Flowers, Gross, Ritter, & Werner, 1982), and the Mundurukú (Frikel, 1957). Always, the taller plants such as bananas are in the outermost ring and the lower ones in the innermost; that is, the field is roughly funnel-shaped.

Several peoples practice zonation without a definite geometric form, which nevertheless largely segregates different species and varieties in their own areas within a single field. This tendency is exemplified by the Ye'kuana (Hames, 1983) and to some degree the Yanomamo (Hames, 1983; Lizot, 1971) and by various groups along the Mirití-Paraná River made up mostly of Yukuna speakers (von Hildebrand, 1975).

Still other peoples practice intercropping, a system in which various species are planted in the same area, so that any plant's nearest neighbor is likely to be of a different species. This system is characteristic of the Siona–Secoya (Vickers, 1983a), whose main crops are manioc, maize, and plantains; of the Cocamilla (Stocks, 1983), who have the same staples; and of the Amahuaca (Carneiro, 1964), who grow primarily maize and manioc. In general, interplanting seems to be associated with maize, and fields intercropped with maize and manioc (the most common combination of the two staples) become essentially manioc monocrops after the earlier-maturing maize is harvested.

Field Size

Because the essence of Amazonian swidden agriculture is the regrowth of a productive field to tall forest, the size of individual fields is important. A field might be so small that much of it was overshaded and invaded by the surrounding jungle while it was in production, or so large that its eventual regrowth in tall timber would be seriously hampered by a lack of in-migrating nutrients and propagules.

Barí fields range from about 0.3 ha to 2.0 ha, this last case being unique and associated with mission life. Mean size is around 0.4 ha, an area close to the apparent mean for all the greater Amazonian peoples (Table 3-1).[2] Traditional swidden fields are quite small by the standards of many agricultural systems, the plurality of them running between 0.4 and 0.6 ha. The ubiquity of this figure suggests that it may represent something of an optimum for unshaded agricultural growth combined with successful regrowth.

Field Locations

Fields are distributed around a settlement with respect to the location of four features: the settlement itself, the other fields, the watercourses, and the soils. Most swiddeners have a field, commonly the major field, located adjacent to the settlement. The Barí actually have their longhouse set in the middle of the most varied and heavily used field. Such is also the case for Aguaruna Jivaro (Boster, 1980). Even peoples who locate their main fields at some distance from their dwellings usually maintain a "house garden" within a few meters of home (Lathrap, 1977). At the other end of the spectrum, a number of researchers agree that the maximum distance people are willing to walk to a plot is about 7 km (Aspelin, 1976; Carneiro, 1961; Isacsson, 1975). Smole (1976) suggests that the Yanoama of the Parima highlands may walk up to 10 km to a field. Distances may be even greater for fields accessible by canoe (von Hildebrand, 1975).

Proximity to other fields is important because of the need for tall forest as a source of recolonizing plants in an abandoned swidden, and as a source of nutrients for the reenrichment of the soil. The Barí never, in my experience, use one side of a field as the beginning of the next, unless both fields are on alluvial soil. Even then, there is usually a boundary of trees between the two. In *terra firme*, two fields are seldom even close.

This pattern is not universal. Three of the four plantain-staple peoples considered here (Shipibo, Embera, Yanomamo) typically make most or many of their fields adjacent to one another. The Kuikuru also follow this pattern with some (though not all) of their manioc fields (Carneiro, 1983), as do the contemporary Siona–Secoya (Vickers, 1979) with their polycropped fields.[3]

Peoples who live in forested areas are not constrained to put their fields close by watercourses, but often do so anyway.[4] The Barí locate their largest fields on patches of alluvium even though their houses, and often the fields that surround them, are placed on elevated spurs above the flood line; these alluvial fields are also those that are kept under cultivation the longest. This riverside preference is widespread but not universal. The Shipibo (Bergman, 1980) and the Embera (Isacsson, 1975), both plantain-staple peoples, manifest it in a particularly strong way, as do the Maroni River Carib (Kloos, 1971), the Candoshi (Stocks, 1983), and the Aguaruna Jivaro (Boster, 1980) among the manioc-staple peoples. The Candoshi farm only river levees, while the Aguaruna, in their steep area, do not have enough of this kind of land and also farm colluvial soils. The Parima highlands Yanoama, plantain-staple people, avoid rivers in general and farm only *terra firme* (Smole, 1976).

As mentioned, alluvial soils are usually (but not always) of higher fertility than the surrounding nonriverine soils. This tendency is probably enough to explain most of the preference for them, although it is also the case that farm plots adjacent to a river have an inherent transportation advantage for peoples who travel primarily by canoes. Cultivation of nonalluvial soils may be attributed to a number of factors: (1) insufficient alluvium; (2) comparable (or better) soils of nonalluvial origin; (3) reduced danger of flooding; (4) reduced pest loss;[5] and (5) decreased visibility of fields to human enemies traveling by water.

Productive Life of a Swidden Field

The essence of swidden is the impermanence of the field. The length of time it is cultivated is directly connected to the density of the population that can be supported, and the degree to which the environment will recover after cultivation.

Three types of fields can be distinguished among the Barí, according to longevity and location with respect to the longhouse and the rivers. As long as the longhouse is occupied (about 10 years, although with substantial periods in which the residents are all absent, the seasonal round having located them in another of their two to five longhouses), the surrounding field is usually kept in cultivation. Additional fields apparently are abandoned after only about 3 years if they are on colluvium, but may be kept in production for as long as 15 years if they are on good alluvial soil (Beckerman, 1975). They figures for Barí alluvial fields are exceptional for Amazonian swiddeners, and conditioned by the Barí practice of semisedentism, in which a single house and its associated fields are utilized normally for a third of a year or less, on average.

More typical are the comparative figures (Table 3-2), which show a mean field cultivation period of a bit over 3 years. It must be stressed that "time to

TABLE 3-2. Years to abandonment of swidden fields.

Language family	People	Range	Mean
Macro-Chibchan	Barí	3–15	10
	Emberá	5–10	
	Yanomamo		4.3
Jivaroan	Jívaro	3–5	
	Huambisa Jívaro	2–4	
	Aguaruna Jívaro	2–4	
	Yaruro		3
	Candoshi		3
Macro-Caribe	Kuikuru	2–5	3
	Trio		3
	Maroni River Carib		4
	Bora	3–5	
Macro-Tukanoan	Cubeo	3–5	
	Uarina	2–3	
Arawakan	Machiguenga	3–5	
	Gran Pajonal Campa	1.3–3	
	Mirití–Paraná	4–6	
	San Carlos	2–3	
Macro-Pano	Shipibo		20+
	Amahuaca		1
	Matses	2–3	
Tupí	Cocamilla	1–3	
Ge-Bororo-Caraja	Mekranoti		3
Unclassified	Piaroa	3–6	

Sources: Barí (Beckerman, 1975; 1983a); Emberá (Isacsson, 1975); Yanomamo (Hames, 1983); Jívaro (Harner, 1973); Huambisa Jívaro and Aguaruna Jívaro (J. S. Boster, personal communication); Yaruro (Leeds, 1961); Candoshi (Stocks, 1983); Kuikuru (Carneiro, 1983); Trio (Riviere, 1969); Maroni River Carib (Kloos, 1971); Bora (Denevan et al., 1984); Cubeo (Goldman, 1948); Machiguenga (Johnson, 1983); Gran Pajonal Campa (Denevan, 1971); Mirití–Paraná (von Hildebrand, 1975); San Carlos (Clark & Uhl, in press); Shipibo (Bergman, 1980); Amahuaca (Carneiro, 1964); Matses (Romanoff, 1976); Cocamilla (Stocks, 1983a); Mekranoti (Gross, Eiten, Flowers, Leoi, Ritter, & Werner, 1979); Piaroa (Kaplan, 1975; Eden, 1974).

abandonment" usually means the time after which a field is no longer weeded. It may produce substantial quantities of food for long after that. Plaintains and bananas, in particular, will continue putting out new shoots, and new bunches of fruit, until they are choked out. Pejibaye palms (*Bactris gasipaes*) will fruit for decades and are often planted in plots nearing abandonment. Other fruit-bearing trees have similarly long life spans. The Bora manage and harvest their "abandoned" swiddens all the way into the tall forest stage (Denevan, Treacy, Alcorn, Padoch, Denslow, & Flores Paitan, 1984).

In general, maize, if grown, is the first crop to be harvested, after 4 to 6 months; manioc follows, with a minimum time to first harvest varying from less than 6 months to over a year, depending on cultivar and conditions. Since manioc can be left in the ground after maturity, manioc fields are typically harvested bit by bit, and even the first harvest of a manioc field may not end until well over 2 years after planting. Manioc is usually replanted as it is harvested; maize is not. Plantains and bananas take 9 months to a bit over a year to bear fruit; usually, after a bunch is harvested, the pseudostem that bore it is cut down so that the new pseduostem(s) already coming up at its base will have more light. Although bananas cannot be stored on the plant indefinitely like manioc, they are perennial, and a staggered harvest can be obtained by staggering the planting of the original clones. They typically continue to bear long after the last of the manioc has been harvested. Finally, pejibaye, avocado, and other trees may remain in production for 20 or 30 years or more.

The Fallow Period

It is likely that it takes on the order of half a millennium to reestablish true climax forest in an Amazonian swidden plot. However, the growth of forest, the recovery of the soil, and the choking out of weeds occur far earlier.

The Barí prefer primary forest for their fields, although they also fell secondary growth on alluvium. The preference for tall forest is widespread, but the age of that forest is not easy to determine without the sorts of chronosequence studies that are only now beginning to be made (Saldarriaga, 1985). Estimates of forest age at the time of felling range from 70 years for the Kuikuru (Carneiro, 1983) to 15+ years for the Huambisa and Aguaruna Carib (J. S. Boster, 1985, personal communication) and the Maroni River Carib (Kloos, 1971) to a minimum of 8 years for the Uarina (Kramer, 1977). Peoples under some land pressure, such as the Cocamilla (Stocks, 1983) and the Embera (Isacsson, 1975) may reclear after only 2 or 3 years, although they prefer to wait 15 or 20. The peoples of the Mirití–Parana region of Colombia, mostly Yukuna speakers, sometimes replant after 5 years, but then let the field regrow indefinitely after that (von Hildebrand, 1975). Vasey (1979) has argued that any fallow time in excess of 10–15 years is inexplicable by agronomic considerations, a position challenged by the findings of Uhl (1980).

The advantage of reclearing successional forest after only a decade or so is the relative ease of cutting early successional soft woods, such *Cecropia* (trumpet tree) and *Ochroma* (balsa wood tree). It is also possible that the continuing decomposition of unburned wood from the first clearing continues to improve the soil (von Hildebrand, 1975). The disadvantage is that if the trees are easy to cut, the underbrush, until it has been completely shaded out, is thicker in secondary growth. The shrubs and vines are not only an obstruction to reclearing but will reestablish themselves in the reused field as weeds, which will increase cultivation time significantly. Also, insect pests, particularly of corn, are much more likely to be a problem in reused fields.

Swidden Settlements

Traditionally, most Amazonian swiddeners lived in large communal dwellings holding up to 100 people, variously called *malocas, bohios*, and longhouses in the literature. Today they tend to live in single-family houses, largely because of pressure from local whites who scorn the longhouses as "uncivilized" and "immoral." Some of the Amazonians, however, adopted dispersed single-family dwellings long ago—possibly as a defensive tactic in the face of slave raiding, expeditions of extermination, and similar barbarities of the conquest. Barí communal houses traditionally held about 50 people. Each local group owned two to five of them, within a recognized territory, and moved among them around the year (Beckerman, 1983c). At present there are no *bohios* left among the Venezuelan Barí, and only four among the less numerous Colombian Barí.

The largest traditional village maintained by any of the slash-and-burn agriculturalists considered here belongs to the Mekranoti, the entire society living in a single settlement of 285 people (Flowers, Gross, Ritter, & Werner, 1982; Werner, 1983). This size may have been typical around the time of the conquest for such peoples as the Cubeo (Goldman, 1948), the Kuikuru (Carneiro, 1960, 1961, 1983), and the Maroni River Carib (Kloos, 1971), among others. Nowadays, these peoples reach no more than half of that population in a single settlement, and that only when subject to some sort of political or economic coercion from the national society in which they are embedded. The typical contemporary swidden settlement holds fewer than 50 people, and in some societies—the Campa (Denevan, 1971), the Amahuaca (Carneiro, 1964)—fewer than 25. These smaller settlements, often consisting of no more than a single extended family, are better thought of as dispersed farmsteads rather than small villages.

The Overall Swidden Landscape

The human-modified swidden landscape is in a state of permanent change as new fields are created and old ones are allowed to grow back to tall forest. A traditional Barí local group of about 50 people usually had about 9 ha of land in active cultivation, distributed among the several longhouses in its territory of about 150 km^2 (Beckerman, 1975, 1976). The ratio of productive land to all other land was about 1 : 1,700. How much of this uncultivated land was cultivable is an open question, as it is for most of the Amazonian peoples. Additionally, the amount that is considered to be in fallow depends on what figure is chosen for the number of years after abandonment when forest suitable for reclearing can be considered to have reestablished itself—a number that may vary according to the definition of tall forest given by an agronomist, a forest ecologist, or the swidden farmer himself. The Barí prefer "virgin" forest—forest in which they themselves see no evidence of previous cultivation. The preference is often met for *terra firme* (colluvial) fields and often breached for alluvial fields. Similar breaches of preferred fallow times are recorded in the literature for other peoples.

In general, it seems likely that recovery times are related to soil type and size of field as well as to the agricultural practices of the people in question. Also, most of the swiddeners considered here seem to be far above the man/land ratio suggested by Boserup (1965) as the point at which crop production demands shrink the fallow period to years too few to permit full recovery of soil, a buildup of enough woody mass to produce a full load of ash, and freedom from weeds. Such may not have been the case before the European conquest of America, when the riverine peoples especially may have had large, dense populations (cf. Carneiro, 1961; Denevan, 1976; Roosevelt, 1980). These farmers, however, would have been aided by annual flood soil renewal and weed depression on at least some of their land.

Site Selection

The criteria most often mentioned as governing the selection of sites for swidden plots are soil and vegetation qualities. In general, people search for soil with good drainage and appropriate chemical properties. The ethnotaxonomies of soils may be fairly elaborate.

The Barí recognize many different soil types, but prefer a small set of them for agriculture. *Bírida*, a brown alluvial soil with considerable sand, is considered best for all crops. Lacking sufficient *bírida*, bananas and plantains may also be planted in *kungbangbaitana*, a wet, less easily worked soil described as "black" although the color is actually a sort of brick-grey. Manioc will rot in *kungbangbaitana*, however, and the secondary soil choice for that crop is *bongkita*, a reddish, clayey colluvial soil with small stones in it.

The chemical properties of the soil may be judged by its color (Boster, 1980; Johnson, 1983), the vegetation growing on it, or even its taste (Boster, 1980). The mechanical properties of the soil are also an issue; a root crop such as manioc apparently does better in "soft" soils, which do not restrict the expansion of the roots (cf. Hames, 1983; Johnson, 1983). Loose texture is also important for *Musa* spp. In practice this preference often means seeking a somewhat sandy soil.

Some of the considerations taken into account when judging the suitability of the covering vegetation have already been mentioned. The farmer looks for a plot that will not be especially onerous to clear and will not pose severe weeding problems after the field is in crop. As is probably rather common, the Barí prefer tall forest but are willing to settle for an earlier stage in the successional sere if it occurs on good alluvial soil. Convenience to the longhouse is another important Barí criterion in swidden site selection. Conversely, it is likely that the site of a future longhouse is influenced by the location of suitable soils in the vicinity. Such would seem to be the case from the areal photographic record of precontact longhouses in proximity to large patches of alluvial soils.

Ethnoagronomy

An aspect of swidden that deserves a great deal more attention (Posey, 1983) is the knowledge that the swiddeners themselves have of their agricultural systems. I suspect major cultural differences here.

The Barí, for instance, say that burning the field clears the ground; keeps vines and weeds from coming up as fast as they otherwise would (although it does not kill them); and produces ashes, which, after it rains, enter the soil and help the manioc grow. They add that ashes are good for all their cultuvars except yams, and, if they are sprinkled around sugarcane, will keep worms from attacking it.

Overall ethnographic opinion is divided as to the traditional recognition of the value of ash as fertilizer, with some authors declaring an inability to get their informants to produce any statements conceding that burning is anything more than a means of removing the fallen forest (Carneiro, 1983; Yde, 1965), while others give evidence that their informants understand the fertilizing function of ash in essentially the same terms that a Western agronomist would (Johnson, 1983; von Hildebrand, 1975). It is likely that this difference in perception is real, reflecting an underlying difference in the conceptualization of farming masked by surface similarity of behavior.

Labor Costs

There are four stages in the creation of a swidden field that occur only once in a single cycle: brushing out, felling trees, burning, and initial planting. The first three are universally the work of men, while the fourth may also involve women.

Clearing

Among the Barí, the most common time for clearing a field is January, in order to take advantage of February's relatively dry weather for maximum drying and a good burn. Fields can be cut at any time of the year, however, if the farmer(s) accept a poor or mediocre burn. Some other swiddeners clear fields more or less exclusively at the beginning of their dry season; still others show only a tendency toward that scheduling.

Like all the swiddening people in Amazonia, the Barí first go through an area chosen for cultivation with machetes, clearing the brush, the vines, and the saplings. Only when the entire area has been brushed out do they return with axes to fell the large trees. For communal fields, all the men from a longhouse work together in clearing the whole field, although after planting each man and his wife or wives are responsible only for the section of it in which their own crops are planted.

The labor demands of these stages of field preparation are rather heavy in terms of man-hours per hectare (Table 3-3). It must be kept in mind, however, that when averaged over the year for the whole area under cultivation, the number of hours spent in agriculture is very low (see Table 3-8).

The pattern of collaboration in clearing followed by individual responsibility from planting onward is common. The Yukuna-speaking people of the Miriti–Parana, for instance, still hold collective work parties for clearing, even though they now live largely in dispersed extended-family dwellings instead of their former longhouses (von Hildebrand, 1975). Collective work is not universal, however; the Machiguenga generally clear their fields alone (Johnson, 1983), as do all the Jivaroan peoples (Boster, 1980; Harner, 1973) and the contemporary Uarina (Kramer, 1977). Individual or very small group clearing seems to be associated, plausibly enough, with a dispersed farmstead pattern of residence. Even peoples who live in longhouses, however, such as the Barí, may sometimes clear individual fields to supplement their portions of communal fields, or perhaps as a declaration of social distance from the community.

Burning

The firing of the field is the preparatory step in which most variability is shown. The Barí simply set fires here and there in the field; usually, not even all the leaves burn, let alone all the twigs and branches. The Machiguenga, however, spend a great deal of time on the burn (Table 3-3), nursing the fire along and

TABLE 3-3. Labor requirements (man-hours/ha) for clearing and burning fields.

Language family	People	Brushing	Felling	Sum	Burning	Total
Macro-Chibchan	Barí	31	55	86	1	87
Jivaroan	Candoshi					250
Macro-Caribe	Kuikuru				11	
Macro-Tukanoan	Siona-Secoya	47	16	63		
Arawakan	Machiguenga	100	100	200	80	280
	Campa			70		
	Mirití-Paraná	32	88	120	1	121
	San Carlos			126	6	132
Macro-Panoan	Shipibo			91	1	92
	Amahuaca	83	33	116	1	117
Tupí	Cocamilla					143–243

Sources: Barí (Beckerman, 1975, 1983a); Candoshi (Stocks, 1983); Kuikuru (Carneiro, 1983); Machiguenga (Johnson, 1983); Campa (Denevan, 1971); Mirití-Paraná (von Hildebrand, 1975); San Carlos (Clark & Uhl, in press); Shipibo (Bergman, 1980); Amahuaca (Carneiro, 1964); Cocamilla (Stocks, 1983); Siona-Secoya (Vickers, 1979, 1983a).

piling up the unburned debris for a second burn (Johnson, 1983). At the other end of the spectrum, the Emberá, who live in one of the wettest places in the world (Figures 3-1 and 3-3), do not burn at all (Isacsson, 1975). The Uarina, plantain cultivators who brush out, plant, and *then* fell the trees, sometimes also omit the burning (Kramer, 1977).

Planting

The initial planting of a swidden field may take place any time from immediately after its burning to as much as 2 months later. It is at this stage that women usually take up their agricultural labors.

Among the Barí, the planting of a new field follows its burning in short order; the field may indeed still be smoldering in some spots. Women bring baskets packed with manioc stems, which have been cut from plants they already have growing in another field. As the men work through their individual sections of the field, excavating shallow holes with machetes (or occasionally with the traditional digging stick), their wives and children work alongside them, pulling up the vines and other volunteer weeds that sprouted during the drying period between the felling and the burning, piling them on mats, and carrying them to the edge of the field. In each hole the man puts five or six pieces of manioc stem, with the "eyes" facing down, at a shallow angle. Men also plant the bananas and plantains around the circumference of the field.

Table 3-4 shows the comparative data for labor investment per hectare for planting, along with the density of the major crop and the sex of the people who plant it. The gender distribution of planting is congruent with Carneiro's (1983) observation that men tend to plant the first crop except in situations of chronic warfare. It can be added that men seem to plant bananas and plantains whether there is war or not.

Another inference has to do with the difference in labor demands per hectare of manioc and *Musa* spp. The peoples concentrating on manioc average well over 150 man-hours per hectare (hr/ha) in planting, most of this labor involved with manioc itself; the two banana–plantain–staple peoples for whom we have comparable figures are the Shipibo (75 man-hours/ha) and the Emberá (109 man-hours/ha); the figure for the Emberá includes the labor of cutting second growth to mulch the banana corms; the brushing-out phase of clearing is part of planting in this case. In both the *Musa* cases, the whole planting figure is for bananas and plantains only. Possibly here is one of the reasons for the enthusiastic adoption of bananas by the Amazonian peoples; they are cheaper than manioc in terms of planting. Apparently, not all the saving is a consequence of their reduced density.

Weeding

Weeding and harvesting, the latter usually combined with immediate replanting, are not single events in the life cycle of a swidden fields, but take place in episodes often so close as to be almost continuous over the life of the field.

TABLE 3-4. Densities of major cultivars (plantings/ha) and man-hours/ha in planting for the first planting of new field.

Language family	People	Crop density: Manioc	Crop density: Musa	Crop density: Maize	Man-hours/ha
Macro-Chibchan	Barí	♂ 5,500			193
	Emberá		♂ + ♀ 4,000?		109 (includes cutting brush in reused fallow)
	Parima Highlands Yanoama		♂ 2,000		
Jivaroan	Jívaro	♀ 3,000			
	Aguaruna Jívaro	♀ 6,500			
	Huambisa Jívaro	♀ 13,300			
Macro-Caribe	Kuikuru	♂ 2,200			180
	Maroni River Carib	16,000			
	Bora	15,000–40,000			
Macro-Tukanoan	Siona-Secoya				186
	Uarina		♂ 1,000		
Arawakan	Machiguenga	♂ 2,500–3,500		10,000	40
	Mirití-Paraná	♀ 10,000–40,000			
	San Carlos	10,000?			166
	Campa	10,000–25,000			
Macro-Pano	Shipibo		♂ 1,000?		75
	Amahuaca			♀ 5,000	

Note: The density given is for the *major* crop for each people; it is preceded by the sex of the farmer who does the actual planting (accompanying weeding may be done by a spouse). Labor figures are for *all* crops planted and for all participants. In most cases the major crop is dominant enough that the difference is of minor importance, but for the Emberá and Machiguenga, maize is sufficiently important as a secondary crop that the major crop does *not* account for all but a small fraction of planting time. A similar situation obtains for the Amahuaca and the Campa, for whom there are no planting time figures.

Data sources: Barí (Beckerman, 1983a); Emberá (Isacsson, 1975); Jívaro (Harner, 1973); Aguaruna Jívaro (Boster, 1983; personal communication); Parima Highlands Yanoama (Smole, 1976); Kuikuru (Carneiro, 1983); Maroni River Carib (Kloos, 1971); Bora (Denevan et al., 1984); Siona-Secoya (Vickers, 1983a); Uarina (Kramer, 1977); Machiguenga (Johnson, 1983); Mirití-Paraná (von Hildebrand, 1975); San Carlos (Clark & Uhl, in press); Campa (Denevan, 1971); Shipibo (Bergman, 1980); Amahuaca (Carneiro, 1964).

Thus the former must be expressed in labor demands per hectare *per year* in order to be comparable cross-culturally, and it must be remembered that the latter, even though given here without a time dimension, refers to a process that typically lasts for more than a year.

The Barí usually need to weed their manioc fields only once between the initial planting and the maturity of the first crop. After a few months the bushy varieties of manioc they traditionally cultivate are able to shade out competitors quite effectively. Only when the ground cover is broken by the beginning of harvesting at 12–15 months do weeds begin to threaten again. The second weeding is not needed until weeks after harvesting has begun, and then only on the areas of the field that have been harvested. Both men and women weed, and, unusually, men probably do as much weeding as women, although the latter account for a good deal of casual weeding in the areas adjacent to those they are harvesting.

The available comparative figures for weeding among the greater Amazonian peoples (Table 3-5) show that again the plantain–banana people (Shipibo) enjoy a labor advantage. Weeding figures are quite variable, reflecting the fact that some people (Machiguenga) weed all their productive land several times a year, whereas others (Barí) weed land, on average, less than once a year. This difference changes the hectarage that is effectively under cultivation, insofar as the work of weeding is concerned—hence the wide difference in weeding investment. It seems likely that there is a difference in the timing of weeding demands

TABLE 3-5. Labor investments (man-hours/ha/year) for weeding; and (man-hours/ha) for harvesting–replanting.

Language family	People	Weeding	Harvesting–replanting
Macro-Chibchan	Barí	75–150[a]	570
Arawakan	Machiguenga	600–900	350–475
	Mirití-Paraná	270	
	San Carlos	186	419
Macro-Tukanoan	Siona-Secoya	101	634
Macro-Pano	Shipibo	106	176

Sources: Barí (Beckerman, 1975, 1983a); Machiguenga (Johnson, 1983); Mirití-Paraná (von Hildebrand, 1975); San Carlos (Clark & Uhl, in press); Siona-Secoya (Vickers, 1983a); Shipibo (Bergman, 1980).

[a]I have previously published the figure of 400 man-hours/ha for Barí weeding (Beckerman, 1983a). This figure is the average amount of time it takes to weed one hectare completely. However, the Barí do not, on average, weed their fields as often as once a year. The figure given here reflects the proportion of all land under cultivation that is actually weeded over a single year.

between manioc and bananas, the former needing more weeding as the plot grows older (after first harvest) and the latter less.

Weeding is an activity in which children often participate, particularly girls. Among the Machiguenga (Johnson, 1983) it seems to be largely restricted to men, whereas among the Cubeo (Goldman, 1948) and the peoples of the Mirití–Paraná (von Hildebrand, 1975) it is almost exclusively the domain of women and their daughters. Both men and women, and children of both sexes, weed among the Barí.

The Amahuaca and Campa (Carneiro, 1964; Denevan, 1971) seem to do very little weeding, while the Machiguenga (Johnson, 1983) do a great deal. We may see here alternative ways of dealing with a common problem. These three peoples live at altitudes over 400 meters, the Campa considerably above that. The strategies of concentrating heavily on weeding, or abandoning fields and settlements every year or two, may both be solutions to the problems of particularly fast and unpleasant (spiny, urticant, well-rooted) weeds at altitudes above 300 or 400 m. The different emphasis on maize, a particularly vulnerable and soil-taxing crop, may also be important here.

Harvesting

Women are responsible for most manioc harvesting among the Barí, although a women's husband may help her out if they happen to be together. A woman goes to a section of her family's field or field segment and begins chopping away the branches of a cluster of manoic plantings, usually those adjacent to the ones she harvested last. After all but enough stem to allow a good handhold is cut away, she heaves the roots out of the ground, often having to loosen the soil around them with a knife. She may probe in the resulting hole if it seems that some roots are still in the ground. When she has them all, she moves on the next planting, until she has a load of 30 or 40 kg. She then cuts the roots apart and packs them vertically in her basket. Replanting is accomplished casually, by reinserting a few sticks from the uprooted planting in the cavity it leaves. The woman then shuffles flatfooted back to the house, carrying with a bark tumpline a loaded basket that often weighs as much as she does. Among the Barí, men probably harvest bananas as often as women do.

Harvesting of manioc is the domain of women for all the peoples treated here (Table 3-5), although men may help, as they do among the Barí and the Campa. Plantain harvest may be dominated by women, as among the Shipibo (Bergman, 1980), or men may have a substantial hand in it, as among the Yanoama (Smole, 1976). Since replanting accompanies the harvesting of manioc and bananas in most swidden fields (exceptions: Campa, Amahuaca), women do a lot of planting even where they do not put the first crop in the ground.

Swidden Productivity

Land

The productivity of aboriginal swidden fields is in many cases higher than that obtained by *criollo* peasants in the same environment. The Barí get, at a conservative estimate (Beckerman, 1983a), over 18 t/ha/yr from their fields, more than twice the average for neo-Venezuelans in the same state (Montaldo, 1973). The Barí figure is unexceptional for South American tropical forest peoples (Table 3-6).

A great deal of the variation in this table is due to the stage of the field; for many cases, the higher end of the range refers to the first crop in a swidden field and the lower end to the last. Some authors apparently give figures only for the first crop, and others attempt an average. Additional variability is accounted for by soil type (hence the very low figure for the Spodosols around San Carlos del Rio Negro), crop spacing and intermixture, and possible completeness of burn (P. Fearnside, 1982, personal communication). The considerably lower yield of bananas compared to that of manioc is also obvious.

TABLE 3-6. Productivity of swidden fields.

		Edible portion	
Language family	People (staple)	kcal/ha/yr (millions)	kg/ha/yr (millions)
Macro-Chibchan	Barí (manioc)	25.0	16.7
	Parima Highlands Yanoama (banana)	2.5	2.8
Jivaroan	Huambisa Jívaro (manioc)	20.0	13.4
	Aguaruna Jívaro (manioc)	37.0	24.7
Macro-Caribe	Kuikuru (manioc)	7.5–21.0	5.0–14.0
	Ye'kuana (manioc)	6.8–25.0	4.5–16.7
Macro-Tukanoan	Siona-Secoya (manioc)	20.0	13.4
Arawakan	Machiguenga (manioc)	13.0	8.7
	San Carlos (manioc)	5.6	3.7
Macro-Pano	Shipibo (banana)	4.3	4.8
	Amahuaca (maize)	13.0–18.0	–
Ge-Bororo-Caraja	Mekranoti (manioc)	29.0	19.3

Sources: Barí (Beckerman, 1975); Parima Highlands Yanomamo (Smole, 1976); Huambisa Jívaro and Aguaruna Jívaro (J. S. Boster, personal communication); Kuikuru (Carneiro, 1983); Ye'kuana (Hames, 1983); Siona-Secoya (Vickers, 1983b); Machiguenga (Johnson, 1983); San Carlos (Clark & Uhl, in press); Shipibo (Bergman, 1980); Amahuaca (Carneiro, 1964); Mekranoti (Flowers et al., 1982; Gross et al., 1979).

Labor

Productivity per unit land may not be very revealing for peoples such as the Amazonian swiddeners, who have a great deal of land. It is likely that these peoples generally aim to maximize not productivity per unit land but, rather, productivity per unit labor.

Several scholars have published energy measures of the labor efficiency of Amazonian swidden (Table 3-7). Here as elsewhere, the Barí with their 30 metabolizable calories of food returned for every 1 calorie of labor above basal metabolism invested, are unexceptional. The ample range, 1 : 20 to 1 : 56, is remarkably small, considering the variation in crops, climate, and techniques, and the opportunities for sampling and measurement error, particularly in the human labor expenditure column. These figures exclude processing, which in the case of bitter manioc would essentially halve the efficiency rate (Dufour, 1984; Clark & Uhl, in press), putting bitter manioc effectively on a par with bananas and plantains as far as overall efficiency goes. The processing of bitter manioc is exclusively a female realm, occupying most of a woman's day among groups relying mainly on that variety. Thus the question of whether to adopt bitter manioc has enormous repercussions on women's labor, but relatively slight ones on men's. It is doubtful that the allegedly somewhat improved quality of the starch of bitter manioc over that of sweet is adequate to explain its prevalence.

In addition to its variability with crop, labor efficiency also declines with the age of the swidden for two reasons: Productivity decreases because of soil exhaustion (and usually weed competition also), and labor requirements increase because of growing weeding demands. There are no studies that present

TABLE 3-7. Labor productivity of swidden agriculture.

Language family	People	Input–output ratio (kilocalories)
Macro-Chibchan	Barí	1 : 30
	Central Yanomami	1 : 20
Macro-Caribe	Kuikuru	1 : 56
Macro-Tukanoan	Siona-Secoya	1 : 52
Arawakan	Machiguenga	1 : 20
	San Carlos	1 : 30
Macro-Pano	Shipibo	1 : 26
	Nambiquara	1 : 24

Sources: Barí (Beckerman, 1975); Central Yanomami (Lizot, 1978); Kuikuru (Vickers, 1979); Siona-Secoya (Vickers, 1979); Machiguenga (Johnson, 1983); San Carlos (Clark & Uhl, in press); Shipibo (Bergman, 1980; Nambiquara (Aspelin, 1976).

associated measurements of these factors as they change over the life of even a single field.

Limitations in Land Productivity

There are three major classes of constraints on the productivity of swidden as practiced by native Amazonian farmers, all of them obliging the abandonment of a field after a relatively few years, and hence causing the low productivity of this form of agriculture per unit land. As shown earlier, its productivity per unit labor is rather high in energetic terms. (Pimentel and colleagues [1973] estimate that the energy efficiency of maize agriculture in the contemporary United States is around 3 : 1.) These limitations on the productive life of a swidden field are soil exhaustion, weed invasion, and pest invasion.

The relative importance of soil type and weed invasion to the decision to abandon a field (or, rather, the interaction of these two factors) is clarified by the Barí, who are semisedentary and use the fields associated with a given longhouse for only part of the year. The Barí may keep riverside alluvial fields in crop for as long as 15 years, while typically abandoning colluvial fields after only 3 years or so. However, the field surrounding the longhouse, whatever its soil type, is usually kept in production for the life of the longhouse—about 10 years. Longhouse-associated fields receive fertilization in the form of garbage and excretions, and probably also enjoy particularly careful weeding.

Because the Barí harvest any particular field for only part of the year, manioc maintains a fairly complete ground cover, and there is relatively little weeding. It takes an average of about 400 man-hours to weed a hectare, but fields are typically weeded less than once a year (averaging over all land in production), so the figure for *annual* man-hours per hectare in weeding is between 75 and 150.

The Barí evidence shows that soil type (and by extension soil exhaustion) is important in the abandonment of a field. The exceptionally long lives of most Barí fields show, however, that exhaustion depends somewhat on harvest rate, which in turn has a great influence on the rate of weed invasion.

The *obiter dicta* about the causes of field abandonment that sprinkle much of the ethnographic literature on Amazonian swidden sometimes take the position that soil depletion or weed invasion is *the* cause of field abandonment in this form of agriculture. I suggest such statements are fundamentally erroneous. The loss of soil fertility depends, among other things, on the kinds of weeds that invade and the success with which they do so; and the identity and success of weed invasion depends, among other things, on the fertility of the soil. Only at the most proximate level of causation can the two be separated.

A third constraint on the productivity of swidden fields is pest invasion. For the mammalian and avian raiders of swidden fields, this invasion is probably more welcomed than feared in most cultures. Most of the vertebrate field pests are desirable game animals. (The Kuikuru, who disdain practically all hunting, are an exception [Carneiro, 1983].) Invertebrate pests are another matter for the Amazonian swiddener. To the best of my knowledge, only maize is attacked often enough and with sufficiently damaging effects by insects, blights, and so forth, that farmers have a potentially serious problem. Hence, one would expect the peoples who grow substantial amount of maize to abandon their fields more frequently than others, to avoid a potential crop disaster. There is some support for this suggestion (Table 3-5). The two groups in this study who rely most heavily on maize, the Amahuaca and the Campa, have the shortest times to abandonment of their fields, although nitrogen depletion of the soil by this demanding crop may also be important.

Options and Opportunities of Swidden

Work

Prominent among the opportunities provided by swidden agriculture must be the free time it provides to its practitioners (Table 3-8). As is the case for many of the labor demands examined here, the Barí are in the midrange of a fairly broad range of labor inputs, both in man-hours/ha/yr for all family members and in man-hours/yr for adult male farmers. Banana–plantain cultivators are

TABLE 3-8. Overall labor demands of Amazonian swidden.

Language family	People	Man-hours/ha/year	Man-hours/year, adult males
Macro-Chibchan	Barí	800–1,000	400–500
	Emberá	327	203
	Central Yanomami		360
Jivaroan	Candoshi	1,000	547
Macro-Caribe	Kuikuru	900	730
	Ye'kuana	589	
Macro-Tukanoan	Siona–Secoya	600	
Arawakan	Machiguenga	2,600	1,400
	San Carlos	660	
Macro-Pano	Shipibo	188	150
	Nambiquara		113
Tupí	Cocamilla	1,113	1,533
Ge–Bororo–Caraja	Mekranoti	1,641	1,359

Sources: Barí (Beckerman, 1975, 1983a); Emberá (Isacsson, 1975); Central Yanomami (Lizot, 1978); Candoshi (Stocks, 1983); Kuikuru (Carneiro, 1961, 1983); Ye'kuana (Hames, 1983); Siona–Secoya (Vickers, 1983a); Machiguenga (Johnson, 1983); San Carlos (Clark & Uhl, in press); Shipibo (Bergman, 1980); Nambiquara (Aspelin, 1976); Cocamilla (Stocks, 1983); Mekranoti (Gross et al., 1979; Flowers et al., 1982; Werner, 1983; Werner et al., 1979).

lower than any of the other swiddeners in both man-hours/ha/yr and labor demands/farmer/yr. The Nambiquara, who obtain a great deal of their food from gathering and hunting (Aspelin, 1976) are the exception to this pattern. Self-sufficient manioc swiddeners spend 1–4 hr/day in agriculture, banana swiddeners only 20 min–1 hr/day, for men's labor.

It would be interesting to know the demands on women's labor for the various kinds of swidden, but the data are insufficient to support much more than general statements. Bitter manioc imposes much greater demands on women in the processing of agricultural produce (a category excluded from Table 3-8) than does either sweet manioc or bananas–plantains (Clark & Uhl, in press; Dufour, 1983, 1984). It is not clear if there is a significant difference in processing time between sweet manioc and plantains. Because women generally do not plant bananas and plantains (the Emberá are an exception), it may be that the overall labor demand on women is lighter with *Musa* than with sweet manioc.

The timing of labor demands in Amazonian swidden can be thought of as a drizzle punctuated by cloudbursts. Harvesting and weeding, the major components of swidden labor, typically take place in snatches of an hour or two (the exception is the Machiguenga) spread rather evenly over the year. Brushing out and felling, burning and planting, however, are usually performed in more or less continuous stretches of full work days (exception: the contemporary, market-oriented Uarina). These seasonal demands are highly correlated with, and in more than a few cases restricted to, the beginning of the dry season for the former and the end of the dry season for the latter.

Land

Many authors have investigated land ownership and usufruct among Amazonian swiddeners, and have sought conflicts over land use and ownership. One cannot avoid the impression that these inquiries have been largely fruitless. Among the Barí, a local group clears communal fields around its longhouses and elsewhere, as members decide in an easy consensus from which any family or adult individual is free to dissent simply by absenting himself either from the clearing (in the case of the subsidiary field) or the activities of house establishment (in the case of the house-surrounding field) by joining another local group. Individuals also may clear private fields or fields made with a small group of friends or kin. The area planted by an individual becomes his at that time and remains so as long as any crops he planted remain in it. I have no information suggesting any disputes about land.

The Barí situation seems to be typical: People take possession of land by working it, and that possession is respected as long as the land produces. Indeed, the occasional disputes over a patch of land (Kloos [1971] reports a memory of a conflict between two Maroni River Carib men who planned to clear the same spot) may simply be instances of people quarreling over other matters who focus their animosity on what would otherwise be a trivial issue. In general, traditional swiddeners are not (in the ethnographic present) stressed by a need to seek or defend land.

Population Density

It is doubtful that contemporary population densities of the South American swiddening peoples (Table 3-9) have much to do with their farming systems. The Barí, for instance, have probably increased their population density threefold in this century, in response to the loss of 85% of their territory (Lizarralde & Beckerman, 1982), without any major change in their subsistence activities until the last decade or so. In contrast, Yaruro density fell by an even greater factor as a result of a massacre (Leeds, 1961). Both land theft and massacre are common events in the history of the great majority, if not all, of the peoples considered here; devastating epidemics have also been frequent occurrences in the 400-odd years since these peoples came into contact with Old World populations. It would be somewhat of a fool's errand to attempt to derive the exceptionally low population densities at which these people live today from features of their farming system. As shown later, tropical forest swidden does support population densities an order of magnitude greater in other parts of

TABLE 3-9. Population densities of contemporary Amazonian swiddeners.

Language family	People	Density (persons/km²)
Macro-Chibchan	Barí	0.1–0.3
	Yanoama	0.18
	Central Yanomami	0.34
Jivaroan	Jívaro	0.46
	Huambisa Jívaro	0.8
	Aguaruna Jívaro	0.9
	Yaruro	0.13–0.56
Macro-Tukanoan	Siona-Secoya	0.26
	Tatuyo	0.2
Arawakan	Machiguenga	0.3
	Campa	1.0
Macro-Pano	Shipibo	4.0
Ge-Bororo-Caraja	Mekranoti	0.01
Unclassified	Piaroa	0.12

Sources: Barí (Lizarralde & Beckerman, 1982); Yanoama (Smole, 1976); Central Yanomami (Lizot, 1977); Jívaro (Harner, 1973); Huambisa and Aguaruna Jívaro (J. S. Boster, personal communication, 1985); Yaruro (Leeds, 1961); Siona-Secoya (Vickers, 1983b); Tatuyo (Dufour, 1983); Machiguenga (Johnson, 1983); Campa (Denevan, 1971); Shipibo (Bergman, 1980); Mekranoti (Werner, 1983); Piaroa (Kaplan, 1975).

the world, where disease susceptibility is lower and land tenure somewhat more secure.

The instructive feature of these low population densities is that they demonstrate that agriculture can be carried on by peoples who live at densities no higher than those of many hunting and gathering peoples. Despite the considerable reduction in mobility that swidden farming entails, these people manage to travel far enough often enough to find mates, as they usually must, given the small sizes of the settlements in which most of them live.

Directions of Change

The directions of change in Amazonian swidden seems usefully categorized under three rubrics. The first is a disintensification, a decrease in investment in labor and field longevity and a dramatic increase in settlement mobility, combined with a shift in emphasis to gathering and hunting as the primary ways of making a living. The Barí may have experimented briefly with this strategy earlier in this century, when much of their land was taken from them (Lizarralde & Beckerman, 1982) and they were sometimes hunted like animals. It is this retreat that we almost certainly see among the Sirionó (Holmberg, 1969) and probably also among the Amahuaca and the Campa. It is a tactic of survival in the face of slaughter and exile, and it tells us little about swidden as a farming system except that a refugee remnant can make use of it even when the large majority of the society's members are dead and the remainder have been driven from their best lands and must carry on their lives under the constant threat of death or slavery.

The second kind of change is an entrance into cash cropping. The Barí are currently experiencing this change in their farming system, and many individuals now grow manioc, plantains, and even cacao for cash. It is a change similar to what Stocks (1983) shows the Cocamilla to have undergone. Labor investments per unit land and per farmer rise, sedentism tends to increase, and there may be a shift in field architecture from monocropped or zoned to intercropped fields, as among the Cocamilla (Stocks, 1983); or, contrariwise, there may be a shift from intercropping to monocropping, as among the Siona–Secoya (Vickers, 1983a). The Barí simply expand their basically monocropped fields.

These sorts of shifts are also expected with a reduction in the amount of land available as a result of expropriation by outsiders, rather than flight to an area in which land is still freely available. These changes are also influenced by taxation, the presence of a market, and deliberate agents of change (e.g., missionaries, development officers).

The third kind of change is no longer available for ethnographic study. It is the development of more land-productive alternatives to typical swidden under aboriginal conditions of population growth or colonization of a new area. This is the kind of development meticulously traced by Denevan and

Schwerin (1978) for the draining of *morichales* (morichal palm swamps) by the Karinya for use as semipermanent fields, which Roosevelt (1980) suggests for the transition from swidden manioc to floodplain maize agriculture along the Orinoco, and which Denevan and Zucchi (1978) suggest for the origin of the ridged fields of the Venezuelan llanos.

The fact that these sorts of departures from swidden are known from greater Amazonia suggests that the systems we see operating today are for the most part tiny remnants of what was once a much larger system of farmers and fields. Swidden in other tropical forest areas supports much thicker populations. Densities reported by Hinton (1978) for rice swidden in Thailand range from 10 to 40 persons/km^2, with fallow periods of 5–20 years, for a system under severe stress. Stauder (1971) reports a density of 13/km^2 for the maize-sorghum swidden of the Majangir of Ethiopia, with a fallow period of 10+ years (probably much more, on average) in a system that could support a much larger population. Both these regions are climatically comparable to much of Amazonia. Data assembled by Ruthenberg (1976:49) for shifting cultivation systems in forested areas of Borneo and Zaire show per-hectare labor inputs roughly twice as high as those expended by the Amazonians, both for the land-clearing phase of the process and for the total investment in getting a crop out of the ground. Only the Machiguenga, of the Amazonian peoples, approach the Bornean and African labor figures. The fact that some aboriginal Amazonian peoples, before the European conquest, needed to develop more land-productive farming systems than swidden, probably should be read to imply that Amazonian swidden today is far less productive per hectare than it was in the past, with contemporary aboriginal peoples in Amazonia investing as little labor as they can, in the face of a great surplus of land and the constant threat of dispossession by force.

It is possible, and I believe it to be quite likely, that the ethnographic data assembled here on Amazonian swidden capture only one end of the range of ethnohistorical variation in this farming system. If we now see only the lower portion of the spectrum of production per unit land of which Amazonian swidden is capable, and the lower end of the range of population densities it can support, then we still have only a very partial idea of the kinds of societies that may have been built upon this farming system in greater Amazonia, or which may yet come to be built upon it.

Acknowledgments

Initial fieldwork with the Barí was funded by the Ford Foundation, the Smithsonian Institution Urgent Anthropology Fund, Sigma Xi, and the University of New Mexico; later fieldwork was supported by the Instituto Venezolano de Investigaciones Científicas.

Unpublished information was kindly made available by James Boster and Christopher Uhl; Allen Johnson helpfully rechecked a number of outlying figures; Edward Fry provided a useful recent publication; Bernie McGuire did most of the tedious work of preparing Figure 3-3; Hope

Boylston helped with this and other figures. The manuscript was retyped more times than I care to remember by Marie Rauch, and its final corrections were entered by Jean Courter.

I am most grateful to these people and institutions; they do not, of course, share in the guilt for any errors, either of commission or of omission, that I may have made.

Notes

1. I have previously published a description of Barí swidden stating that they have only one variety of manioc (Beckerman, 1983a); more recent fieldwork reveals that I was wrong.

2. The data in this table, like those in the tables that follow, are compiled from direct statements in the sources listed, and from calculations made on related data presented in those sources. The reliability of figures presented in them is variable, and the reader is referred to the sources cited for the evaluation of any particular figure.

3. It is possible that excessively large areas of cutting, rather than fragility of the ecosystem per se, are responsible for the apparently anthropogenic savannahs in Yanomamo territory.

4. As all swidden agriculture is done in forest, peoples who practice this system in savannah areas—Yaruro (Leeds, 1961); Guajibo (Eden, 1974); Pemon (Thomas, 1982); Nambiquara (Aspelin, 1976)—perforce locate their fields in gallery forest or, occasionally, in patches of forest unassociated with rivers but owing their presence to otherwise adequate year-round moisture. Settlements are usually located nearby, although not necessarily in the forest itself.

5. Pest loss is reduced on nonalluvial soils because pacas and capybaras, potential major pests, are riverine animals; and because it is possible that all the mammalian crop predators avoid ridge tops to some extent (Carneiro, 1964).

References

Alvim, P. de T., & Kozlowsky, T. T. (Eds.). *Ecophysiology of tropical crops*. New York: Academic Press, 1977.

Aspelin, P. L. Nambicuara economic dualism: Levi-Strauss in the garden, once again. *Bijdragentot de Tall -, Land-en Vokenkunde. Anthropologica*, 1976, 18, 1-31.

Beckerman, S. *The cultural energetics of the Barí (Motilones Bravos) of northern Colombia*. Unpublished doctoral dissertation, Department of Anthropology, University of New Mexico, Albuquerque, 1975.

Beckerman, S. Los Barí: Sus reacciones frente a la contraccion de sus tierras. In N. S. Friedemann (Ed.), *Tierra, tradicion y poder en Colombia* (pp. 65-83). Bogotá, Colombia: Instituto Colombiano de Cultura, 1976.

Beckerman, S. The use of palms by the Barí Indians of the Maracaibo basin. *Principes*, 21, 143-154.

Beckerman, S. The abundance of protein in Amazonia: A reply to Gross. *American Anthropologist*, 1979, 81, 533-560.

Beckerman, S. Barí swidden gardens: Crop segregation patterns. *Human Ecology*, 1983a, 11, 85-102.

Beckerman, S. Does the swidden ape the jungle? *Human Ecology*, 1983b, 11, 1-12.

Beckerman, S. Optimal foraging group size for a human population: The case of Barí fishing. *American Zoologist*, 1983c, 23, 283-290.

Beckerman, S. A note on ringed fields. *Human Ecology*, 1984, 12, 203-206.

Bergman, R. W. *Amazon economics: The simplicity of Shipibo Indian wealth*. Dellplain Latin American Studies (Vol. 6). Syracuse, NY: Syracuse University, Department of Geography, 1980.

Boserup, E. *The conditions of agricultural growth: The economics of agrarian change under population pressure*. Chicago: Aldine, 1965.

Boster, J. S. How the exceptions prove the rule: Analysis of informant disagreement in Aguaruna manioc identification. Unpublished doctoral dissertation, Department of Anthropology, University of California, Berkeley, 1980.

Boster, J. S. A comparison of the diversity of Jivaroan gardens with that of the tropical forest. *Human Ecology*, 1983, 11, 47–68.

Carneiro, R. L. Slash-and-burn agriculture: A closer look at its implications for settlement patterns. In A. F. C. Wallace (Ed.), *Men and cultures: Selected papers of the Fifth International Congress of Anthropolitical and Ethnographic Sciences*. Philadelphia: University of Pennsylvania Press, 1960.

Carneiro, R. L. Slash-and-burn cultivation among Kuikuru and its implications for cultural development in the Amazon basin. In J. Wilbert (Ed.), *The evolution of horticultural systems in native South America: Causes and consequences—A symposium*. Caracas, Venezuela: Sociedad de Ciencias Naturales La Salle, 1961.

Carneiro, R. L. (1964). Shifting cultivation among the Amahuaca of eastern Peru. *Volkerkand Liche Abhandlungen*, 1964, 1, 9–18.

Carneiro, R. L. The cultivation of manioc among the Kuikuru of the upper Xingu. In R. B. Hames & W. T. Vickers (Eds.), *Adaptive responses of native Amazonians*. New York: Academic Press, 1983.

Clark, K. E., & Uhl, C. Decline of the traditional subsistence life at San Carlos de Rio Negro. In *Structure and function of Amazonian forest ecosystems in the upper Rio Negro*. Athens: University of Georgia Institute of Ecology, in press.

Cochrane, T. T., & Sanchez, P. A. Land resources, soils and their management in the Amazon region: A state of knowledge report. In S. B. Hecht (Ed.), *Amazonia: Agriculture and land use research*. Cali, Colombia: Centro Internacional de Agricultura Tropical, 1982.

Conklin, H. C. The study of shifting cultivation. *Current Anthropology*, 1961, 2, 27–67.

Denevan, W. M. Campa subsistence in the Gran Pajonal of eastern Peru. *The Geographical Review*, 1971, 4, 496–518.

Denevan, W. M. The aboriginal population of Amazonia. In W. M. Denevan (Ed.), *The native population of the Americas in 1942*. Madison: The University of Wisconsin Press, 1976.

Denevan, W. M., & Schwerin, K. H. Adaptive strategies in Karinya subsistence, Venezuelan llanos. *Antropologica*, 1978, 50, 3–91.

Denevan, W. M., Treacy, J. M., Alcorn, J. B., Padoch, C., Denslow, J., & Flores Paitan, S. Indigenous agroforestry in the Peruvian Amazon: Bora Indian management of swidden fallow. *Interciencia*, 1984, 9, 346–357.

Denevan, W. M., & Zucchi, A. Ridged field excavations in the central Orinoco Llanos, Venezuela. In D. Browman (Ed.), *Advances in Andean archaeology*. The Hague: Mouton, 1978.

Dufour, D. L. Nutrition in the northwest Amazon: Household dietary intake and time–energy expenditure. In R. B. Hames & W. T. Vickers (Eds.), *Adaptive responses of native Amazonians*. New York: Academic Press, 1983.

Dufour, D. L. The time and energy expenditure of indigenous women horiculturalists in the northwest Amazon. *American Journal of Physical Anthropology*, 1984, 65, 37–46.

Eden, M. J. Ecological aspects of development among Piaroa and Guahibo Indians of the upper Orinoco basin. *Antropologica*, 1974, 39, 25–56.

Ericksson, E. Air borne salts and the chemical composition of river waters. *Tellus*, 1955, 7, 243–250.

Falesi, I. C. Soils of the Brazilian Amazon. In C. Wagley (Ed.), *Man in the Amazon*. Gainesville: The University Presses of Florida, 1974.

Flowers, N. M., Gross, D. R., Ritter, M. L., & Werner, D. W. Variation in swidden practices in four central Brazilian Indian societies. *Human Ecology*, 1982, 10, 203–217.

Frikel, P. Agricultura dos indios munduruku. *Boletim do museu Paraense Emilio Goeldi, Antropologia* 4, Belem, 1957.

Geertz, C. *Agricultural involution*. Berkeley: University of California Press, 1963.
Geisler, R., Knoppel, H. A., & Sioli, H. The ecology of freshwater fishes in Amazonia: Present status and future tasks for research. *Applied Sciences and Development*, 1973, 2, 144–162.
Goldman, I. Tribes of the Vaupes–Caqueta region. In J. H. Steward (Ed.), *Handbook of South American Indians, Vol. 3: The tropical forest tribes*. Washington, DC: U.S. Government Printing Office, 1948.
Gross, D. R. Protein capture and cultural development in the Amazon basin. *American Anthropologist*, 1975, 77, 526–549.
Gross, D. R., Eiten, G., Flowers, N. M., Leoi, F. M., Ritter, M. L., & Werner, D. W. Ecology and acculturation among native peoples of central Brazil. *Science*, 1979, 206, 1043–1050.
Hames, R. Monoculture, polyculture, and polyvariety in tropical forest swidden cultivations. *Human Ecology*, 1983, 11, 13–34.
Harner, M. S. *The Jívaro: People of the sacred waterfalls*. New York: Anchor/Doubleday, 1973.
Harris, D. R. The ecology of swidden cultivation in the upper Orinoco rain forest, Venezuela. *Geographical Review*, 1971, 61, 475–495.
Herrera, R. A. *Nutrient distribution and cycling in an Amazon caatinga forest on Spodosols in southern Venezuela*. Unpublished PhD thesis, Department of Soil Science, University of Reading, Reading, England, 1979.
Hinton, P. Declining production among sedentary swidden cultivators: The case of the Pwo Karen. In P. Kunstadter, E. C. Chapman, & S. Sabharsri (Eds.), *Farmers in the forest: Economic development and marginal agriculture in northwestern Thailand*. Honolulu: The University Press of Hawaii, 1978.
Holmberg, A. R. *Nomads of the long bow: The Siriono of eastern Bolivia*. New York: The Natural History Press, 1969.
Irvine, M. D. Rain forest adaptations: Patch management through succession. Paper presented at the 80th annual meeting of the American Anthropological Association, Washington, DC, 1981.
Isacsson, S. E. Observations on Choco slash-mulch agriculture: Work diary and dietary of an Embera domestic group in mid-eastern Choco, Colombia. *Goteborgs Ethnografiska Museum, Annual Report for 1975*, 21–45.
Johnson, A. Machiguenga gardens. In R. B. Hames & W. T. Vickers (Eds.), *Adaptive responses of native Amazonians*. New York: Academic Press, 1983.
Kaplan, J. O. *The Piaroa, a people of the Orinoco basin: A study in kinship and marriage*. Oxford: Clarendon Press, 1975.
Kass, D. C. Polyculture cropping system: Review and analysis. *Cornell International Agriculture Bulletin*, 1978, 32, 1–68.
Klinge, H., & Rodrigues, A. Litter production in an area of Amazonian *terra firme* forest. Part 1: Litter-fall, organic carbon and total nitrogen contents of litter. *Amazoniana*, 1968a, 1, 287–302.
Klinge, H. & Rodrigues, A. Litter production in an area of Amazonian *terra firme* forest. Part 2: Mineral nutrient content of the litter. *Amazoniana*, 1968b, 1, 303–310.
Klinge, H., & Rodrigues, W. A. Phytomass estimation in a central Amazonian rain forest. In H. E. Young (Ed.), *IUFRO biomass studies*. Orono: University of Maine Press, 1974.
Kloos, P. *The Maroni River Caribs of Surinam*. Assen, The Netherlands: Kroninklijke Van Gorrum & Co., N.V., 1971.
Kramer, B. J. Implicaciones ecologicas de la agricultura Urarina. *Amazonia Peruana*, 1977, 1, 75–86.
Lathrap, D. The "hunting" economies of the tropical forest zone of South America: An attempt at historical perspective. In R. B. Lee & I. DeVore (Eds.), *Man the hunter*. Chicago: Aldine, 1968.
Lathrap, D. W. Our father the cayman, our mother the gourd: Spinden revisited, or a unitary model for the emergence of agriculture in the New World. In C. A. Reed (Ed.), *Origins of agriculture*. The Hague: Mouton, 1977.

Leeds, A. Yaruro incipient tropical forest horticulture—Possibilities and limits. In J. Wilbert (Ed.), *The evolution of horticulture systems in native South America: Causes and consequences—A symposium*. Caracas, Venezuela: Sociedad de Ciencias Naturales La Salle, 1961.

Lieth, H. Primary production of the major vegetation units of the world. In H. Lieth & R. H. Whittaker (Eds.), *Primary productivity of the biosphere*. New York: Springer-Verlag, 1975.

Linares, O. F. "Garden hunting" in the American tropics. *Human Ecology*, 1976, 4, 331–349.

Lizarralde, R., & Beckerman, S. Historia contemporanea de los Barí. *Antropologica*, 1982, 58, 3–52.

Lizot, J. Economie ou société? Quelques thèmes à propos de l'étude d'une communaute d'Amerindiens. *Journal de la Société des Americanistes*, 1971, 60, 137–175.

Lizot, J. Population, resources and warfare among the Yanomami. *Man*, 1977, 12, 497–517.

Lizot, J. Economie primitive et subsistance. *Libre*, 1978, 4, 69–113.

Manshard, W. *Tropical agriculture*. New York: Longman, 1974.

Mayobre, F., San Jose, J. J., Orihuela, B. E., & Acosta, J. Influencia del nivel de fertilización y riego sobre el crecimiento de *Manihot esculenta* Crantz var. Cubana, 1982a, 171–196.

Mayobre, F., San Jose, J. J., Orihuela, B. E., & Acosta, J. Caracteristicas morfológicas, anatómicas y fisiológicas que influyen sobre el crecimiento de una comunidad de *Manihot esculenta* crantz, var. Cubana. *Alcance: Revista de la Facultad de Agronomia, Universidad Central de Venezuela*, 1982b, 31, 197–206.

Meggers, B. J. *Amazonia: Man and culture in a counterfeit paradise*. Chicago: Aldine-Atherton, 1971.

Montaldo, A. Importancia de la yuca en el mundo actual con especial referencia a Venezuela. *Revista de la Facultad de Agronomia de la Universidad Central de Venezuela*, 1973, 22, 17–40.

Moran, E. *Developing the Amazon*. Bloomington: Indiana University Press, 1981.

Patino, V. M. *Plantas cultivadas y animales domesticos en America equinoccial* (6 vol.). Cali, Colombia: Imprenta Departmental, 1963–1974.

Pimentel, D., Hurd, L. E., Bellotti, A. C., Forster, M. J., Oka, I. N., Sholes, O. D., & Whitman, R. J. Food production and the energy crisis. *Science*, 1973, 182, 443–449.

Pires, J. The forest ecosystems of the Brazilian Amazon: Description, functioning, and research need. In *Tropical forest ecosystems: a state-of-knowledge report*. Paris: UNESC, 1978.

Posey, D. A. Indigenous ecological knowledge and development of the Amazon. In E. F. Moran (Ed.), *The dilemma of Amazonian development*. Boulder, CO: Westview Press, 1983.

Pottier, B. *America Latina en sus lenguas indígenas*. Caracas, Venezuela: UNESCO–Monte Avila, 1983.

Purseglove, J. W. *Tropical crops: Dicotyledons*. London: Longmans, 1968.

Purseglove, J. W. *Tropical crops: Monocotyledons*. London: Longmans, 1972.

Riviere, P. *Marriage among the Trio*. Oxford: Clarendon Press, 1969.

Rogers, D. J., & Fleming, H. S. A monograph of *Manihot esculenta* with an explanation of the taximetric methods used. *Economic Botany*, 1973, 27, 1–113.

Romanoff, S. Informe sobre el uso de la tierra por los Matses, en la selva baja peruana. *Amazonia Peruana*, 1976, 1, 97–130.

Roosevelt, A. C. *Parmana: Prehistoric maize and manioc subsistence along the Amazon and Orinoco*. New York: Academic Press, 1980.

Ruthenberg, H. *Farming systems in the tropics*. Oxford: Clarendon Press, 1976.

Salati, E., & Vose, P. B. Amazon basin: A system in equilibrium. *Science*, 1984, 225, 129–138.

Saldarriaga, J. *An 80 year chronosequence of forest regeneration following slash and burn agriculture in upper Rio Negro region of Amazonia*. Unpublished doctoral dissertation, Department of Biology, University of Tennessee, Knoxville, 1985.

Sanchez, P. A. Multiple cropping: An appraisal of present knowledge and future needs. In M. Stelly (Ed.), *Multiple cropping*, American Society of Agronomy special publication no. 27, Madison, WI 1976.

Schubart, H. O. R., & Salati, E. Natural resources for land use in the Amazon region: The natural systems. In S. B. Hecht (Ed.), *Amazonia: Agriculture and land use research*. Cali, Colombia: Centro Internacional de Agricultura Tropical, 1982.

Schwerin, K. H. Apuntes sobre la yuca y sus orígenes. *Tropical Root and Tuber Crops Newsletter*, 1970, 3, 4–12.

Sigaut, F. *L'agriculture et le feu: Rôle et place du feu dans les techniques de préparation du champ de l'ancienne agriculture européenne*. Paris: Mouton, 1975.

Smith, N. J. H. Anthrosols and human carrying capacity in Amazonia. *Annals of the Association of American Geographers*, 1980, 70, 553–556.

Smith, N. J. H. *Rainforest corridors: The transamazon colonization scheme*. Berkeley: University of California Press, 1982.

Smole, W. J. *The Yanoama Indians: A cultural geography*. Austin: University of Texas Press, 1976.

Sombroek, W. G. *Amazon soils: A reconnaissance of the soil of the Brazilian Amazon region*. Wageningen: Centre for Agricultural Publications and Documentation, 1966.

Stark, N. Nutrient cycling pathways and litter. *Bioscience*, 1972, 22, 355–360.

Stauder, J. *The Majangir: Ecology and society of a southwest Ethiopian people*. Cambridge: Cambridge University Press, 1971.

Stelly, M. (Ed.). *Multiple cropping*. American Society of Agronomy special publication no. 27. Madison WI, 1976.

Sternberg, H. O. *The Amazon river of Brazil*. New York: Springer-Verlag, 1975.

Stocks, A. Candoshi and Cocamilla swiddens in eastern Peru. *Human Ecology*, 1983, 11, 69–84.

Tai, E. A. Banana. In P. Alvim & T. T. Kozlowski (Eds.), *Ecophysiology of tropical crops*. New York: Academic Press, 1977.

Thomas, D. J. *Order without government: The society of the Pemon Indians of Venezuela*. Urbana: University of Illinois Press, 1982.

Uhl, C. *Studies of forest agricultural and successional environments in the upper Rio Negro region of the Amazon basin*. Unpublished doctoral dissertation, Department of Botany, Michigan State University, East Lansing, 1980.

Uhl, C., & Jordan, C. F. Structure and composition of the *tierra firme* forest on oxisols at San Carlos de Rio Negro: Static and dynamic perspectives. In *Structure and function of Amazonian forest ecosystems in the upper Rio Negro*. Athens: University of Georgia Institute of Ecology, in press.

Ungemach, H. Chemical rain water studies in the Amazon region. In J. M. Idrobo (Ed.), *II Simposio y foro de biologia tropical Amazonica*. Bogotá, Colombia: Asociación Pro-Biologia Tropical, 1970.

Vasey, D. E. Population and agricultural intensity in the humid tropics. *Human Ecology*, 1979, 7, 269–283.

Vickers, W. T. Native Amazonian subsistence in diverse habitats: The Siona–Secoya of Ecuador. *Changing Agricultural Systems in Latin America*, 1979, 1, 6–36.

Vickers, W. T. Tropical forest mimicry in swiddens: A reassessment of Geertz's model with Amazonian data. *Human Ecology*, 1983a, 11, 35–46.

Vickers, W. T. The territorial dimensions of Siona–Secoya and Encabellado adaptation. In R. B. Hames & W. T. Vickers (Eds.), *Adaptive responses of native Amazonians*. New York: Academic Press, 1983b.

von Hildebrand, P. Observaciones preliminares sobre utilización de tierras y fauna por los indígenas del rio Mirití–Paraná. *Revista Colombiana de Antropología*, 1975, 18, 183–292.

Walter, H. *Vegetation of the earth*. New York: Springer-Verlag, 1973.

Walter, H., & Lieth, H. *Klimadiagramm-Weltatlas*. Jena: VEB Gustav Fischer Verlag, 1967.

Went, F. W., & Stark, N. The biological and mechanical role of soil fungi. *Proceedings of the National Academy of Science*, 1968, 60, 497–504.

Werner, D. Why do the Mekranoti trek? In R. B. Hames & W. T. Vickers (Eds.), *Adaptive responses of native Amazonians*. New York: Academic Press, 1983.

Werner, D., Flowers, N. M., Ritter, M. L., & Gross, D. R. Subsistence productivity and hunting effort in native South America. *Human Ecology*, 1979, 2, 303–315.

Wiley, R. W. Intercropping – Its importance and research needs. *Field Crop Abstracts*, 1979, 32, 1–10, 73–85.

Yde, J. *Material culture of the Waiwai*. Copenhagen: National Museum of Copenhagen, 1965.

Young, A. *Tropical soils and soil survey*. Cambridge: Cambridge University Press, 1976.

4

Milpa in Yucatán: A Long-Fallow Maize System and Its Alternatives in the Maya Peasant Economy

PETER T. EWELL AND DEBORAH MERRILL-SANDS

The majority of Mexican farmers are peasants, members of rural communities with small holdings and limited resources beyond the labor of their own families. Although they produce a significant proportion of the nation's food supply, they have shared very unequally in the benefits of economic development. Many have been relegated to agroclimatically marginal regions, and many have faced unfavorable exchange relationships in the input, product, and labor markets. On the basis of empirical experience accumulated over generations, they practice a variety of traditional agricultural systems suited to particular conditions in the heterogeneous landscape. To meet the needs of their households, most peasant farmers have developed flexible strategies of diversification. They simultaneously grow a portion of their direct food requirements, produce crops and animals for sale, and work off the farm on a seasonal or intermittent basis.

The State of Yucatán is a semiarid limestone plain in the lowland tropics of southeastern Mexico (Figure 4-1). Although it was a major site of pre-Hispanic Maya civilization and has been continuously inhabited for thousands of years, the region is marginal for conventional forms of agricultural development. The soils are thin and rocky, the rainfall patterns are highly irregular, and there are no permanent sources of surface water. Since the Spanish conquest over 450 years ago, maize (*Zea mays*) and other food crops have been grown almost exclusively in small communities of Maya peasants scattered through the scrubby deciduous forest. The *milpa*, a traditional slash-and-burn system of temporary cultivation and continuous rotation through forest fallow, is adapted to the patchy and difficult environment. Nevertheless, average yields today are among the lowest in Mexico, and the capacity of the system to provide a secure food supply or an adequate income to most peasant families is deteriorating. This chapter outlines the characteristics of the *milpa* system, the

Peter T. Ewell. Social Science Department, International Potato Center (CIP), Lima, Peru.

Deborah Merrill-Sands. Research Fellow. International Services for National Agricultural Research, The Hague, The Netherlands.

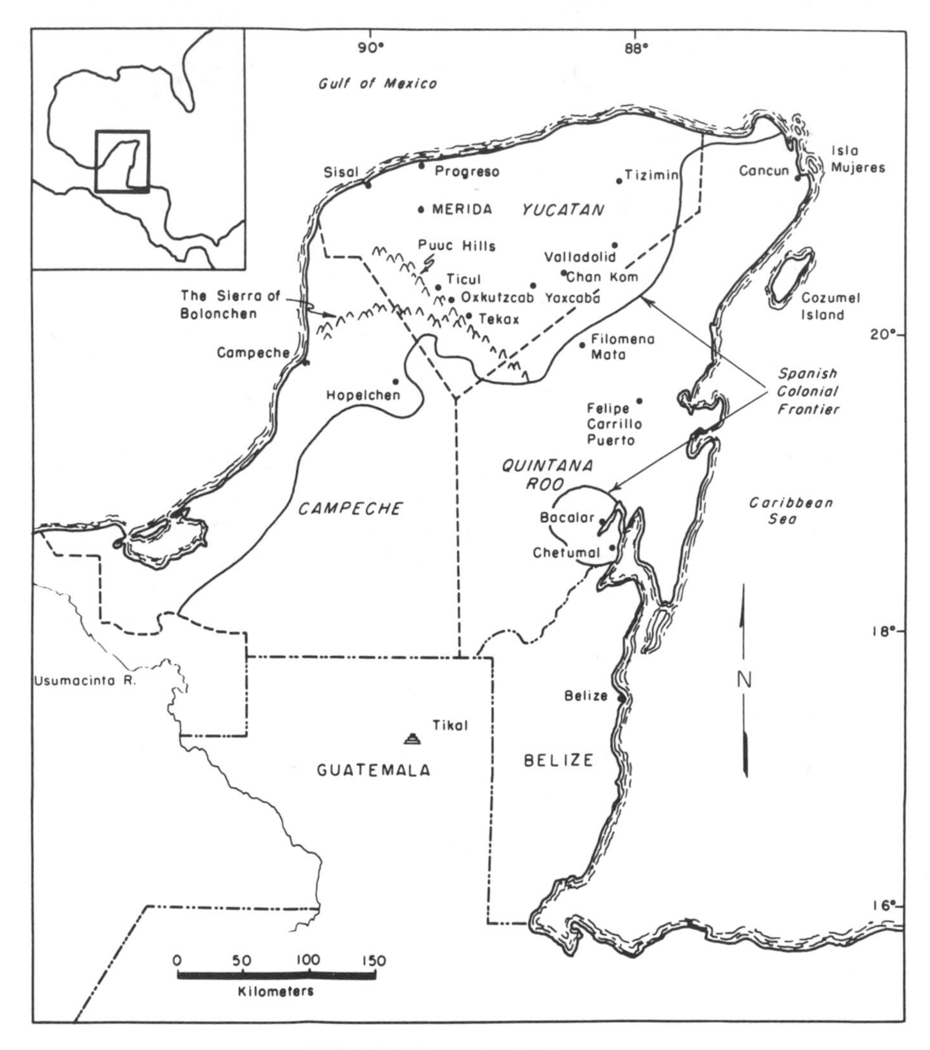

FIG. 4-1. The peninsula of Yucatán.

increasingly serious dilemma faced by the Maya, and two alternative agricultural enterprises that have been developed within the organizational and decision-making framework of the peasant economy in response to these pressures.

Since the late 1960s, approximately 8,000 Maya peasant families have become small-scale beekeepers producing honey for the export market, using European honey bees, *Apis mellifera*, and the introduced technology of the movable frame hive. They produce 5–10% of the honey traded internationally. Their apiaries are scattered throughout the forest so that the bees can forage on the abundant flowering trees and shrubs. The majority exploit commerical beekeeping as a complementary activity to subsistence-oriented production of the *milpas*. The land-use patterns and calendars of labor tasks do not compete. Income from the sale of honey helps finance the next *milpa*, carries the family through the period of food shortages before the next harvest, and usually covers the basic cash needs of the household. This combination of commercial and subsistence production activities increases the economic stability of the peasant household and buffers it against fluctuations both in the market economy and in agroclimatic conditions. The mixed system predominates in the central, eastern, and southern parts of the state.

A region bordering on the Puuc hills in southern Yucatán has become a center for the production of fruits and vegetables for the regional market. Over the past 40 years, approximately 3,500 Maya families have shifted from *milpa* to more intensive production systems. Small-scale irrigation units have been built by the federal government, but the peasant farmers have maintained the autonomy to develop their parcels gradually, with incremental investments of labor time and cash earned in migratory labor. To provide a more or less even flow of income through the year, to minimize risk in an uncertain market, and to accumulate equity in the long-term security of their households, they have planted diverse combinations of annual, semiperennial, and perennial crops. They have adapted techniques from the *milpa* and from the traditional backyard tree–gardens, or *solares*, and they have adopted selected elements of modern technology.

The peasant economy is organized to attempt to meet the subsistence needs and cash income requirements of the household, and, if possible, to build up equity in its long-term security. Unlike commercial farmers, peasants do not necessarily attempt to maximize direct monetary returns to a given unit of land or labor time. The labor of their families is simultaneously the principal resource that they command and an overhead cost. Their consumption requirements include food and an increasing number of goods and services that must be purchased with cash income. Although *milpa* yields are low and uncertain, the system persists because it gives each household partial control over its food security by providing a crucial portion of its basic consumption needs. The *milpa* has become internalized into the cultural values of the Maya; in the communities, a family that is forced to buy maize is considered poor.

Although these two alternative enterprises are characteristic of different agricultural regions in Yucatán, they developed in response to stresses within the *milpa* economy and share certain characteristics. They do not conflict with the production calendar of the *milpa*, and the labor involved is less onerous. They do not require a large block of capital to get started, and the production units can be built up gradually over a period of years. They permit incremental investments of labor time, the most abundant resource in the peasant economy, and require only moderate amounts of cash, which is chronically scarce. They do not expose the household to excessive risk, and the level of management can be adjusted in response to climatic variations and the fluctuating fortunes of the producers' families. The yields and net returns to these peasant enterprises are modest. They have, however, permitted many households to take advantage of the improved infrastructure and new markets that have become available in the regional economy without sacrificing control over their resources or threatening the stability of the family economy.

The Milpa System in the Physical Environment of Yucatán

Making *milpa* is the principal activity of approximately 40,000 Maya Indian families who live in small rural communities throughout Yucatán. The crops and the sequence of cultural operations in contemporary *milpa* cultivation recall those reported in Spanish chronicles of the 16th century (Tozzer, 1941). The system remains virtually the only source of food crops in the state and is the central element in a combination of resource uses in the traditional Maya peasant economy. Although average maize yields of less than 1 t/ha are among the lowest in Mexico, the system is flexibly adapted to the variable, patchy, and difficult agroclimatic characteristics of the region.

Patterns of Access to Groundwater

The peninsula of Yucatán (Figure 4-1), is a massive limestone shelf that was slightly uplifted from the sea in relatively recent geological time, from the Pleistocene to the Eocene eras (Isphording, 1975; Wilson, 1980). It is divided into three states, although since colonial times most of the population has been concentrated within what is now the state of Yucatán. The rural settlement pattern of small, dispersed villages can be largely explained by the irregularly distributed points of access to groundwater.

There is no surface water whatever in the state—no rivers, streams, or lakes. Rainfall percolates very rapidly through the porous limestone, forming a lens of fresh water with a mean thickness of 70 meters, which floats on intruded seawater (Doering & Butler, 1974; Lesser, 1976). Within 100–150 meters of the north coast, where the groundwater is less than 30 meters below the surface sinkholes called (*cenotes*) and hand-dug wells provide access to it. Most of the ancient Maya ruins and contemporary villages are located at these *cenotes*, which are formed when the roofs of underground caverns collapse.

In the southwest, the Puuc hills rise to 100 meters in an abrupt escarpment, and the depth to the groundwater drops to over 150 meters within a few kilometers. Only in recent years have deep wells been drilled in any numbers. The lands to the south have been a sparsely populated forest frontier since before the Spanish conquest (Roys, 1957). Nevertheless, a few scattered seasonal sources of water can be found. Rain accumulates in surface depressions in the rocks (*sartenejas*). In a few areas, seasonally charged water tables can be tapped with shallow wells (*chenes*). Infiltration basins where internal drainage has become blocked (*aguadas*) were improved by the ancient Maya as reservoirs and are still used.

The eastern frontier of the settled regions of Yucatán does not correspond to any natural boundary and has fluctuated through the course of history.

Irregular Rainfall Patterns

The climate of the northern Yucatán Peninsula is characterized by high temperature and relative humidity throughout the year, and by a rainfall regime that is irregular in both time and space (Contreras, 1958; García, 1973). The peninsula is located in the intertropical zone. The prevailing trade winds blow across it in the summer months, bringing warm, moist air from the South Atlantic and the Caribbean. The landscape presents no topographical relief sufficient to trigger orographic precipitation. Local, convectional phenomena provoke unpredictable, scattered downpours. The agricultural cycle is controlled by a marked seasonal pattern, with a rainy season that runs from late May or early June through October, punctuated with a relatively drier period in August, called the *canícula* in Spanish. Then cooler air pushes down from the north, and the precipitation drops off sharply. Most of the native trees lose their leaves during the dry season, and there is too little water to support more than one crop a year except under very special circumstances.

The average annual rainfall decreases gradually and irregularly from 1,500 millimeters on the southeastern coast of Quintana Roo (south of our area of study) to 900 millimeters at the city of Mérida. The annual weather patterns in any location are so variable that the concept of an "average" year has little meaning. The probability of the rainfall achieving the mean in a given month and place is never as high as .5, and varies between .31 in March and .49 in September (García, 1973). The highly variable rainfall is the principal source of risk for peasant agriculture in Yucatán. Maya farmers recognize this, and, despite the attenuation of many agricultural rituals in recent years, the elaborate 3-day ceremony in which the farmers make a collective appeal to the gods for continued rain after the planting, called the *chac-chac*, is still prevalent in Maya communities. The peasant farmer adjusts his calendar of agricultural activities according to his assessment of the probability of a favorable sequence of first rains, a midseason drought, and other subtle factors. Nevertheless, *milpa* yields are uncertain and highly variable.

Patchy Associations of Shallow Soils

In the flat plains of Yucatán, young, generally alkaline, and shallow soils have been formed in complex, patchy associations controlled by the patterns of microrelief (Flores Mata, 1977; Hernández X., 1958; Wright, 1967). They are not easily classified according to standard international taxonomies, and almost everyone who works in Yucatán, including the government research and extension services, uses Maya names for them.

Maya peasants recognize a wide range of soil types. As in indigenous systems elsewhere in Mexico, they classify soils according to their use characteristics, including the vegetation they support and their physical properties (Williams & Ortiz-Solorio, 1981). Significant soil variations over short distances are crucial to the *milpa* system, and the farmers are sensitive to various indicators of natural fertility, including the presence of indicator plants.

In the central part of the state, where *milpa* production is most concentrated, the topography is a patchy association of low stone hummocks and flat depressions. The soil intermixed with stones on the hillocks is called *tsek'el* in Maya; the depressions are filled with a different red soil called *kankab*. A single plot of 3–4 hectares is a patchwork of both types. Less common are *kakab*, a darker, more fertile soil with higher concentrations of organic matter, associated with sites of previous human occupation.

More varied and complex soil associations are found along the margins of the Puuc hills. Further to the south, the frontier lands are dotted with ephemeral lakes and seasonally inundated lands called *bajos*. There is increasing evidence that the heavy hydromorphic soils that have formed in them, called *akalche'* in Maya, were intensively managed by the ancient Maya with raised beds and other techniques (Turner, 1983; Turner & Harrison, 1983). They are little used by peasants today, and recent government projects that have attempted to introduce mechanized rice and maize production into the *bajos* have encountered serious difficulties (Argaez & Maldonado, 1980).

The soils in most parts of the Yucatán are too thin and rocky to permit mechanization. There is evidence of significant potential to increase yields with fertilizers and other improvements within the context of traditional management practices (Pool, 1980). Nevertheless, agronomic research has been limited, and many agricultural development projects have failed because they have introduced technology inappropriate to the local conditions (Argaez & Montañez, 1975; Ewell, 1984; Warman, 1972).

The Milpa System

Milpa management practices are based on detailed, empirical knowledge of the dynamics of forest regeneration and of particular local conditions. The calendar of the principal tasks is controlled by the seasonal weather patterns. Up to 20 months elapse between the time a new plot is selected for clearing and the time when the last of the crop is harvested. The timing of some operations, such as burning and planting, is critical. Others, such as tree felling and harvesting, are carried out a little at a time, a few days a week, over periods of several months. Key junctures are celebrated with syncretic religious ceremonies that combine Catholic ritual with traditional Maya practices (Redfield & Villa Rojas, 1962). Figure 4-2 outlines the labor tasks of a typical *milpa*.

A plot in its first year of cultivation is called the *milpa roza*. An average of approximately 60% of its area is recleared and planted a second time, when it is called the *milpa caña*. The land is then allowed to revert to fallow; third- and fourth-year *milpas* are uncommon. In a given year, therefore, a peasant family is working two plots with a modal total area of approximately 4 hectares. In a good year, an operation this size should provide enough food for household consumption and a surplus to store, feed to animals, or sell on the market.

A new site is selected in August or September according to two principal criteria: its distance from and convenience of access to the village, and the age of the secondary vegetation. Although only a fraction of the forest land of any one community is cleared and cultivated in a given year, aerial photos reveal a nearly continuous checkerboard of plots in various stages of regrowth. The patterns are more dispersed in less densely populated frontier areas, but the *milpa* is a system of continuous fallow rotation. Unlike slash-and-burn practices in many parts of the tropics, it is not a brief transitional phase between forest and completely different land uses such as extensive pasture. It does not lead to serious erosion in the northern Yucatán Peninsula because the land is flat and because internal soil drainage is rapid.

The trees are felled with axes between August and January, depending on the age of the stand. Large trunks are easier to cut at the end of the rainy season, when the sap is running, and require many months to dry. The timing is less important if the vegetation is small. The trees are cut at waist height, and many resprout from their original root systems after a plot is left in fallow, which accelerates the sequence of secondary succession in old fields (Hernández X., 1958; Miranda, 1958). There is a detailed Maya terminology for the different stages of secondary succession (Steggerda, 1941), which vary significantly over short distances depending on the underlying soil associations (Illsley Granich & Hernándex X., 1980).

The management of the *milpa* is focused on the single day when the plot is burned. A good burn consumes all the trunks and branches, which have previously been spread evenly over the ground. It kills many weed seeds and leaves a thick cushion of ash, which washes into the soil, providing phosphorus and other nutrients. Some organic matter is destroyed, and nitrogen is volatilized, but the fire passes too quickly to damage the soil structure significantly. A poor burn leaves a plot littered with trash, and weed competition with the crops becomes a serious problem.

Maize seed is physically mixed with beans (*Phaeseolus vulgaris, P. lunatus*) and squash (*Cucurbita* spp.), and they are planted together in holes made with a steel-tipped stick.[1] The seed ratio in the mixture is adjusted in relation to the variations in soil type within every parcel. In general, the average density of the companion crops is low (Arias Reyes, 1980). There are real advantages to planting early. If one waits too long, competition from weeds in the first 2 weeks after germination will be severe. Furthermore, yields can be significantly reduced if the maize "tassels" during the *canícula*, the midseason drought, or if it is not entirely mature before the heavy rains of September. The rains are often

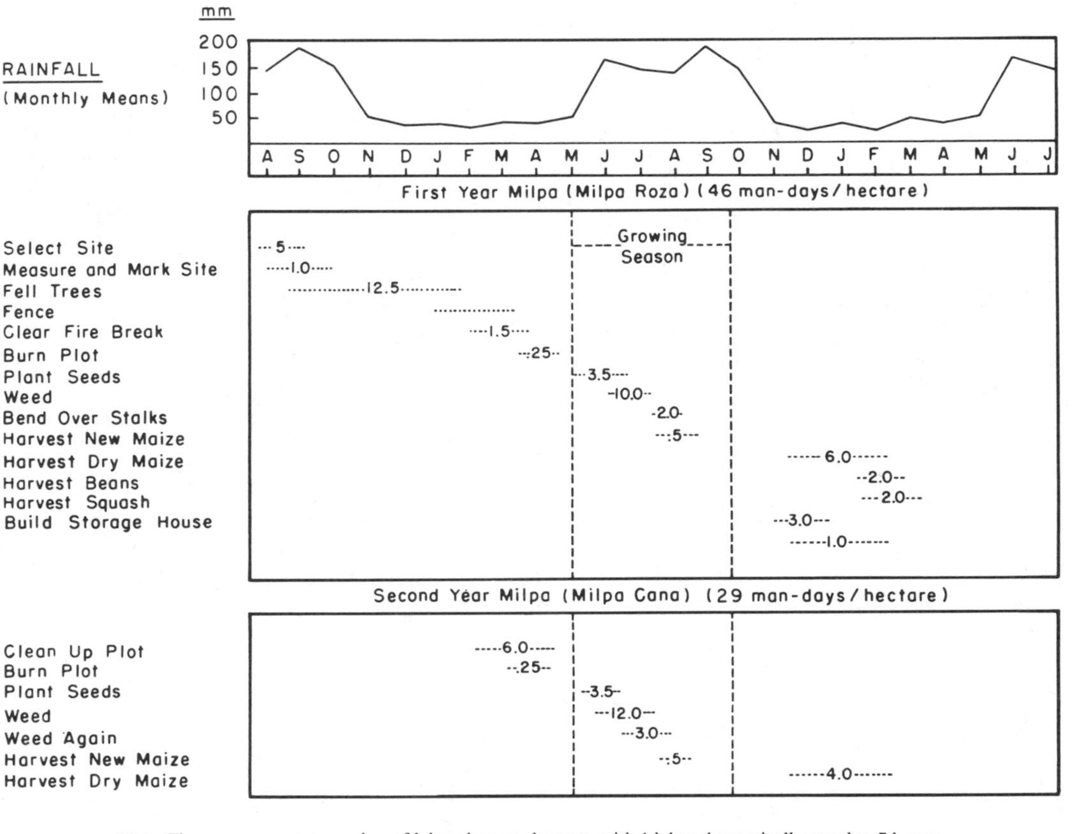

Note: Figures are average number of labor days per hectare, with 1 labor day typically equal to 7 hours.

FIG. 4-2. The traditional *milpa* system: Distribution of the principal cultural operations.

associated with microclimatic factors causing fungal disease (Tec Poot, 1978). On the other hand, if the seeds are planted too early, before the rains become well established, a single downpour may stimulate them to germinate. If there is a pause before it rains again, the seedlings will wither and die, and the farmer will have to replant at considerable expense in seed and time.

A first-year *milpa roza* is normally weeded once in July; a *milpa caña* is often weeded a second time. The work is carried out with either a machete or a small curved knife called a *loché*, which is used to uproot persistent weeds. Except for clearing the plot in the first place, this is the most time-consuming operation. Plots are almost always left to revert to fallow after the second year, principally because the labor required to weed becomes too great; the fertility of the young soils is not so quickly exhausted (Emerson, 1953; Hernández X., 1958; Pool Novelo, 1980; Steggerda, 1941).

The harvest begins in late August or early September with the new maize, which is the occasion for celebration and a ritual called *pibinal*. The majority of the crop is left in the field to dry. It is harvested, stored in rustic structures in the field, and hauled to the villages little by little over the course of the next dry season. These tasks overlap the beginning of the next year's cycle.

A number of minor crops are grown on a small scale, including sweet potatoes (*Ipomoea batatas*), cassava (*Manihot esculenta*), and other root crops; pineapples, sugarcane, and tobacco (Perez Toro, 1942; Redfield & Villa Rojas, 1962). Small fertile patches within the *milpas*, called *pach pa'kaal* in Maya, are intensively cultivated with tomatoes, watermelons, and other vegetables. The walled yards surrounding the houses in the villages, called *solares*, are planted with a great variety of fruits and other tree crops for household use, as well as with short-season maize (Barrera, 1981; Vara Moran, 1980). Delicate seedlings, herbs, spices, and some vegetables are grown in platforms raised on poles, called *kanché*, in order to protect them from the pigs and poultry confined in the *solares* for fattening. The small animals are important to the household economy because they provide a way to convert damaged or surplus *milpa* products into flexible household assets. Animals can be slaughtered for family celebrations or sold for cash in emergencies. Before the advent of modern beekeeping, hives (made from hollowed trunks) of indigenous bees (*Meliponae beecheii*) were also commonly found in the *solares*. Their honey was consumed in the household for ritual, medicinal, and sweetening purposes (Merrill-Sands, 1984; Weaver & Weaver, 1981). Deer, paca (*tepisquintle*), and other animals are hunted in the forest, where construction materials and other useful products are gathered. Wild "famine foods" have been collected and eaten in periods of hardship for centuries (Marcus, 1982).

One of the most important sources of stability in Maya peasant agriculture is the long tradition of empirical knowledge on which it is based. Nevertheless, it would be a great mistake to conclude that the people live in a static, autarkic, traditional equilibrium, or that pressures for change are only recent. On the contrary, they have lived in an uneasy, asymmetrical relationship with the dominant economy since their communities were forcibly reorganized in the first years of the Spanish colony.

A Brief History of the Maya Peasant Economy

Yucatán is a forest densely dotted with ruins: of the ancient and late Maya, of colonial and 19th-century *haciendas*, and, less conspicuously, of repeated cycles and abandonment of Maya villages. All of the contemporary towns and most of even the smallest hamlets can be located on 16th-century maps. Some have grown into prosperous commercial centers; others have faded into little groups of wattle-and-thatch houses clustered around the massive remains of churches and agricultural processing factories. The relationship between settlement patterns, peasant food production, commercial crop production, and the fluctuating forest frontier is the key to understanding the agricultural history of the region.

Pre-Hispanic Maya Agriculture

Recent archaeological studies have found increasing evidence of sophisticated hydrological and intensive agricultural systems of the ancient Maya throughout the Yucatán Peninsula, but mostly south of our study zone. These include irrigation, raised fields, reservoirs, cisterns, and other types of infrastructure (Harrison & Turner, 1978; Matheny, 1982; Neugebaur, 1983; Turner, 1983; Turner & Harrison, 1983). Many scholars have suggested that the ancient culture was based on elaborate techniques of land use and plant cultivation, including agrosilviculture and multiple cropping systems (Barrera, Gómez-Pompa, & Vásquez-Yánes, 1977; Turner & Miksicek, 1984). Although the *milpa* as it is practiced today has roots that extend back for centuries, pre-Hispanic systems were both more varied and more intensive (Tozzer, 1941).

The Colonial Economy

The Spanish conquered Yucatán in 1546, finally crushing a loose confederation of Maya states which had resisted fiercely for 18 years (Roys, 1957). They established direct administrative control over the northwestern part of the peninsula, where groundwater was easily accessible. The survivors of the wars and of catastrophic epidemics, which drastically reduced the population, were forcibly concentrated into villages dominated by a church (Cook & Borah, 1974; Gerhard, 1979). The population of the peninsula did not reachieve its preconquest numbers until the 1950s (Figure 4-3). The *conquistadores* were granted licenses to collect tribute from Indian communities, called *encomiendas*. The Europeans lived in the cities and did not practice agriculture at all. Maize and other food crops, as well as cotton and other tribute goods, were produced by the subject Maya population using simplified forms of their traditional technology, particularly the *milpa*.

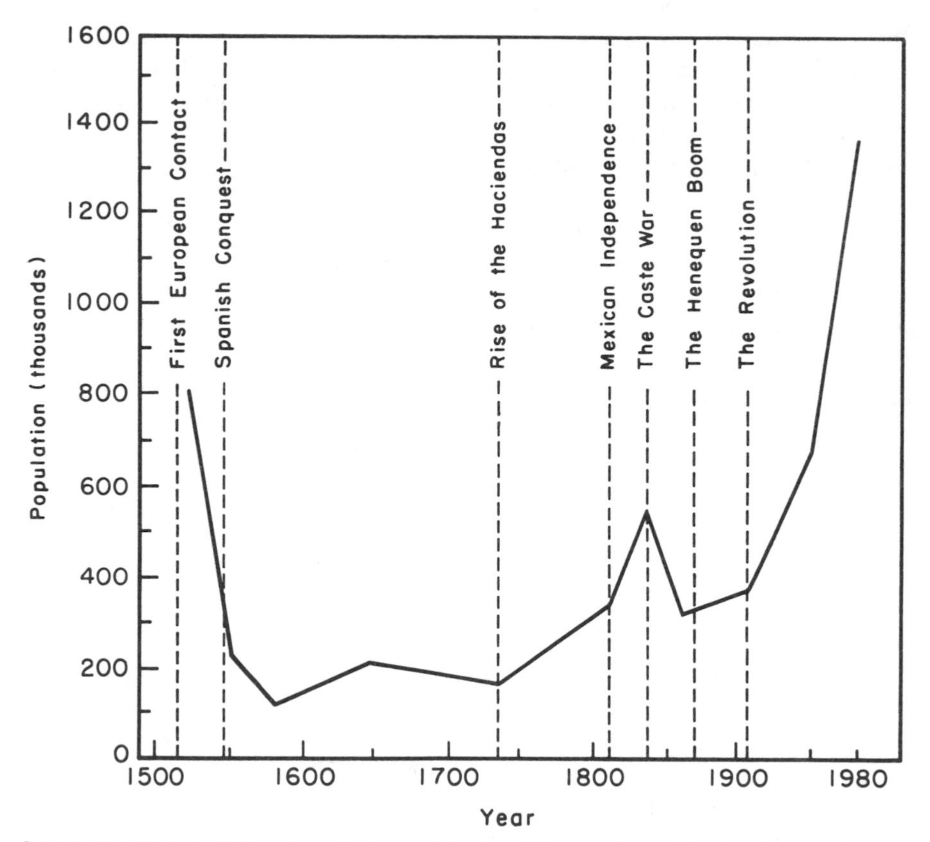

Sources: Data for 1517–1960 are from Sherburne F. Cook and Woodrow Borah, "The Population of Yucatan, 1517–1960," *Essays in Population History: Mexico and the Caribbean* (Berkeley: University of California Press, 1974), Vol. 2, p. 122. Figures for 1970 and 1980 are taken from the Mexican national population census for those years.

FIG. 4-3. Population estimates for the peninsula of Yucatan, 1517–1980.

The peasants were formally protected against slavery and other abuses, but these laws were enforced only erratically. The forest lands beyond the pale of Spanish control provided a sanctuary for Indians who could escape to live in small hamlets. These outlaw settlements were periodically raided by Spanish military expeditions to bring the fugitives back to communities subject to tribute (Cook & Borah, 1974; Patch, 1979).

The semiarid peninsula of Yucatán was a provincial backwater and, in contrast to the rest of Mexico, the *hacienda*, or landed estate, did not become an important institution until the late 18th century, when colonial restrictions on direct trade with Spain and Cuba were relaxed. At that time, large areas of land and significant numbers of workers were organized into *haciendas* to produce hides, tallow, honey, and beeswax for export. The estates developed a symbiotic, if asymmetrical, relationship with the Maya peasant communities, which continued to base their economy on the production of food. They secured a core permanent labor force by granting parcels rent-free to their peons, and through other mechanisms. Additional seasonal labor requirements were obtained from associated, but independent, villages, which also sold surplus maize and other commercial products to the cities (Patch, 1979). They depended on the *haciendas* for access to water points and other services, as well as for part-time employment (Strickon, 1965). The expansion of the *hacienda* system was a dramatic change, but it was limited to the historical area of Spanish domination. The autonomy of the peasant communities was partially preserved, and their economy remained centered on food production.

The Caste War

After Mexico became independent from Spain in 1821, a new class of landowners expanded their holdings by declaring undocumented land "idle," among other strategems (Patch, 1979). New production technology and markets became available, and the cultivation of sugarcane and other commercial crops expanded into the southern and eastern frontiers, where the agroclimatic conditions were relatively favorable (Cline, 1947, 1948; Suárez Molina, 1977). The increasing demand for land and labor conflicted with the requirements of the *milpa*, disrupted food production, and broke down the symbiotic relationship that had existed between the *haciendas* and the peasant communities. These pressures on resources, combined with an attempt to control and tax Maya villages in what had formerly been loosely administered border areas, provoked the bloody rebellion known as the Caste War in 1847. The rebels, many of whom had been armed and trained to participate in a series of civil skirmishes between the Mexican central government and various federalist factions in Yucatán, massacred the white population of entire regions (González Navarro, 1970; Reed, 1964). In 1848 they controlled 80% of the peninsula and came to within 17 kilometers of the city of Mérida before the tide was turned. A fifth of the entire population of Yucatán had been killed or had emigrated by the early 1860s (Cook & Borah, 1974). Large areas of the state were abandoned for over 75 years.

The Henequen Boom

In the last third of the 19th century, Yucatán's economy was reconsolidated and dramatically expanded in the secure northwestern corner surrounding Mérida on the basis of the rapid development of henequen plantations.[2] Sisal (*Agave* spp.) fiber became a major export crop as the raw material for binder twine after the invention of the mechanical reaper by Cyrus McCormick in 1875, which revolutionized agriculture in the Great Plains of the United States. During its brief boom, Yucatán was the world's predominant sisal supplier and the richest state in Mexico. Owing in part to secret price fixing by the International Harvester Trust, the value of fiber was maintained at artificially low levels

(Joseph, 1978). The plantation owners kept their profits up by keeping labor costs at an absolute minimum through a system of debt peonage. In Yucatán, the Maya still refer to this period as the "era of slavery." Peasants were brought into servitude with some small debt; the wage rates and the prices that they were forced to pay at the *tiendas de raya*, or plantation stores, were manipulated so that they could never work their way back to freedom. Workers were openly bought and sold between plantations for prices unrelated to this ostensible debt, and the whip was used regularly to maintain discipline (Turner, 1912).

Outside the henequen zone, which was confined to a tightly integrated area within 100 kilometers of Mérida, *haciendas* organized to produce sugarcane, cattle, and maize. These and other commercial products also expanded at the expense of the peasant communities. It has been estimated that 75–90% of the rural population had lost individual access to land by the Revolution of 1915 (Joseph, 1978:53; Tannenbaum, 1929:33). The region lost its capacity to feed itself and has been dependent on imports ever since.

The Revolution and the Resurgence of the Traditional Peasant Economy

Despite the terrible social conditions, the Mexican Revolution did not reach Yucatán until the state was occupied by federal troops in 1915, and reforms were imposed (Joseph, 1978). The "slaves" were freed by decree, and large tracts of forest land were divided into *ejidos* in a comprehensive program of agrarian reform.[3] The *ejido* is a community land grant administered by elected representatives. Households in the communities are given usufruct rights to land. Eligibility requirements vary among communities. In Yucatán, the average *ejidal* grant is 3,000 hectares, theoretically enough to permit a *milpa* fallow cycle of 15–20 years. The haciendas were not physically destroyed, but the henequen industry lost its low-cost labor force and faced increasing competition in the world market. It was nationalized in stages over the next decades but has been in protracted decline since the 1920s.

The Maya peasant economy, on the other hand, experienced a dramatic resurgence. Drawing upon their cultural traditions of resource use and *milpa* production, the peasants redeveloped a subsistence-oriented economy. They fanned out from the ex-haciendas into the forest hinterlands (Redfield & Villa Rojas, 1962; Revel-Mouroz, 1980; Ryder, 1977). Villages that had been virtually depopulated came back to life, and new frontier settlements were formed in a process that Redfield calls "hiving off" (Redfield, 1955:28). Partly to escape social conflict and partially to find *monte alto* – forested land that had been fallowed for a long time – a group of families would establish an isolated camp in the hinterlands. These satellite hamlets maintained political and ceremonial ties with their parent communities for some years, after which many became fully independent.

In the first two decades after the Revolution, *monte alto* was freely available. Informants recall maize yields of between 2.0 and 4.5 tons per hectare, or between two and four times the current regional average (Merrill-Sands, 1984; Rosales, 1980b). In part because of this initial flush of productivity and in part because of a relatively favorable economic and institutional environment, most families were able to meet both their subsistence and their surplus production needs from their *milpas* in most years. Data collected in the principal central *milpa* region in the 1930s by Redfield and Villa Rojas (1962) and Steggerda (1941) show that 40–50% of the maize produced was available for storage in the event of a shortfall the following year, or for sale or exchange for manufactured goods and foodstuffs not produced directly by the household.

The cash needs of most families were modest. Basic commodities such as salt, sugar, coffee, cocoa, clothing, sandals, shoes, soap, machetes, bullets, thread, and candles were obtained in the small village stores or from itinerant peddlers. Meat was generally obtained from hunting or household production. Corn and eggs were the principal medium of exchange in local transactions and as payments for labor. The use of currency was restricted to large items such as cattle or gold chains for women—common forms of saving surplus wealth—or for purchases made in large towns (Redfield & Villa Rojas, 1962).

The importance of maize sales increased as the linkages between the peasant communities and the regional economy expanded. In the southern zone, merchants and relatively prosperous peasants produced maize in large commercial *milpas* worked with hired labor. In the central area of the state, modest surpluses were produced more widely. In 1930, one-quarter of the families in Chan Kom planted an area larger than necessary to meet their households' subsistence and exchange requirements (Redfield & Villa Rojas, 1962:53). The surplus was sold directly or used to fatten swine, which were sold at a good profit to traders from the larger towns in the region. By the 1940s, hogs had become an important business in communities located on improved roads (Redfield, 1950).

The Milpa Economy in Recent Decades

Beginning with a disastrous series of droughts and locust plagues in the 1940s, which brought many villagers to the brink of famine, the fission of the peasant population into *monte alto* has been gradually and unevenly reversed. It is not true in any absolute sense that the population has exceeded the carrying capacity of the land; large areas of the state have fewer than 10 persons/km^2, and numerous small hamlets have been abandoned.

The rural people have clustered into relatively large villages where employment opportunities and other services are available, particularly schools. Many regard education and a knowledge of Spanish as the principal means by which their children may escape poverty and improve their standard of living. This has led to the concentration of *milpas* within smaller areas, a reduction in fallow periods, and a simplification of management practices as people take care of *milpas* in combination with other activities. Although the *ejidal* land of most Maya communities has never been divided into individual lots, the peasants have gradually been restricted to *milpa* sites within their own *ejidos* and to areas not controlled by the local elite or designated for government projects. The rural population as a whole increased by only 20% between 1940 and 1960, but pockets of population pressure have put the traditional *milpa* system under stress.

Data from the village of Chan Kom reveal the changes that have occurred within the *milpa* system over the past 50 years resulting from the more concentrated settlement pattern. Between 1932 and 1936, when Chan Kom was an isolated frontier settlement, an average of 29 households worked plots of an average size of 4.8 hectares and harvested a combined average total of 140 hectares. The mean distance of the *milpas* from the community was 8.5 kilometers, and the households had access to unclaimed national lands outside of their official *ejido* land grant (Steggerda, 1941:112; Redfield, 1941:45).

Today, the average size and the distance of the *milpas* has decreased, while the total area planted and the number of households working *milpa* in the community has approximately doubled. Between 1980 and 1982, an average of 68 households planted *milpas* of an average size of 3.8 hectares. The average distance of the milpas from the village was only 4.3 kilometers (Merrill-Sands, 1984). The more concentrated land use results in a shortened fallow period. In the 1930s the average fallow period in the same region was approximately 20 years (Steggerda, 1941), while today in Chan Kom it is only 11 years (Merrill-Sands, 1984).[4] The average for the state as a whole has fallen below 10 years (Gallegos, 1981).

Sustained yields depend on the extensive use of land and on long fallow periods for forest regeneration. The contemporary peasants constantly lament that "the *milpa* doesn't produce"—that their harvests are small and uncertain and the returns to their labor very low. This perception in part reflects a decline in the relative value, or purchasing power, of maize. Nevertheless, although continuous time-series data are not available, there is good evidence that both fallow periods and average yields have declined significantly since the 1930s. At that time, the standard rule of thumb of what a *milpa* should yield in an average year was 25 "loads," or approximately 1 t/ha (Redfield & Villa Rojas, 1962; Steggerda, 1941). Since the 1940s, informants have reported expectations only half as great (Baraona & Montalvo, 1981; Merrill-Sands, 1984; Redfield, 1950).

After World War II, and particularly in the decade between 1963 and 1973, the Mexican government maintained the price of maize low relative to other goods to keep down urban wage rates and to encourage industrial development. In constant 1960 pesos, the guaranteed price fell from 800 to 537 pesos per ton between 1960 and 1973, an average annual decline of 4.6% (Montañez & Aburto, 1979:63). In Yucatán, where yields were low, commercial maize production with hired labor was drastically reduced. Peasant households received less value for their modest surplus at the same time as their needs for cash and manufactured goods were increasing as a result of the integration of the rural villages into the regional economy (Palerm, 1980; Stavenhagen, 1976; Warman, 1972). They found themselves working ever harder while the fruits of their labor yielded less and less. Maize prices were increased in the late 1970s, but only to a value of 700 pesos per ton in 1960 pesos, or less than the prices of 20 years before (Hall & Price, 1982:308).

Milpa yields are too low and too variable to assure either a secure food supply or an adequate income to most Maya peasant families. A recent study in an isolated village in Quintana Roo found that a family of the average size of seven members and of median economic status used 2,600 kilograms of maize in a year—they ate 2,200 in various forms and fed 400 to a small collection of backyard animals (Baraona & Montalvo, 1981). They also consumed approximately 100 kilograms of beans. A 4-hectare *milpa* can supply these needs in a year with average weather conditions (Table 4-1), but the production that any one family obtains varies widely depending on local weather conditions, the size of their plot, and other variables. Many households must sell maize at the harvest to pay debts or to meet emergency needs, only to buy it back later at higher prices. Food shortages in the months before the harvest are widespread

TABLE 4-1. Apparent economic returns to *milpa* production.

Crop	Area[a]	Yield per hectare (kg)			Production (kg)			Price per kilogram (pesos)	Value (pesos)		
		Low average	Mean	High average	Low average	Mean	High average		Low average	Mean	High average
Maize	4	400	700	1,000	1,600	2,800	4,000	6.50[b]	10,400	18,000	26,000
Beans	4	10	25	55	40	100	220	25.00[c]	1,000	2,500	5,500
Squash	4	24	36	45	96	144	180	40.00[d]	3,840	5,760	7,200
Total	4								15,250	26,260	38,700
Estimated cost of 158 days of family labor, valued at the federal daily minimum wage of 175 pesos (U.S.$6.75/day)[e]									27,650	27,650	27,650
Apparent economic returns									–12,400	–1,400	11,000
								(U.S.$)	(–475)	(–50)	(425)

[a]This accounting exercise is based on an operation with 2.5 hectares of first-year *milpa roza* and 1.5 hectares of second-year *milpa cana*. Beans and squash are planted in limited areas within the *milpa*. The yields of these crops are calculated as total output over the entire area, and are much lower than the actual agronomic yields of the patches where they are planted.

[b]Both the official wholesale guarantee price and the average retail price of maize were approximately 6.50 pesos per kilo in 1981. Both prices fluctuated as much as 25%, depending on the location of a particular village and the time of year.

[c]The average retail price of various types of beans was 25 pesos.

[d]Squash seed is grown primarily for sale, and the wholesale price fluctuates through the year depending on supply conditions.

[e]The federal minimum wage is adjusted periodically to account for inflation in the cost of living, and is used as a very imperfect measure of what it costs to support a family. The actual wages that a *milpa* producer could earn off the farm vary considerably.

in many years, and the government organizes emergency distribution programs.

The returns to labor from a typical subsistence-oriented *milpa* can be illustrated using prices from the 1981–1982 cropping season (Table 4-1). It is assumed that all the work tasks (see Figure 4-2) are done with family labor, which is valued at the federal minimum wage in the region. In fact, at least some hired help would be used in peak periods. Rural retail prices are used for maize and beans, to indicate what it would cost to buy the food if it were not produced. Squash is grown for its dried seed, which is sold to middlemen at local wholesale prices. On the basis of these simplified calculations, the apparent economic returns are negative except in unusually good years, and most households must supplement their incomes from other sources.

The development of the regional economy of Yucatán has led to the integration of the peasant sector, as producers of both staple and commercial agricultural products, as providers of labor on a part-time and intermittent basis, and as consumers. An extensive road-building project in the late 1960s and 1970s augmented the network of linkages between the peasant communities and the urban centers. The new roads stimulated the development of commercial enterprises in the peasant sector, because farmers could more easily get their products to market, but they were also strategic in expanding the distribution of consumer goods. They provided easier access to migratory wage labor opportunities in urban and tourist centers, where the peasants have been exposed to new values.

Rural development, including improved education and other services, has led to increases in the families' cash requirements and to changes in the criteria for a desirable standard of living (Elmendorf & Merrill, 1977). The traditional marks of status—filled graineries, backyards crowded with poultry and hogs, herds of cattle, and flashy gold chains on the women—are being replaced by consumer goods such as transistor radios, televisions, fancy manufactured furniture, and refrigerators.

The Maya peasant households have adapted to these changed conditions by reducing their dependence on the *milpa* and by diversifying into alternative cash-generating activities—both commercial agricultural enterprises and migratory wage labor. The kind and number of economic activities exploited by a household depends largely on its access to resources and the factors of production—land, labor, and capital. Despite the *ejido* land tenure system, access to community resources is not equal. Key family groups in the communities have consolidated political and economic power and they control access to resources both within the *ejidos* and through their linkages with merchants and powerful figures in government (Domínquez, 1979; Elmendorf & Merrill, 1977; Goldkind, 1965, 1966; Merrill-Sands, 1984; Pérez Ruiz, 1980).

In the village of Chan Kom, for example, three households of the local elite have fenced for their exclusive use 200 hectares of some of the best *ejido* lands, favored by cenotes, wells, fertile *ka'kab* soils, and proximity to the village. Furthermore, they have used their political power to circumvent the state law prohibiting the fencing of cattle in the *ejidos*. Not only do they maintain approximately 75 head of cattle on *ejido* lands, but they have also seeded 50 hectares in improved grasses, effectively precluding this land's future use for *milpa* agriculture (Merrill-Sands, 1984). Clearly, these elite families have greater flexibility in their economic alternatives than poorer households do. The latter, with limited access to resources and factors of production other than household labor, are constrained to making *milpa* and to wage labor (Table 4-2).

A large number of peasant households in Yucatán have developed both beekeeping and small-scale fruit and vegetable production as diversification strategies in response to the attenuation of the *milpa* economy. In the first case, the farmers have maintained food production at a level sufficient to meet their subsistence requirements, and have invested in honey production as a commercial enterprise to meet their cash needs. In the more fertile Puuc region in the southwestern part of the state, farmers have gradually abandoned *milpa* as they have established intensive production systems in small, irrigated parcels.

Beekeeping and the Mixed Subsistence–Commercial Production System

Over the past two decades, commercial beekeeping with European bees (*A. mellifera*) and the introduced technology of the movable frame hive has expanded dramatically within the peasant sector. Whereas previously honey production had been dominated by large-scale, capital-intensive enterprises owned by urban entrepreneurs, 80% of the estimated 3,000 beekeepers in the state

TABLE 4-2. Participation by socioeconomic groups in principal production strategies, Chan Kom, Yucatan, 1976 and 1981.

	Strategy									
	Apiculture		Migration		Milpa		Cattle		Commerce	
Socioeconomic[a] group	1976	1981	1976	1981	1976	1981	1976	1981	1976	1981
I (10 HH)	80%	90%	20%	50%	70%	80%	70%	80%	40%	70%
II (16 HH)	44	56	31	63	75	75	25	37	–	–
III (42 HH)	31	55	28	69	81	67	23	26	–	–
IV (31 HH)	–	13	19	58	95	71	–	–	–	–
Total (99 HH)	30%	46%	25%	62%	82%	73%	22%	25%	04%	07%

[a]The ranking method was based on Dewalt (1983). Three informants were asked to divide all the households in the village into socioeconomic groups according to their own criteria. Two informants used four groups, and one used five. The rankings were summed and then the households were divided according to the natural breaks in the distribution. There was a strong correlation between the three rankings ($r = .76$, significant at the .01 level).

were small-scale peasant producers by 1970 (Calkins, 1974). The widespread process of adoption has continued; in 1982 it was estimated that at least 20% of the rural households in Yucatán kept bees (Merrill-Sands, 1984).

The combination of beekeeping with *milpa* is a model example of the successful integration of subsistence and commercial strategies into a mixed peasant production system:

1. The strategies are complementary with respect to their production goals, their demands on the factors of production, and their production calendars.
2. The sources of risk for the strategies are distinct.
3. Commercial beekeeping is a viable and lucrative cash strategy within the context of the peasant economy.

Complementarity

In the mixed system, the goal of *milpa* production is primarily to meet the consumption needs of the household—to generate *use value*. Commercial beekeeping, on the other hand, generates the *exchange value* needed to procure merchandise and services. In many households, it has supplanted the commercial role of *milpa* production as it generates considerably more exchange value with less onerous work. This is not to say that *milpa* products are never sold, but beekeeping has become the principal means of obtaining cash.

Honey production protects and supplements the household's subsistence. It permits the maize harvest to be retained in full for home consumption. Locally grown maize is much preferred by the Maya peasants for its taste and cooking qualities—an important consideration with respect to a staple food consumed on a daily basis. Beekeeping also makes the household less vulnerable to capricious *milpa* yields. In the event of a shortfall, honey can usually generate sufficient income to permit the family to purchase maize in the months before the next harvest. In turn, the subsistence strategy protects the household from too much dependence on the market economy, which is volatile and often highly exploitative of the peasantry.

Beekeeping and *milpa* production can be successfully integrated within a diversified production system because they are complementary with respect to their demands on land, labor, and capital. *Milpa* production relies primarily on land and labor, whereas beekeeping depends on labor and capital (Table 4-3). The *milpa* system of permanent rotation through long-term forest fallow maintains the resource base for the bees. They forage primarily on the prodigious floration of the natural forest vegetation in various stages of regrowth, but also on the flowers of the crops. For example, pollen from tasseling corn is the primary stimulus for renewed colony development after the dearth period during the rainy season.

Beekeeping and the *milpa* are complementary in their use of labor. With

TABLE 4-3. Model of subsistence-commercial mixed production system integrating *milpa* and beekeeping, 1981.

	Milpa	Beekeeping
Production unit	4 ha	45 hives
Household labor investment	136	110
Total labor investment	160	118
Percentage of paid labor	15%	7%
Period of most intense labor demands	June–July, October–March	March–May
Value of assets	Hand tools	31,000 (US$1,190)
Production costs:		
Inputs	700	3,900
Transport		1,600
Labor	2,500	900
Equipment replacement	200	2,100[a]
Total cash expenditures	3,400 (U.S.$130)	8,500 (U.S.$335)
Products sold	100 kg pepita 250 kg corn	1,700 kg honey
Income from sales	5,400 (U.S.$208)	23,800[b] (U.S.$915)
Net cash income from sales	2,000 (U.S.$77)	15,300 (U.S.$589)
Products allocated for household consumption	3,000 kg corn, 50 kg beans, 40 kg pepita	10 kg honey
Value of consumption goods	22,300[c] (U.S.$858)	140 (U.S.$5.00)
Value produced per household day	178 (U.S.$6.84)	139[b] (U.S.$5.40)

Note: All values expressed in 1981 Mexican pesos. The figures reflect 1981 production and marketing conditions.

[a]Equipment replacement expenditures below normal because of 1981 drop in honey price. Producers delayed expenses.

[b]Abnormally low, reflecting the drop in price of honey.

[c]Calculated using the purchasing price of corn—that is, what households would have to pay if they did not produce it themselves.

the exception of the harvest season, beekeeping entails light, but continuous, labor throughout the year. In contrast, the *milpa* requires sporadic periods of intense, taxing labor, which the peasants say saps them of their strength.

The timing of the strategies' respective demands on labor are also highly compatible. The periods of lowest activity in the *milpa* are those of highest labor use in beekeeping, as is evident in the pattern of labor investment in both beekeeping and *milpa* for the 1981 production cycle in Chan Kom (Figure 4-4). The period of greatest labor use in beekeeping is during the dry season, particularly April–May, when the harvests occur and more frequent trips to the apiaries are required to take water and arrange the hives to accommodate the honey flow.

In the *milpa*, on the other hand, the dry season is a time of greater flexibility and less drudgery. Its most intense period of labor activity involves planting and weeding during the rainy season (June–July). These activities must be performed on schedule and as rapidly as possible. In beekeeping, these are flexible months of lower labor investment. The required tasks can be accommodated to the more rigid schedule of work in the *milpa*.

The analysis of labor cycles demonstrates that beekeeping is a highly suitable cash strategy to supplement the *milpa*. Beekeeping permits households to intensify production through the use of surplus labor because it does not compete for labor at crucial periods in the production cycle of the *milpa*, nor does it overtax the labor power of the household.

Finally, beekeeping and *milpa* are complementary in their demands on the scarcest factor of production in the peasant economy—capital. *Milpa* production relies primarily on labor and land, requiring cash investment only for seed and hand tools. The minimal competition for capital makes commercial beekeeping's requirement for capital investment in hives and equipment feasible within the household economy.

Beekeeping relies much more heavily on cash inputs than does the *milpa*.

FIG. 4-4. Labor investment in beekeeping and *milpa* production, Chan Kom, 1981.

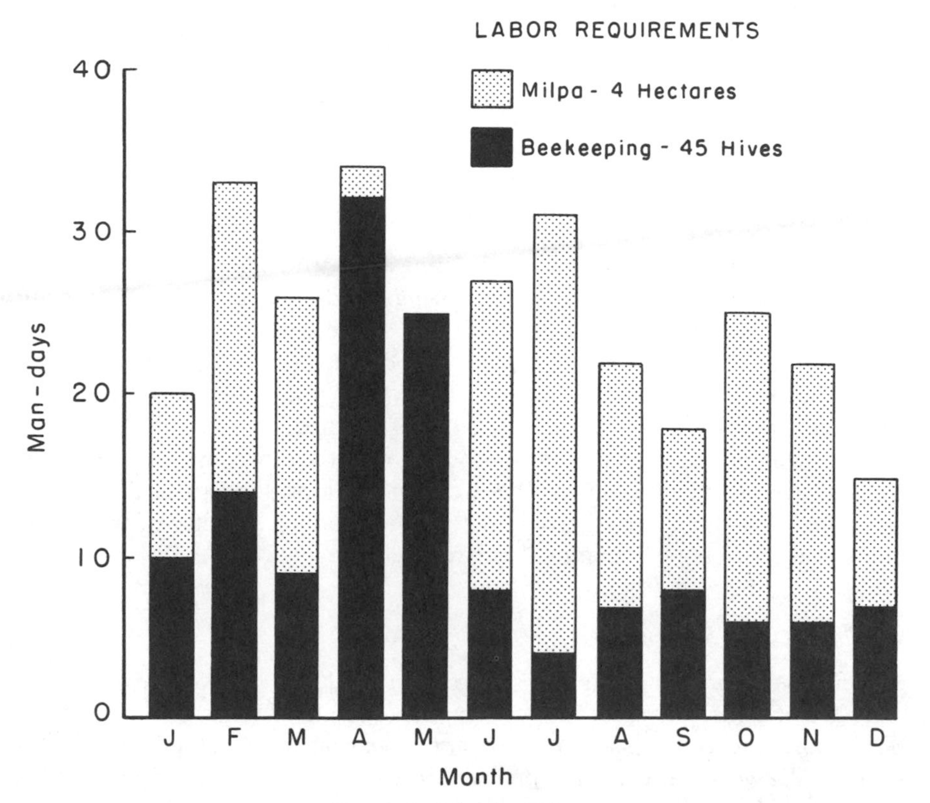

Note: Typical labor day = 7 hours.

The peasant household strives to maximize cash production through labor investment in its commercial strategies, and regulates the level of cash investment in beekeeping in accord with the price and yield of honey as well as with the extent of cash demands of the household. In some years, the immediate cash needs of the household undermine its ability to maintain an adequate level of inputs for beekeeping—to the detriment of the following year's production. Income from the sale of honey provides cash at critical points in the milpa cycle—at periods when households must often hire additional labor, and when they must often purchase maize or *tortillas* as stocks run low before the first harvest.

The inputs and outputs of a model subsistence–commercial mixed system of 4 hectares of *milpa* and a developed production base of 45 hives has been developed (Table 4-3). The figures are derived from consolidated production data from the village of Chan Kom for the 1980–1982 production cycle. Although cash income from beekeeping is lower than normal due to the drop in the 1981 honey price, the figures reveal the respective strategies' complementary use of the factors of production and their distinct economic roles within the household economy. The higher relative value produced per household labor day for *milpa* should be interpreted in light of the low honey price in 1981, the above-average yields for *milpa*, and the fact that the value is computed using the rural retail prices for *milpa* products, or what peasants would have to pay if they did not produce their own food. If the *milpa* products had simply been sold for cash, the value produced per labor day would have been considerably lower.

Risk Independence

The sources of risk for *milpa* and beekeeping are distinct. Good honey production depends on intermittent showers during the dry season, which stimulate the trees and shrubs in the forest to flower; successful *milpa* production depends on the timing and abundance of rain during the wet season. Although yields and prices are erratic in both activities, their integration into the diversified production system helps to reduce the risk. Since they are not interdependent, a failure in one does not harm production in the other; in fact, they often compensate for one another in the event of a shortfall or price drop. This distribution of risk is very important for the economic stability of a peasant household operating in a marginal environment.

Viability as Cash Strategy

Modern beekeeping is an appropriate commercial strategy for the economic and natural conditions under which peasants operate in Yucatán. Commercial beekeeping reached the peasant sector through a process of technological diffusion from large-scale, capitalist production enterprises that rented *ejido* lands and hired peasant farmers. These operations demonstrated that there was a viable market for honey, and that commercial production using *A. mellifera* bees and the introduced technology of the movable frame hive could be lucrative. Peasants learned basic management techniques by observing and working for the commercial enterprises in the *ejidos* without having to assume any risk. The capitalist operations also provided access to the means of production. Through pilfering, capturing escaped swarms, and purchasing secondhand equipment, peasant farmers were able to begin small-scale honey production units with minimal cost.

In contrast to many other commercial enterprises, which entail a large outlay of capital or burdensome credit obligations, the production base in beekeeping can be built up through the incremental investment of surplus wealth. The cash returns are immediate while, at the same time, long-term equity is accumulated within the household. A producer can begin with one or two hives, or even with swarms captured in the forest, and then gradually expand by reinvesting the proceeds from honey sales in equipment and by forming new colonies with divisions from the original hives. Beekeepers usually expand at a rate of 50% in the first years, funneling all their proceeds into equipment.

The market for honey is relatively stable and accessible to peasant producers. Improved infrastructure and the establishment of regional collection centers for honey throughout the state have lowered the cost and reduced the difficulties confronted by small products trying to get the product to market. Two export cooperatives buy all the honey produced at a fixed base price. Any profits that accrue after processing and marketing expenses are subtracted are distributed among the producers at the end of the year.

Finally, commercial beekeeping is a lucrative cash strategy. Through the union of small savings and their own labor, peasant households can produce a substantial income. The amount of surplus honey available for exchange from an average-sized production unit of 35 hives in the central zone has been calculated, subtracting maintenance and equipment replacement costs. These figures are compared to the amount of honey needed by the household for sale to meet its minimal cash needs as expressed in terms of a standardized market basket of food items and manufactured goods (Figure 4-5).[5]

The cash income, however, is erratic because of variable yields and fluctuations in the world price for honey and the exchange rate (Figure 4-5). For example, in 1980 and 1981 lower world prices, compounded by spiraling national inflation in Mexico, significantly eroded the purchasing power of honey. The devaluation of the Mexican peso in 1982 increased the value of the harvest, but it was one of the worst production seasons in recent history. Peasant beekeepers regarded the poor conditions as temporary and strove to maintain the production base so that they would be in a position to reap future benefits.

Thirty to forty hives represent a significant asset in the context of peasant economy. It embodies a large share of a household's savings and represents the

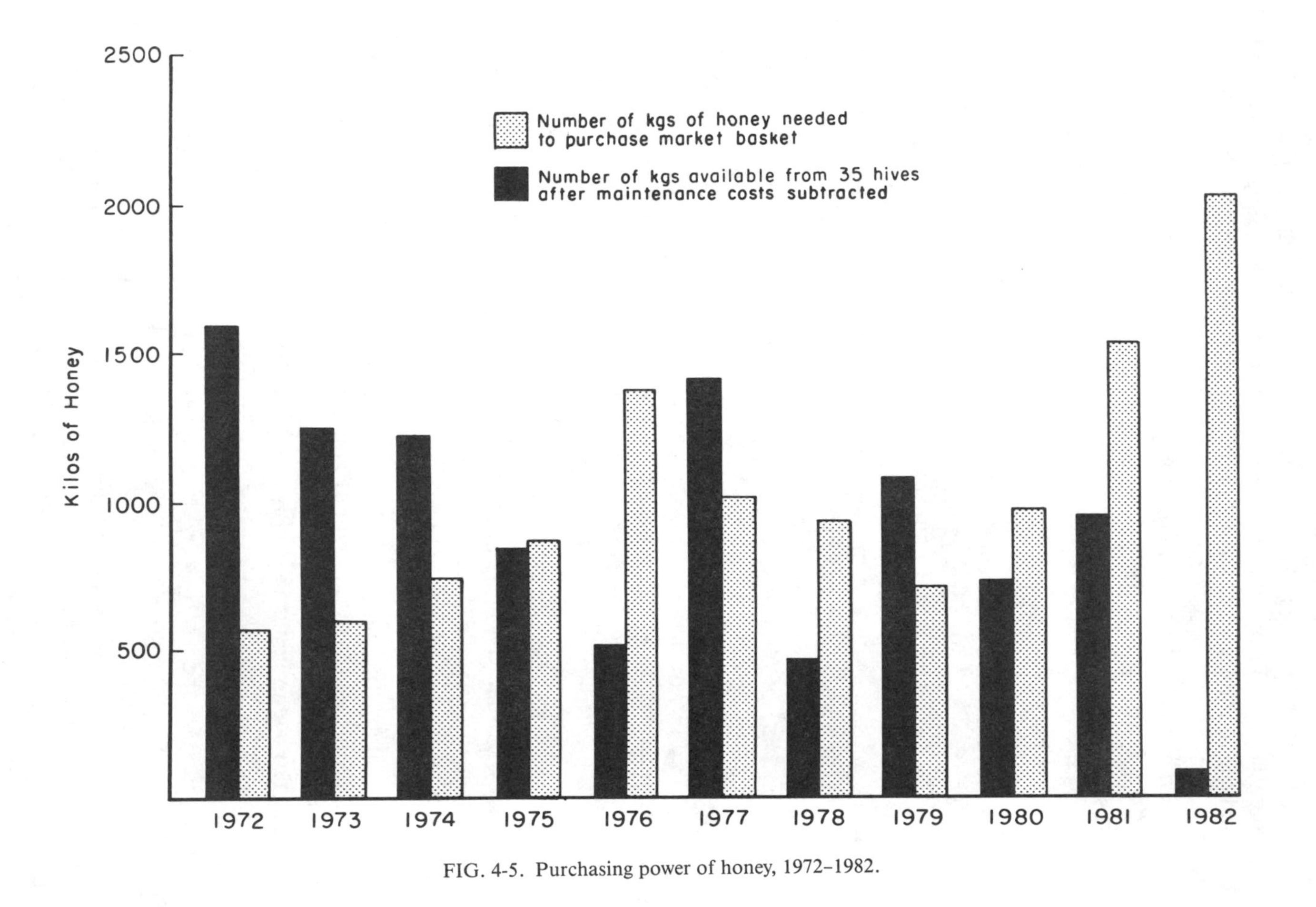

FIG. 4-5. Purchasing power of honey, 1972–1982.

children's future inheritance—not something to be lightly sacrificed because of a market slump. During lean years in beekeeping, or as a supplementary cash strategy when they have surplus labor power, beekeeping households often exploit migrant wage labor on an intermittent basis. In Chan Kom, for example, during the 1980–1982 period of low returns in beekeeping, 37% of the households adopted migrant wage labor to compensate for their cash deficit. Few households, however, are willing to sacrifice their control over resources, decision making, and the future equity of their households for the short-term benefits of wages. Full dependence on migrant wage labor is viewed as risky, unprofitable in the long term, and for some even degrading. It represents a loss of freedom and harbors the threat of returning peasants to the era of slavery of the henequen boom.

Intensive Peasant Production Systems in the Puuc Region

The Puuc region in southern Yucatán is an irrigated oasis where Maya peasants have intensified their traditional management practices to produce fruits and vegetables for the urban market. Over the past 40 years, approximately 3,500 families have accumulated resources in irrigated parcels, which were at first only one of several activities within their diversified household economies. They have gradually abandoned *milpa*, although many continue to grow at least some maize for home consumption. The majority have financed the development of their parcels with long periods of migratory labor throughout the Yucatán Peninsula and further afield. The median number of years that the members of a sample of farmers worked outside of their home communities was 14 (Ewell, 1984:129). Access to improved land has been secured through membership in *ejido* communities and through participation in organized groups that have petitioned the federal government for the construction of small-scale irrigation units. Most of their operations are small, between 0.5 and 4–5 hectares, often divided between two or more parcels at different stages of development. The farmers share similar backgrounds and work histories. Almost all of them are Maya speakers with strong ties to the culture, and almost none of them are more than a generation away from the traditional, *milpa*-based economy.

The Fruit Zone of Yucatán

Fruits and vegetables are relatively minor crops in Yucatán, representing only 10% of the total value of agricultural output in 1981 (Secretaria de Programación y Presupuesto, 1982). Fruit is widely grown on a small scale for household consumption in traditional backyard tree gardens, or *solares*, in rural villages. Commercial demand is concentrated in urban areas, particularly Mérida, where nearly 40% of the state's population lives. The region is supplied with fresh produce from other parts of Mexico, but small farmers in Yucatán are able to compete for a residual share of the market for some crops. Two-thirds of the state's commercial fruit output, particularly citrus, is produced in an area centered on the market town of Oxkutzcab, which is 100 kilometers from Mérida.

The peasants of this and the neighboring communities along the margin of the Puuc hills have access to a variety of agroecological zones with distinct soil types and cropping capacities within walking distance. The area was one of the most important sources of Mérida's maize supply during the colonial period (Patch, 1979). It was the center of the stage's sugar industry in the 19th century and has produced a wide variety of commodities, including citrus fruit, tobacco, castor oil, and others, on a smaller scale (Suárez Molina, 1977). As a trading center, it has concentrated on the products of the sparsely populated forest frontier lands to the south. A railroad was opened in 1901, and the region is located on a major highway linking Mérida with Quintana Roo.

At the time of the Revolution, the peasant population was highly concentrated around the haciendas (Rejón Patron, 1981). After the liberation of the "slaves" and the agrarian reform, the ex-peons fanned out and reestablished autonomous family enterprises based on the *milpa*. Many produced vegetables for sale in the market in *conucos*, a small-scale, intensive adaptation of traditional slash-and-burn techniques to the production of horticultural crops through an extended growing season. Trade was taken over by small shopkeepers, who provided household necessities and credit to the peasant farmers and marketed their surplus production. Some of these merchants were financed by brokers in Mérida to outfit and supply chicle camps in the southern frontier in the 1920s and 1930s. To feed these workers, and to provide an additional source of commercial maize for the market, they hired peasants to work in large *milpas* in the frontier. Oxkutzcab became an important transshipment point as carters brought diverse products to warehouses along the rail line (Domínguez, n.d.; Rosales Gonzales, 1980a).

The successful merchants, most of them Maya-speaking natives of the area, invested their profits in agricultural enterprises based directly in the town. For most of them, the first step was to buy cattle, which were released to browse in the Puuc hills at night and were confined in a corral with a well during the day. The expansion of these operations required increased water-pumping capacity. Starting in the 1920s, small diesel engines gradually replaced windmills and provided enough water to establish irrigated fruit and vegetable operations. The products were shipped to Mérida on the railroad. By the late 1930s the Oxkutzcab area was producing a significant proportion of the fresh fruit sold in the urban market, and had established its reputation as "the garden of Yucatán," renowned for its lush vegetation and for the cultivation of many kinds of crops (Pérez Toro, 1972).

At that time, the federal government was interested in diversifying the agricultural economy of the state away from henequen and maize, with preference for the *ejidal* sector. A program of small-scale irrigation was initiated, which gave the peasant population access to irrigated parcels. Over the next 40 years, 135 units were built as part of several distinct development programs. In all, they were designed to provide water to over 6,000 hectares and to benefit approximately 3,500 families (Secretaría de Agricultura y Recursos Hidráulicos, 1982). The most common type are called the *unidades antiguas de riego*, or old-type units, to distinguish them from the sprinkler systems that were introduced later. They are designed for gravity water distribution and are divided into individually assigned parcels with an average size of less than 1 hectare. In the 1940s and 1950s the users participated directly in the construction of the main canals and administered the systems themselves through elected representatives. Since the early 1960s, the federal government has taken over almost all administrative functions and has provided technical assistance at erratic intervals. There has been little continuity in these programs, and the accumulation of the resources necessary for the development of the parcels, as well as their management, has been the autonomous responsibility of the participant farmers.

Unlike other regions of Mexico where peasant agriculture predominates, such as Oaxaca, Yucatán does not have a well-established system of peasant markets. The Maya communities are small and widely scattered and produce essentially the same set of subsistence crops. Oxkutzcab is a notable exception: a bustling trade center, which has evolved as a wholesale assembly point where buyers from the urban markets come with trucks to buy small lots of diverse products from many small peasant farmers. Most of the sellers are Maya women, the wives of the producers. The prices are unpredicatable, vary over a wide range depending on both seasonal and short-term supply conditions, and are subject to manipulation by powerful merchants. Nevertheless, the market has permitted the intensification of agriculture.

Production Systems

All the parcels in the old-type irrigation units are cultivated using simple hand tools and production techniques derived from those used in traditional systems. In the traditional, rainfed, backyard tree gardens, a few individuals of many kinds of trees are planted to meet the direct needs of the peasant household. The plots mature into highly diverse plant communities (Barrera, 1981; Vara Moran, 1980). The development of irrigation and the evolution of the wholesale assembly market have permitted the Puuc region to become relatively specialized in oranges, which account for 60% of the trees in production (Comisión Nacional de Fruticultura, 1980). The operation and maintenance of the irrigation systems is administered by a federal irrigation district, and other services have been provided by the government at erratic intervals. Nevertheless, the peasants have maintained the autonomy to develop production systems gradually. In a pattern that closely resembles traditional citrus groves in the Mediterranean countries of Europe (Burke, 1951), the individual parcels are diversified, with between 4 and 20 species of fruit, as well as with annual and semiperennial crops when the trees are small. They form complex, multi-storied arrangements, crowded into both the vertical and the horizontal space. The most frequent companion species are avocados (*Persea americana*) and mangos (*Manglifera indica* L.)—large trees with delicate, high-value fruit—and mandarin oranges (*Citrus reticulata* or *C. nobilis*), a small species with growth characteristics that can be combined conveniently with oranges in the architecture of the parcels. In addition to three or four principal crops, most farmers continue to plant a few individuals of sapotes (*Manilkara zapota*), anonas (*Annona* spp.), and many other tropical fruits of the region.

The development costs of a parcel are concentrated in the first few years. If there were no revenue until the fruit trees came into production 5 to 7 years later, the farmer would be forced to find other sources of income to support his family at the same time as he irrigated, weeded, and cared for the young seedlings in a period when careful management is crucial to future yields. Within a given year, the production of a variety of crops ensures that there is a more or less even flow of income for the household and to pay for water, labor, and other cash costs of production when they are required in the annual cycle. Diversification spreads the risks of fluctuations, both in the yields of individual crops and in unpredictable market prices. The sequence permits the gradual accumulation of equity in the long-term stability of the household economy.

A common development pattern is as follows. A group of peasants organize themselves to solicit the construction of an irrigation unit. After a number of years the project is built, and each farmer clears a parcel and builds a network of irrigation ditches connected to the main canal. Maize and other *milpa* crops are planted for 2 years before grafted fruit tree seedlings are laid out. Vegetables and other short-season crops are grown in the open space between the young trees for 4 or 5 years. These are gradually overlapped with semiperennials such as papaya (*Carica papaya*), cassava (*Manihot esculenta*), and bananas (*Musa* cvs.). As the predominant trees come into production and the shade canopy closes over, other perennials are planted, usually shorter or taller species to take advantage of the vertical space.

Vegetables and other short-season crops require frequent irrigation and weeding. Pest and disease problems have become serious in recent years, and many farmers spray with a "cocktail" mixture of insecticides, fungicides, and foliar fertilizer as often as every 4 or 5 days. Increasing shade cover suppresses weed growth. Mature trees are fertilized only occasionally and require irrigation less often. The total output of the parcel gradually decreases. An older peasant farmer is assured of a diversified, relatively secure source of income for progressively less labor and cost, and has something tangible to pass on to his heirs. A young family with the means and work force to do so acquires additional land on which to begin the sequence again. In any particular case, the combination of crops, the intensity of labor and input use, and the net economic returns depend on the relative importance of fruit and vegetable production in the family economy, on the stage of parcel development, and on the resources that the farmer has been able to accumulate in the course of his life.

Types of Management

The impact of these variations is illustrated by the apparent net economic returns of six farmers (Table 4-4).[6] Part-time farmers operate small parcels in combination with other sources of income with which they pay a portion of the costs of production. Some, such as Mario Hau, have backyard gardens with a function similar to that of traditional *solares*. His principal occupation is as an independent mason and construction worker, but he pays help to produce fruit for home consumption and maintains the parcel for the long-term security that it provides should his business flag. Manuel Gongora is just getting started, and his highly negative apparent net income reflects the fact that he is building an investment in the future of his household.

The majority of the producers in the region are family-labor farmers. Transito Us operates a highly diversified parcel, which he has built up gradually

TABLE 4-4. Irrigated fruit and vegetable operations in Oxkutzcab and Akil, Yucatán: Summary of receipts, estimated cash costs of production, estimated value of family labor, and estimated apparent net returns, standardized on a per-hectare basis, of six selected farms, 1981–1982.

Farms	Size of parcel (*hectares*)	Number of crops sold	Value of receipts (U.S.$/ha[a])	Estimated cash costs (U.S.$/ha)	Estimated value of family labor (U.S.$/ha)	Apparent net returns (U.S.$/ha)
Type I: Part-time farms						
1. Mario Hau	0.5	7	925	675	310	−60
2. Manuel Gongora	0.5	3	380	475	3,180	−3,275
Type II: Family-labor farms						
3. Transito Us	1.2	15	3,130	400	2,375	355
4. Juan Pech	1.9	7	8,075	3,210	1,460	3,405
Type III: Hired-labor farms						
5. Alberto Suarez	3.0	8	1,675	625	60	990
6. Francisco Castro	2.8	5	3,230	1,110	0	2,120

Source: Ewell, 1984, appendix tables B6–B11.

[a]All monetary figures have been converted at the exchange rate of 26 pesos per dollar.

over a period of many years. He spends very little on inputs, and his apparent net income is modest. Nevertheless, he was able to support three children in school and to invest in a bull calf, a local measure of relative prosperity. Juan Pech has recently returned from 14 years of labor overseas and was investing his accumulated savings in the redevelopment of several parcels. He has planted high-value avocado seedlings, and relies heavily on purchased inputs and hired labor to produce vegetables intensively. He plans to reduce his cash costs as soon as his parcels are well established and his children are through school.

A small but influential group are the hired-labor farmers, who do not work their parcels themselves. They are merchants or have other sources of capital, which they have invested in fruit production. They were the first to develop irrigated parcels in the region, and they have introduced and tested new crop varieties, new inputs, and new techniques, which have later been adopted by their less prosperous neighbors. They are not necessarily the most efficient producers, however, because their parcels are not the principal source of income on which the security of their families depends. Alberto Suarez, for example, developed his 3-hectare citrus parcel over 20 years earlier with his income as the manager of a large commercial beekeeping operation. He maintains it with a minimum of cash expenditure as he helps his son establish a cattle ranch in Quintana Roo. At 68 years of age, he regards the modest income it earns as a kind of pension. Francisco Castro is a fruit merchant who produces choice avocados, a delicate fruit with high input costs, as a source of supply and for the prestige that the cultivar confers in an agricultural community.

Most of the fruit producers have had very limited resources beyond the labor of their own families. They have built up their parcels gradually, with small, incremental investments of labor time and cash earned in migratory employment. They have planted diverse combinations of crops, in a variety of arrangements and sequences, to provide income for consumption and to pay for production costs throughout the year. With very little assistance from the government beyond the construction of the units themselves, they have adjusted their traditional production practices and have adopted modern inputs selectively. Yields and net returns are modest by commercial standards, but the development of parcels has stablized the economy of many families.

Comments

Both commercial beekeeping and small-scale irrigated fruit production have developed as peasant management systems. They are strategies of adaptation to a difficult environment, to shifting household consumption requirements, and to dynamic conditions in the larger economy. Both have become viable complements or alternatives to the *milpa*. The government has provided infrastructure and marketing services, and the peasant families have responded to regional economic opportunities while maintaining control over their resources.

These two cases of development of peasant agriculture contrast sharply with the results of many large, formal government development programs that have attempted to develop modern agriculture in the coastal lowlands of Mexico, an area seen as having underexploited potential (Barkin, 1978; Ewell & Poleman, 1980). The majority of these schemes have neither achieved their goals nor provided a stable livelihood to the participating peasant families. They have been based on untested technical assumptions and inadequate evaluation of the capacity of the natural resources. They have been financed through credit systems that have not provided sufficient or timely working capital. They have centralized decision making into the hands of a diffuse and poorly coordinated bureaucracy. The officials have been under heavy pressure to elaborate schemes that conform to ambitious national policies and rigid guidelines, and they have often attempted to fit the peasant into a Procrustean bed.

A valuable lesson for agricultural development can be drawn from the changes in peasant agriculture described here for Yucatán. The relative success of the mixed *milpa*-commercial beekeeping and fruit production systems has been due to the fact that each family has retained the autonomy to intensify management practices at a rate determined by its own resources and fluctuating fortunes. Yields and net returns of fruit production and of the mixed *milpa*-commercial beekeeping system could be increased further if better technical services were provided and if the markets were more stable. Yet, as this case study of the evolution of the *milpa* and its alternatives illustrates, the challenge that faces the agricultural development in Yucatán specifically and in Mexico in general is to meet two basic criteria: The household unit must retain control over the decisions which are crucial to the stability of its economy, and any new enterprises must be sufficiently flexible to adapt to a variable natural and economic environment.

Notes

1. The native crop varieties are diverse, and the Maya terminology distinguishes a large number of local variants. A study of the native races of maize in Mexico has classified those found in Yucatán into three families according to their genetic characteristics (Wellhausen, Roberts, & Hernández X., 1952). The peasants distinguish three broad types according to the length of the vegetative cycle and the size of the ears, characteristics that are directly correlated with yield (Hernández X., 1959:27). The earliest, *nalt'el*, which produces tiny ears in as few as 50 days, is now grown only rarely. *Xmehenal* is an intermediate type with a growing cycle of approximately 3 months, which is usually grown in backyard *solares* and in other especially fertile areas. The common *milpa* maize is called *xnuknal*. It matures in approximately 4 months, is resistant to climatic fluctuations, and stores relatively well in the husk. There are white and yellow varieties of all these types. *Xk'olibu'ul* and *tsama'* are black beans of a species of *Phaseolus vulgaris*. *Ib*, which are lima or seiva beans (*P. lunatus*), as well as *Xpelon*, which are cowpeas (*Vigna sinensis*) are often eaten immature as vegetables. A number of squash (*Cucurbita* spp.) species and varieties are produced, primarily for their dried seeds. The problem of the erosion of genetic diversity has not been studied in detail. The *campesinos* lament it, and select what seed is locally available.

2. Two of the several fiber-bearing *Agave* species that are native to Yucatán are cultivated commercially. *Agave sisalana*, which is called *ya'axki* in Maya, is true sisal. It has been established

in East Africa, Brazil, Haiti, and other parts of the world. Most of the commercial crop in Yucatán is *Agave fourcroydes, Sak-ki* in Maya, because it is adapted to drier conditions. Its fiber is slightly inferior. Although it is also commonly called *sisal*, it is correctly identified as *henequen* in the specialized crop and trade literature.

3. The *ejido* is a complex institution unique to Mexico. The word, referring to the common lands in medieval Spanish villages, was applied in a general way to communal holdings in the colonial period. After the Revolution it was embodied in a series of laws as a specific institution with several variants. In the most common type, the land that had been despoiled of a village by a *hacienda* was restored to it as an *ejido*. The henequen plantations were not expropriated until 1937, when they were converted into collective *ejidos* to maintain economies of scale in the vertically integrated production and processing system. They are effectively managed by the government as nationalized enterprises.

4. Data collected in an intensive study on *milpa* production in the community of Yaxcabá, also in the maize-producing zone, indicates a similar pattern. The average *milpa* size in 1978 was approximately 4 hectares, the average distance was 4.2 kilometers, and the average fallow was 16 years (Arias Reyes, 1980).

5. The amount of honey available for exchange was calculated from average annual yields, the selling price of honey, and the cost of a standardized partial maintenance package for each year from 1972 to 1982. The market basket was constructed from detailed consumption data collected in a Maya community in 1980 (Baraona & Montalvo, 1981). The prices used for the consumption items were those charged by the principal rural store in Chan Kom over the same period. It was assumed that the household meets all its maize and bean requirements, and half of its egg consumption, from home production. Villagers consume meat fairly regularly, but this item was not included in the market basket because demand is highly elastic and price information is incomplete. A conservative estimate of spending on manufactured goods is based on observations of buying patterns and on interviews with store owners in Chan Kom. The list includes the kinds of items typically purchased with honey money, but it is not intended to be a complete budget of the actual expenditures of a peasant household.

6. The partial budgets are based on a uniform accounting procedure. The receipts from the sale of all crops were calculated using a weighted average price after the harvesting and transportation costs were deducted. The cash costs of production were limited to a uniform set of factors: irrigation water, fertilizer, spray chemicals, and hired labor. The value of family labor was estimated using average wage rates in the region. The time of the women in the market was counted at the same rate as that of the men, and the labor of children under 12 was put in at half-price. The total represents approximately what the farmer would have had to pay to hire in the work. The bottom line, the apparent net income in a single year, is a useful standard for comparison. Nevertheless, it does not include the value of the family's accumulated assets or the other sources of income.

References

Argaez, I., & Maldonado, O. Hacia una nueva estrategia para la apertura de areas con suelos mechanizables. In *Quintana Roo: Problemática y perspectiva*. Cancun, Mexico: Centro de Investigaciones Cientifícas de Quintana Roo, 1980.

Argaez, I., & Montañez, C. *Yucatán: las condiciones del desarrollo de la agricultura de subsistencia*. Mérida, Mexico: Escuela de Economía, Universidad de Yucatán, 1975.

Arias Reyes, L. M. La producción actual en Yaxcabá, Yucatán. In Efraím Hernández X. (Ed.), *Seminario sobre producción agrícola en Yucatán*. Mérida, Mexico: Gobierno del Estado de Yucatán, 1980.

Baraona, M., & Montalvo M. C. Filomena Mata: Pequeña sociedad y alimentación. Mexico City: Centro de Investigaciones Para el Desarrollo Rural (CIDER), 1981. (Mimeographed.)

Barkin, D. *Desarrollo regional y reorganización campesina*. Mexico City: Centro de Ecodesarrollo Nueva Imagen, 1978.

Barrera, A. Sobre la unidad de habitación tradicional campesina y el manejo de recursos bioticos en el area maya Yucatanense. *Biotica* (Xalapa, México), 1981, 5, 115–129.

Barrera, A., Gómez-Pompa, A., & Vásquez-Yánes, C. El manejo de las selvas por los mayas. *Biotica* (Xalapa, Mexico), 1977, 2, 44–61.

Burke, J. Henry. *A study of the citrus industry of Italy*. United States Department of Agriculture, Foreign Agriculture Report, 1951, 59.

Calkins, C. *Beekeeping in Yucatán: A study in historical cultural zoogeography*. Unpublished PhD dissertation, University of Nebraska, 1974.

Cline, H. F. The "Aurora Yucateca" and the spirit of enterprise in Yucatán: 1821–1847. *Hispanic American Historical Review*, 1947, 27, 30–60.

Cline, H. F. The sugar episode in Yucatán, 1825–1850. *Inter-American Economic Affairs*, 1948, 2, 79–100.

Comisión Nacional de Fruticultura, México. Proyecto Citrícola del Banrural Peninsular, Censo de cítricos en el sur de Yucatán. (Mimeographed, Mexico City, 1980).

Contreras Arias, A. Bosquejo Climatológico. In E. Beltrán (Ed.). *Los recursos naturales del sureste y su aprovechamiento*. Mexico City: Instituto Mexicano de Recursos Naturales Renovables, 1958.

Cook, S. F., & Borah, W. The Population of Yucatán, 1517–1960. *Essays in population history: México and the Caribbean* (Vol. 2). Berkeley: University of California Press, 1974.

Dewalt, B. R. Inequalities in wealth, adoption of technology, and production in a Mexican ejido, *American Ethnologist*, 1983, 2, 149–168.

Doering, D. O. & Butler, J. H. Hydrogeologic constraints on Yucatán's Development. *Science*, 1974, 186, 591–595.

Domínguez, J. L. *La formación de la burguesia en Oxkutzcab*. Mérida, Mexico: Instituto Nacional de Antropología e Historia, n.d. (Mimeographed).

Elmendorf, M., & Merrill, D. *Socioeconomic impact of development in Chan Kom, 1971–1976: Rural women participate in change*. Unpublished manuscript prepared for the World Bank, Population and Human Resources Division, Development Economics Department, 1977.

Emerson, R. A. A preliminary survey of the milpa system of maize culture as practiced by the Maya indians of the northern part of the Yucatán peninsula. *Annals of the Missouri Botanical Garden*, 1953, 40, 51–62.

Ewell, P. T. *Intensification of peasant agriculture in Yucatán*. Ithaca, NY: Cornell/International Agricultural Economics Study, A.E. Research 84-4, 1984.

Ewell, P. T., & Poleman, T. T. *Uxpanapa: Agricultural development in the Mexican tropics*. New York: Pergamon Press, 1980.

Flores Mata, G. *Los suelos de la peninsula de Yucatán y sus posibilidades agropecuarios*. Mexico City: Secretaría de Agricultura y Recursos Hidráulicos, Subsecretaría de Planeación, 1977. (Mimeographed.)

Gallegos, G. de la C. E. *"La milpa," sistema tradicional de producción de maiz asociado en la peninsula de Yucatán*. Mérida, Mexico: Instituto de Investigaciones Agrícolas (INIA), 1981. (Mimeographed.)

García, E. (Ed.) *Precipitación y probabilidad de la lluvia en la republica Mexicana y su evaluación: Campeche, Yucatán, y Quintana Roo*. Mexico City: México, Secretaría de la Presidencia, Comisión de Estudios Sobre el Territorio Nacional, 1973.

Gerhard, P. *The southeastern frontier of New Spain*. Princeton, NJ: Princeton University Press, 1979.

Goldkind, V. Social stratification in the peasant community: Redfield's Chan Kom reinterpreted. *American Anthropologist*, 1965, 67, 863–884.

Goldkind, V. Class conflict and cacique in Chan Kom. *Southwestern Journal of Anthropology*, 1966, 22, 325–345.

González Navarro, M. *Raza y tierra: La guerra de castas y el henequén*. Mexico City: El Colegio de México, 1970.

Hall, L. L., & Price, T. Price policies and the SAM. *Food Policy*, 1982, 7, 302–314.

Harrison, P. D., & Turner, B. L., II (Eds.). *Pre-historic Maya agriculture*. Albuquerque: University of New Mexico Press, 1978.

Hernández X., E. La agricultura. In E. Beltrán (Ed.), *Los recursos naturales del sureste y su aprovechamiento* (Vol. 3). Mexico City: Instituto Mexicano de Rucursos Naturales Renovables, 1959.

Illsley Granich, C., & Hernández X., E. La vegetación en relación a la producción en el ejido de Yaxcabá, Yucatán. In E. Hernández X. (Ed.), *Seminario sobre la producción agrícola en Yucatán*. Mérida, Mexico: Gobierno del Estado de Yucatán, 1980.

Isphording, W. C. The physical geography of Yucatán. *Transactions of the Gulf Coast Association of Geological Sciences*, 1975, 25, 231–262.

Joseph, G. M. *Revolution from without: The Mexican revolution in Yucatán, 1915–1924*. Unpublished PhD dissertation, Yale University, 1978.

Lesser Illades, J. M. *Estudio hidrogeológico e hidroquimico de la Peninsula de Yucatán*. Mexico City: Secretaria de Recursos Hidráulicos, 1976. (Mimeographed.)

Matheny, R. T. Ancient lowland and highland Maya water and soil conservation strategies. In K. V. Flannery (Ed.), *Maya subsistence* (pp. 157–178). New York: Academic Press, 1982.

Marcus, J. The plant world of the sixteenth and seventeenth century lowland Maya. In K. V. Flannery (Ed.), *Maya subsistence*. New York: Academic Press, 1982.

Merrill-Sands, D. *Commercial beekeeping or migrant wage labor: Alternative cash strategies in the peasant economy of Chan Kom, Yucatán, México*. Paper presented at the Society for Applied Anthropology, San Diego, 1983.

Merrill-Sands, D. *The mixed subsistence-commercial production system in the peasant economy of Yucatán, México: An anthropological study in commercial beekeeping*. PhD dissertation, Cornell University, 1984.

Miranda, F. Estudios acerca de la vegetación. In E. Beltrán (Ed.), *Los recursos naturales del sureste y su aprovechamiento* (Vol. 2). Mexico City: Instituto Mexicano de Recursos Naturales Renovables, 1958.

Montañez, C., and Aburto, H. *Maíz: Política institucional y crisis agrícola*. Mexico City: CIDER/Nueva Imagen, 1979.

Neugebaur, B. Watershed management by the Maya civilization of central Yucatan, Mexico. *Vierteljahresberichte* (Germany), 1983, 94, 395–409.

Palerm, A. *Antropologia y marxismo*. Mexico City: CISINAH/Nueva Imagen, 1980.

Patch, R. W. *A colonial regime: Maya and Spaniard in Yucatán*. Unpublished PhD dissertation, Princeton University, 1979.

Pérez Ruiz, M. L. Organización del trabajo y toma de decisiones en la familia campesina milpera. In E. Hernández X. (Ed), *Seminario sobre la producción agrícola en Yucatán*. Mérida, Mexico: Gobierno del Estado de Yucatán, 1980.

Pérez Toro, A. *La milpa*. Mérida, Mexico: Gobierno de Yucatán, 1942.

Pérez Toro, A. La fruticultura en los suelos pedregósos de Yucatán. Mexico City: Comisión Nacional de Fruticultura, Folleto No. 5, 1972.

Pool Novelo, L. El estudio de los suelos calcimórficos con relación a la producción maicera. In E. Hernández X. (Ed.), *Seminario sobre producción agrícola en Yucatán*. Mérida, Mexico: Gobierno del Estado de Yucatán, 1980.

Redfield, R. *The folk culture of Yucatán*. Chicago: University of Chicago Press, 1941.

Redfield, R. *A village that chose progress: Chan Kom revisited*. Chicago: University of Chicago Press, 1950.

Redfield, R. *The little community: Viewpoints for the study of the human whole*. Uppsala: Almqvist and Wiksells, 1955.

Redfield, R., & Villa Rojas, A. *Chan Kom: a Maya village*. Chicago: University of Chicago Press, 1962.

Reed, N. *The caste war of Yucatán*. Stanford, CA: Stanford University Press, 1964.

Rejón Patrón, L. Tabi, una hacienda azucarera de Yucatán a fines del siglo XIX. In B. Gonzalez R. (Ed.), *Yucatán, peonaje y liberación*. Mérida, Mexico: FONAPAS, 1981.

Revel-Mouroz, J. *Aprovechamiento y colonización del trópico húmedo Mexicano*. Mexico City: Fondo de Cultura Económica, 1980.

Rosales Gonzáles, M. Comerciantes en Oxkutzcab, 1900–1950. *Yucatán: Historia y economía* (Mérida, Mexico), 1980a, 17, 64–73.

Rosales Gonzáles, M. Etapas en el desarrollo regional del Puuc, Yucatán. *Yucatán: Historia y economía* (Mérida, México), 1980b, 18, 41–53.

Roys, R. L. *The political geography of the Yucatecan Maya*. Publication 613. Washington, DC: Carnegie Institution of Washington, 1957.

Ryder, J. W. Internal migration in Yucatán: Interpretation of historical demography and current patterns. In G. D. Jones (Ed.), *Anthropology and history of Yucatán*. Austin: University of Texas Press, 1977.

Secretaría de Agricultura y Recursos Hidráulicos (SARH), México. *Dirección general de distritos y unidades de riego: La operación y el desarrollo en el distrito de riego -48*. Ticul, Yucatán, Mexico, 1982. (Mimeographed.)

Secretaría de Programación y Presupuesto (SPP), México. *Informe económico 1981: Yucatán*. Yucatán, Mexico: SPP, 1982.

Stavenhagen, R. *Capitalismo y campesinado en México*. Mexico City: INAH, 1976.

Steggerda, M. *Maya Indians of Yucatán*. Publication 531. Washington, DC: Carnegie Institution of Washington, 1941.

Strickon, A. Hacienda and plantation in Yucatán. *América Indígena*, 1965, 25, 42–57.

Suárez Molina, V. *La evolución económica de Yucatán a través del Siglo XIX*. Mérida, Mexico: Universidad de Yucatán, 1977.

Tannenbaum, F. *The Mexican agrarian revolution*. New York: MacMillan, 1929.

Tec Poot, J. El K'ankubul-ha. *Boletín de la escuela de ciencias antropológicas de la universidad de Yucatán* (Mérida, Mexico), 1978, 32, 30–35.

Tozzer, A. M. *Landa's relacion de las cosas de Yucatán*. Papers of the Peabody Museum of Archaeology and Ethnology, No. 18. Cambridge, MA: Harvard University Press, 1941.

Turner, B. L., II. *Once beneath the forest: Prehistoric terracing in the Rio Bec region of the Maya lowlands*. Dellplain Latin American Series, No. 13. Boulder, CO: Westview Press, 1983.

Turner, B. L. II, & Harrison, P. D. (Eds.). *Pulltrouser Swamp: Ancient Maya habitat, agriculture, and settlement in northern Belize*. Austin: University of Texas Press, 1983.

Turner, B. L., II, & Miksicek, C. H. Economic plant species associated with prehistoric agriculture in the Maya lowlands. *Economic Botany*, 1984, 38, 179–193.

Turner, J. K. *Barbarous Mexico: An indictment of a cruel and corrupt system*. London: Cassell & Company, 1912.

Vara Moran, A. La Dinámica de la milpa en Yucatán: El solar. In E. Hernández X. (Ed.), *Seminario sobre la producción agrícola en Yucatán*. Mérida, Mexico: Gobierno del Estado de Yucatán, 1980.

Warman, A. *Los campesinos, hijos predelictos del régimen*. Mexico City: Editorial Nuestro Tiempo, 1972.

Weaver, E. C., & Weaver, N. Beekeeping with the stingless bee (*Melipona beecheii*) by the Yucatecan Maya. *Bee World*, 1981, 62, 7–19.

Wellhausen, E. J., Roberts, L. M., & Hernández X., E. *Races of maize in Mexico: Their origin, characteristics, and distribution*. Jamaica Plain, MA: Bussey Institution of Harvard University, 1952.

Williams, B. J., & Ortiz-Solorio, C. Middle American folk soil toxonomy. *Annals of the Association of American Geographers*, 1981, 71, 335–358.

Wilson, E. M. Physical geography of the Yucatán peninsula. In E. H. Moseley & E. D. Terry (Eds.), *Yucatán: a world apart*. University: University of Alabama Press, 1980.

Wright, A. C. S. *El reconocimiento de los suelos de la peninsula de Yucatán, México: Informe final*. Chapingo, Mexico, Escuela Nacional de Agricultura, 1967. (Mimeographed.)

5

Swiddeners in Transition: Lua' Farmers in Northern Thailand

PETER KUNSTADTER

Farmers in Southeast Asia have practiced swidden agriculture for millennia. It has served as the primary system for many subsistence farmers. As a supplement to irrigated agriculture it occurs as a first step in clearing and preparing land for permanent field farming.[1] Except in more isolated areas such as the foothills and mountains of northern Thailand, swidden farming declined in importance as a primary source of subsistence in the late 19th and early 20th centuries after large-scale irrigation projects were developed in the major river valleys.

By the 1960s, four basic types of swiddens were found in Thailand. Each type was more or less associated with specific sets of environmental, economic conditions, and ethnic groups (Table 5-1). The long cultivation–very long fallow type is practiced by the Hmong and other so-called hill tribes. It has served as the model for the negative public stereotype for all hill tribe swiddeners, who were believed to cut and burn primary forests for temporary fields, which they then abandoned after a short period of cultivation. Such negative perceptions have guided the policy for highland development, which is directed at settling shifting cultivators and reforesting abandoned (or fallow) swiddens.

The short cultivation–short fallow type of swidden practiced in the foothills by Northern Thai has been the most widespread in terms of population use and land area. The short cultivation–long fallow type followed in the middle elevations by Lua' and Karen has been the most stable and probably the most widespread among non-Thai minorities. The short cultivation–variable fallow type has been used by Northern Thai and some minorities as a supplement to tea production and has been the least extensive of the four types.

Swidden systems have come under increasing pressure since the 1960s because of population increase, loss of forest, and increasing governmental efforts to modify traditional land-use systems in the highlands (Kunstadter, Sabhasri, Aksornkoae, Chunkao, & Wacharakitti, forthcoming). Average an-

Peter Kunstadter. Medical Anthropology Program, Department of Epidemiology and International Health, and Institute for Health Policy Studies, University of California, San Francisco, California.

TABLE 5-1. Types of cultivation systems in the Northern Thai highlands in the 1960s.

Cultivation to fallow ratio (years:years)	Altitude range (m)	Method and depth of tilling	Vegetation/ regrowth	Ethnic groups	Land tenture	Village stability	Economic importance	Major crops
Swidden								
Short cultivation–short fallow (1:3–5)	200–500	Hoe, plow; shallow, stumps removed	Bush/bush	Northern Thai	Traditional individual use rights for recultivation	Permanent	Supplement to wet rice	Rice
Short cultivation–long fallow (1:10 ±)	700–1,400	Weeding tool; shallow, stumps left for regrowth	Secondary forest/ secondary forest	Karen, Lua'	Traditional community ownership, household rights for recultivation	Permanent	Primary subsistence	Rice
Short cultivation–variable (1:3–10)	1,000–1,500	Hoe; shallow	Primary or secondary forest/ secondary forest	Northern Thai and others	Variable	Permanent	Supplement to tea	Rice
Long cultivation–very long fallow (3 ± :50 ±)	1,000 +	Hoe; deep, stumps removed	Primary forest/grass	Akha, Hmong, Lahu,	Household use rights only while actively cultivated	Usually temporary	Primary for cash and subsistence crops	Rice, maize and opium
Irrigated								
Annual or multicrop dry season or no fallow	200–1,500	Plow, harrow; deep, stumps removed	Secondary forest/rice	Northern Thai, Karen, Lua'	Individual legal ownership	Permanent	Primary subsistence and cash crops	Rice

nual natural increases in the highland population have been around 3%, resulting in the doubling of the highland population in a little over 20 years (Kunstadter 1984a, 1984b), and migration of hill dwellers from Burma and Laos further increased the rate of growth of the highland population. (Recent demographic research (e.g., Kamnuansilpa, Kunstadter & Auamkul, 1987) confirms the fact that population growth rates of different swiddening groups vary, with Lua' and Karen annual rate of natural increase under 3%, and Hmong 5% or more. Because of population increase and recent changes in implementation of government policies, as of 1987 few Hmong still practiced the traditional long cultivation–very long fallow swidden system in mature high-altitude forests. In association with suppression of poppy cultivation, control of cutting forests, and resettlement of highland villages, many Hmong were farming on a short cultivation–short fallow cycle; some Hmong villages had been relocated to lower elevations (300–700 m); and some Hmong farmers were growing maize as a cash crop, with extensive use of farm machinery on permanent upland fields.) The highland population increased at a time when there was virtually no net migration into the lowlands and when the rate of natural increase in the lowland population was declining rapidly because of widespread use of family planning. From 1963 to 1982 Thailand lost approximately 43% of its remaining forest, primarily as a result of the expansion of the area devoted to upland crops (Klankamsorn & Charuppat, 1983; Wacharakitti, Boonnarm, Sanguantan, Boonsaner, Silapatong, & Songmai, 1979). The loss of forest in the northern region (where most of the highland population is located) was at a slightly lower rate than the national average, but most of the loss in the north was of fully mature forests at the highest elevations.

Short cultivation–long fallow subsistence swiddeners (Karen, Lua') have lived for many hundreds of years in permanent villages at the middle elevations. Since the 1960s they have continued their conservative swidden practices in secondary forests. Because of the decline in per capita amounts of land, many have now been forced to reduce the length of fallow, with consequent declines in yield. Thus their swidden system is coming to resemble the short cultivation–short fallow system of the lowland Northern Thai. Where possible, many have switched the emphasis of their farming to terracing and permanent cultivation of irrigated fields.

Groups such as the Hmong began migrating into Thailand from Laos and Burma in the last decades of the 19th century. In Thailand they lived in temporary villages where they practiced long cultivation–very long fallow swidden. They expanded rapidly, both in population and in land area. They competed for land with the more conservative short cultivation–long fallow swiddeners, and also cleared and farmed large areas of mature forest at the higher elevations. As forest was replaced by grassland, which required much more labor for cultivation, those who could do so continued to advance southward into as yet unexploited highland forests.

Starting in the 1970s an extensive road network was built into the highlands, which accelerated the pace of change and was associated with a variety of projects that, along with increased access, often encouraged the migration of lowlanders into the highlands and further increased the pressure on swidden land. Governmental authorities became alarmed at the rapid rate of forest destruction and began large-scale reforestation projects, commonly on fallow swiddens at the middle elevations. This further reduced the amount of land available for swiddening by the more conservative middle-elevation groups.

Government development projects were aimed at stabilizing the settlements of the highest elevation groups and promoting nonnarcotic cash crops. These were often used as supplements rather than substitutes for opium poppy (*Papaver somniferum*) and had the effect of further increasing the demands for land.

This chapter describes the short cultivation–long fallow swidden system as it was practiced in the 1960s in the middle elevations by Lua' villagers in Mae Sariang District, Mae Hongson Province, and by their Skaw Karen neighbors. Some of the causes and consequences of changes in their farming system are examined as they existed through the early 1980s, a time of rapid population growth and economic development in Thailand. The description is based on fieldwork beginning in 1963, with periodic visits to the research area during the following two decades.[2]

The Lua' Swiddeners of Pa Pae in Northwest Thailand: Village and Environment

The Lua' are a Mon-Khmer–speaking group who are the autochthonous population of northwestern Thailand. They have lived in the area since the time of the earliest historical records, and are also recognized in the folk history as the original settlers and rulers of both the lowlands and the valleys in the northwestern part of the country. In the early 1960s a number of Lua' villages in the mountains still retained their ethnic identity and followed traditional land-use customs. By that time the lowland Lua' ancestors were only a memory of the inhabitants of former Lua' villages ("our grandparents were Lua', but we are Northern Thai").

The system described here is the one practiced by people of Pa Pae village (Figure 5-1). Pa Pae is a Lua' village located at an altitude of about 700 m, in the hills about 25 km (a long day's walk) northeast of Mae Sariang District town. Similar systems have been used by Lua' and other highlanders in the hills of northern and western Thailand, as well as neighboring parts of Burma and Laos. The hill country around Pa Pae ranges from about 700 to 1,200 m. Valleys are relatively narrow, and hill slopes are generally not more than 50 degrees. Soils in the area are derived from granite and are highly variable in terms of fertility, water-holding capacity, erodability, and other qualities, in association with microgeological conditions, and with the above-surface vegetation and recent history of land use.

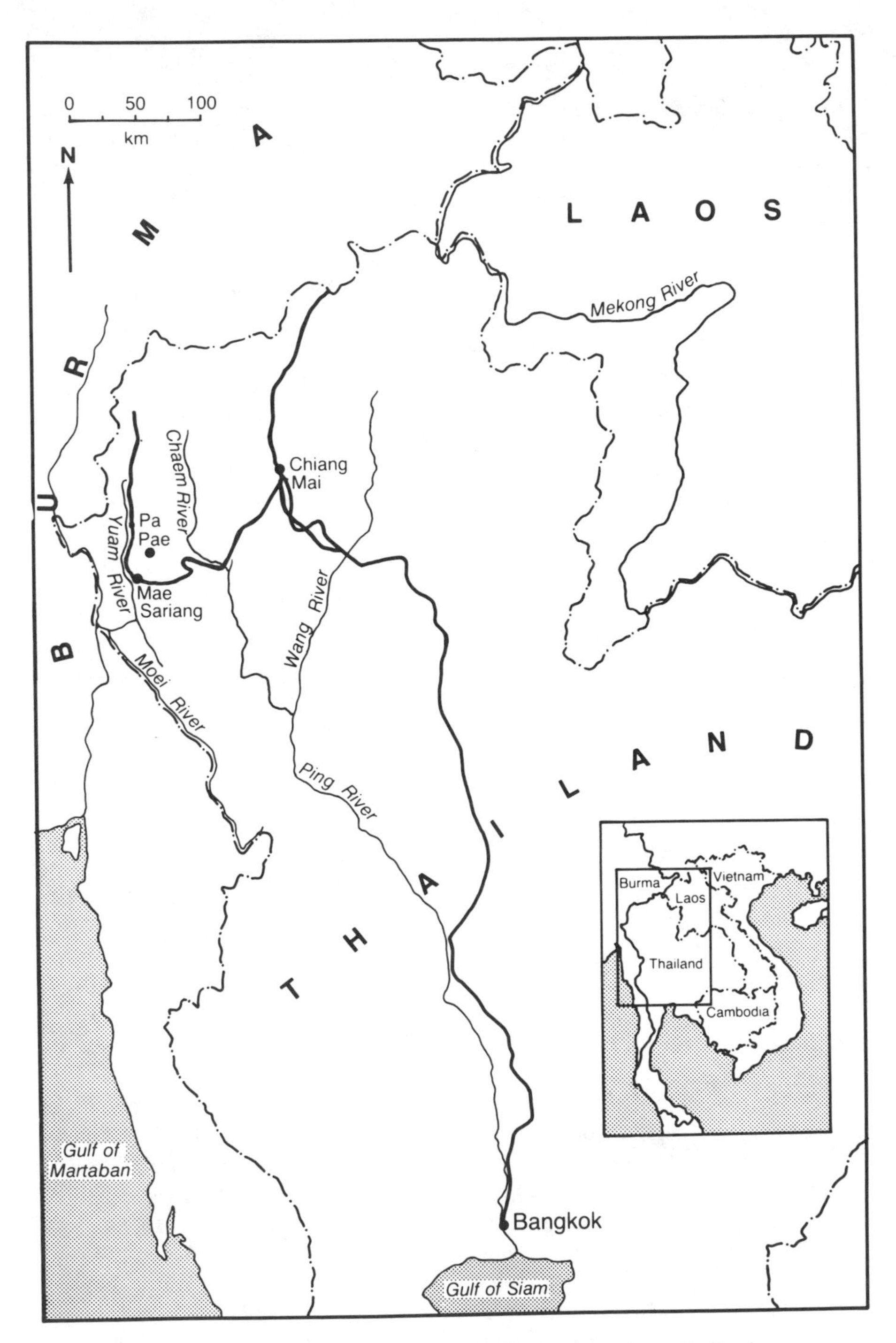

FIG. 5-1. Location of Pa Pae and Mae Sariang in northern Thailand.

Climate in this area is monsoonal, with a dry season running from November through April. Onset of the rains varies from early in May to as late as July. Almost all rain falls between June and October, with annual rainfall measured at Pa Pae in 1967–1968 of 1.43 m. The weather is clear and cool in November through January and then warms up in February through April, with relief from the heat provided by monsoon cloudiness. Temperature measured in a cleared swidden in 1968–1969, elevation 700 m, ranged from a low of 2.2°C (January 20, 1968) to a high of 37.2° (April 25, 1969), as compared with a low of 5.6° (January 4, 1969) and a high of 34.1° (April 25, 1969) in an uncleared forest, elevation about 800 m. Relative humidity in the cleared swidden ranged from 15% toward the end of the dry season (March 20, 1968) and 27% on the same date in the old forest, to 100% during much of the rainy season, when cloud cover was virtually continuous and the temperature hovered in the low 20s (Kunstadter, 1978:127–131).

Natural vegetation in the Pa Pae area has been replaced almost entirely by secondary forest in various stages of regrowth following swidden cultivation. The original vegetation probably ranged from Mixed Deciduous Dipterocarp and Dry Evergreen Forest at the lower elevations to Hill Evergreen on the higher slopes. The true Deciduous Dipterocarp forest forms the lower boundary of this system.

In 1964 Pa Pae had a population of 196. The old man who was recognized by the government as Pa Pae's village headman was also the head of the *samang* or priestly lineage, descended from the ancient Lua' royalty. He was illiterate but had as his assistant a man then of about 40 who had taught himself to read and write during his younger days when he worked away from the village in the mines. The village was surrounded by a small mature forest, which was preserved as a home for the spirits, and also to keep the village cool. Anyone who violated the rule against cutting in this sacred grove was sentenced by the headman to pay a fine for this offense against the spirits. A school staffed by Border Patrol Police had been established about 1960, but there were no other government facilities in the village.

In the 1930s Pa Pae villagers employed lowlanders to teach them how to terrace and farm irrigated fields. In the mid-1960s irrigated fields made up less than 3% of the agricultural land in use (allowing for all the land used in the 10-year swidden cycle), but accounted for about 40% of the rice produced annually by the villagers. Despite the economic importance of irrigated farming, swidden activities dominated the organization of the village.

Annual Cycle of the Lua' Swidden System at Ban Pa Pae

Through the mid-1960s village religion and the annual cycle of village activities revolved around swidden activities. Although many of the organizational features have changed in the past two decades, the basic swidden techniques remain the same and, for convenience, are described in the present tense.

Site Selection

The new year's farming activities start in January, when village elders inspect the site that is next in the regular rotation of swidden fields. In general, all villagers cut their swiddens in a contiguous block, following a regular round in a 10-year cycle. If the site appears suitable (vigorous secondary forest regrowth, no evidence of recent fire) the elders take an omen by examining the gallbladder of a chicken or pig. If conditions are not favorable, or the omen is bad (gallbladder not full and shiny), the elders look for a suitable site in blocks that are scheduled to come up later in the cycle. Once the general location has been decided, each household head visits the plot his household had cultivated during the previous cycle and sacrifices an animal to check the gallbladder to confirm that he will be successful in farming there again. In the event of a bad omen or conditions that do not seem suitable, the field owner might try to exchange sites with someone else or to find an unclaimed portion of the year's block. The headman, as chief *samang* priest, and his religious assistants (*lam*) may take precedence in claiming any piece of land they want, but apparently they very rarely claim land over which another household has prior use rights.

The *samang* chief has the authority to resolve disputes between villagers over access to particular pieces of land. If a dispute cannot be resolved in any other way, he can simply forbid them to cultivate. He also regulates the timing of burning and planting of swiddens. Swiddens are burned on a day he sets in consultation with his religious assistants, at a time when all the elders join in performing the necessary ceremony; no one is allowed to plant a field until he has begun. The chief *samang* derives his authority from his relationship with spirits of the land. In addition to ultimate control over land use, he is entitled to receive a leg of any large game animal that the villagers might kill.

Clearing

Beginning usually in February, each household sets about to cut the forest that covers its field. Working uphill, so the trees they cut knock down the vegetation below them, they cut the smaller trees about 75 cm above the ground. They also cut down some of the larger trees, but for the most part these are only trimmed, so as to prevent shading of the growing crops. The only tool for this work is a heavy knife with a long steel blade (30 cm or more), sharpened to a razor's edge with a fine-grained stone. Men do most of the cutting, especially if it requires climbing trees, but women also cut trees, especially in households where there is only one adult male. At this time they also harvest any bamboo poles growing on their swidden site, and drag them down to the stream at the bottom of the swidden, to save them for fencing after the fire. Generally they leave the tops of ridges uncut and do not cut the vegetation along streambeds or in very narrow ravines. Ridge tops are not favored for swidden farming because the soil is shallow and sandy, and the slopes are often steep. If this type of land is farmed, they may plant a particular rice variety that tolerates the dry soil conditions.

Trees left standing in the fields or growing on ridge tops serve as seed sources for regenerating the forest after the swidden has been left to fallow. Ravines and stream beds are left uncut in order to reduce the danger of erosion. Cutting the swiddens early in the dry season, when the weather is still cool, has the effect of conserving soil moisture by greatly reducing transpiration from the vegetation, and by putting a mulch of branches and leaves over the surface, so as to reduce direct loss of moisture by evaporation.

Firebreaks

Once the forest has been felled, the farmers allow the slash to dry for about 6 weeks. They watch carefully in order to prevent or extinguish accidental fires before the swidden fires are set. If a fire breaks out that threatens the drying swidden or fallow fields, each household is obliged to send one member immediately to fight the fire. Failing this, they are assessed a fine at the end of the year for each day of communal work they have missed. Before the date set for burning the swidden, each household contributes labor to the task of cutting and sweeping a firebreak all around the swidden block, to ensure that the fire will not burn into uncut forest, as that would reduce forest growth and soil fertility in blocks to be cut in future years.

Burning

The day picked by the headman for burning is in the hottest time of the year. If another village is farming an adjacent swidden block, the headmen of the two villages try to arrange to burn on the same day. If blocks are not adjacent, neighboring villages try to arrange to burn on different days so there will be enough people to patrol the village and fields to put out fires started by flying embers. In recent years headmen have lost their ability to regulate the time of burning, and accidental spread of swidden fires has led to disputes within and between villages over responsibility for damages from escaped fires. About noon, at the hottest, driest time of day (about 30–35°C, relative humidity about 10–20%), after the older women of the village have made small sacrifices to the ancestors, and while the older men make sacrifices to the myriad spirits of the forest, the younger men go to the swiddens. After spraying rice liquor on bamboo torches, with an invocation for fire to burn the fields as liquor burns the mouth, they light the torches and touch off the tinder-dry slash, running first along the tops of the fields, then along the sides, and finally along the bottoms, before seeking shelter from the smoke and flames in a cleared valley or streambed. The flames burn upward and toward the center of the fields, eventually creating an indraft of 40–65 km/hr and firestorms that quickly consume the most flammable parts of the fuel. Flames may leap as high as

100m, and an updraft carries the smoke up to 3,000–6,000 m. Cumulus clouds form above the updraft, and occasionally the firestorm triggers thunder, lightning, and showers. The area burned at Pa Pae per season is about 100 ha, consuming fuel with an oven-dried weight of about 55,000 mt.

Temperatures reach 600°C at the soil surface, but only about 70–150° (depending on soil moisture) at a depth of 2 cm. After the fire, the surface of the soil is blackened and covered with a thin ash. The top 2–3 cm of the soil are very dry and loose, but the heat of the fire does not penetrate more than about 5 cm. Below that level the soil remains cool and moist. Insects in the aboveground vegetation and in the top layer of soil are killed, but those living in burrows (e.g., ants, termites, worms), which are important for maintenance of soil structure, survive the fire. Fire-resistant plant species also survive. Within a few weeks most of the trees begin sending out shoots and leaves from their burned-over stumps. By the end of the growing season these form the basis for rapid regeneration of a secondary forest, composed largely of fire-resistant species (Zinke, Sabhasri, & Kunstadter, 1978).

Planting Minor Crops

While the fire still smolders in the larger logs, field owners walk across the ashes, pray, and place a *taleo* woven of split bamboo near where they plan to build their field shelter. The *taleo* is a sign to people and to spirits that the farmer has claimed the field, temporarily, from the spirits who normally dwell in the forest. The claim is reinforced by scattering the seeds of a few flowers, including marigolds (*Tagetes* spp.), cosmos (*Cosmos sulphureum*), and especially red and yellow *Celosia argentea*, which the spirits are believed to shun, around the shelter site. The same afternoon they plant maize seeds (*Zea mays* L.) in shallow holes which they make with a short metal-tipped digging stick. They use the same implement to plant a few yams (*Dioscorea alata*) in the lower, better-watered portion of their swiddens, and they flick cotton seeds (*Gossypium* sp.) over the surface of their swiddens with a short, flat piece of bamboo.

Reburning and Fencing

During the next 2 weeks the farmers lay charred logs along the contours of the hill slopes in order to retard soil erosion, and they also collect unburned poles for fencing or firewood. They pile and reburn the rest of the debris left by the fire. If their field contains deep crevices subject to erosion, they often build pole frameworks at this time for squash or other viny plants (*Cucurbita pepo*, *Lagenaria leucantha*, *Luffa acutangula*, *Tricosanthes* sp.), which also serve to reduce erosion.

Also before rice planting begins, the villagers attempt to fence the swidden block to keep out cows and water buffaloes, which generally are turned loose to graze in fallow fields. Lua' villagers make fences of posts and poles, carefully tied closely enough together to keep out large animals. Once this is done, anyone whose cattle break into the area that is being farmed is liable to a fine, payable to the field owner for any damage done.

Planting Rice

The villagers usually start planting swidden rice (*Oryza* spp.) about the same time as the Northern Thai New Year water-throwing ceremony (April 14). The actual date is set by the headman, who coordinates the date with the phase of the moon, which is believed to influence the success of the crop (depending whether the moon was waxing or waning). Before the planting actually begins, each household, together with the owners of the neighboring swidden, marks the boundaries between the swiddens with a line of charred logs. Usually, once the swidden fire has burned out, the line left from the previous cycle is still visible as a ridge of soil raised by the termites when they consumed the earlier boundary markers. These lines are further marked by planting tall growing (3–4 m) sorghum (*Sorghum vulgare*) along them, so that villagers will not inadvertently stray into a neighbor's field. Trespassers are liable to pay for any damage or loss, and must give one chicken to the rightful owner of the trespassed field. The owner has to sacrifice the chicken to his field spirits, who would be offended by any trespass, even the inadvertent cutting of a single rice stalk.

Fields are planted with the aid of long (3–4 m) bamboo planting poles, tipped with an iron point. The upper end of these poles is fashioned so that they sound like a gong each time they are thrust into the soil. Younger men sing as they wield these poles, making shallow (3–4 cm) depressions in the soil, about 14 holes/m^2. They are followed by women, children and older men, who throw a few seeds at each hole. They make no attempt to bury the seeds, but the soil is so friable that by the end of the planting it has usually slipped enough to cover most of the seeds. Each household attempts to arrange exchange labor with relatives and friends and to time their planting so that they will have a large crew and can finish quickly—on a single day if possible.

Upland (dry) nonglutinous rice is the main crop and forms the dietary staple, but glutinous rice is required for some ceremonies and each household plants many different kinds of rice (depending on their taste preferences, the soil conditions on the swidden, and the availability of seed). Farmers who expect to run out of rice before harvest often plant early-ripening (but lower-yielding) varieties; farmers who like glutinous rice (*O. glutinosa*) plant more than those who do not. Farmers also plant a large number of other crops in their swiddens, including beans (*Dolichos lablab*, *Dolichos* sp., *Glycine max*, *Phaseolus* sp., *Psophcarpus tetragonolobus*); squash (*Cucurbita pepo*); chilis (*Capsicum frutescens*); okra (*Hibiscus esculentus*); sesame (*Sesamum indi-*

cum); millet (*Setaria italica*); mustard (*Brassica* sp.); taro (*Colocasia antiquorum*); Job's tears (*Coix lachryma-jobi*); and others (see Kunstadter, Sabhasri, & Smitinand, 1978, for listing of species planted in swiddens).

Timing of Swidden Planting and the Rains

Timing of swidden planting is crucial for success, but the most important variable, timing of the rains, cannot be controlled. Planting must be finished before the rains start (seed placed in wet soil will rot, not germinate), but not too long before the rains are well established (or the young plants will die from lack of moisture). The onset of the monsoon is usually in mid-May, but rains may come several weeks earlier or later than usual, with disastrous results. Because there is still some soil moisture after the swidden fire, if the planting and rains are timed correctly, the rice will germinate and will have enough roots to prevent being washed out if the early rains are heavy. As insurance against washouts, some of the farmers plant seed extra thick on the shallow slopes at the bottom of their fields, and transplant the extra seedlings into bare spots they missed in planting or that were swept clear by heavy early rains. If everything goes well, the rice germinates and is 5–10 cm high before the heavy rains begin.

Weeding

Weeds are one of the chief constraints on crop yields. The fields should be weeded as soon as it is possible to distinguish rice and other cultigens from weeds. Weeding (mostly the work of women and children) continues through the rainy season until a few weeks before the harvest. The farmers attempt to complete three or four rounds of weeding over their entire swidden field before harvest.

The rainy season weather is cloudy and cool (about 20°C, day and night) but very damp (relative humidity near 100% at all times). The weeders traditionally shielded themselves from the downpour with a leaf-lined basketry framework resembling a turtle shell. Now they wrap themselves with a sheet of plastic from the market. Bending at the waist, they use a short-handed tool with an L-shaped iron blade, sharpened on the lower side and end. This is the only implement for weeding. They cut and scrape the weeds, disturbing only the upper few centimeters of the soil surface and leaving the debris on tree stumps, where it decays. Weeding is the most time-consuming and uncomfortable task of swiddening, lasting from mid-May throughout the rainy season, until September, when the rice crop begins to ripen.

In the meantime various other crops start to yield—first mustard greens, then beans, squash greens, maize, peppers, and others.

Pest Control

The growing crops are subject to loss from insects, rodents, birds, deer, wild pigs and bears, domestic animals, and humans. Losses may be major, especially if pests come in great numbers or if large animals get into the fields and trample the crops. Traditional techniques for control of insect pests include swidden burning (which kills insects and some rodents), limitation of cropping to a single year per swidden block (which prevents buildup of pest populations over more than a single year), separation of blocks for swiddens cut in successive years (which restricts spread of pests from previous years' field blocks), and religious rituals (which the people hope will protect their crops against the spirits that influence the crop pests). Rodents in the fields are controlled by deadfalls and snares. Birds are shot at or scared by noisemakers; large mammals are hunted or scared away by noisemakers, including wind- and water-powered clappers or gongs, and by blowing on horns and conversing, or, in recent years, by playing radios or tape recorders loudly in the shelters at night; domestic animals are fenced out, and their owners, if identified, are fined. Villagers attempt to limit human thieves by staying in their fields when the crops begin to ripen.

Postharvest losses may also be major. Villagers attempt to limit loss to rodents by putting rat shields, made either of kerosene tins or of slippery leaves, around the posts supporting their rice barns. Prayers and small gifts of food are also offered to the soul of the rice to prevent the rice from vanishing and to make it last until the next harvest.

The great variety of crops grown in each field, plus the diversity of rice strains with different ripening times, probably protects the farmers against the total disaster that might strike if only a single variety were planted. Each household has its own preferences for rice in terms of taste and appearance, and each household generally preserves its own seed stock, carefully selected at harvest time from the panicles that appear to be the best. A very great number of different strains of rice have evolved as a result of this selection process. The resulting genetic diversity, preserved in each household, probably offers some protection against crop pests.

Harvest

Some of the poorer families, who have been out of rice for several months, begin harvesting their early-ripening rice in September when it is still in the "milk" stage, before the seeds have fully formed and hardened. The main harvest does not begin until mid-October, when the heavy rains stop and the late rice ripens, first on the lower parts of the fields and then on the upper slopes. Men and women use a short-handled sickle, with a curved, serrated blade about 20 cm long, to cut the stalks of ripened rice. They rest these gently to dry for a day or two on the stubble, and hope they can complete the threshing without the interruption of a late-season rainstorm. They carry big bundles of the dried sheaves to a threshing platform, which has been leveled and lined with mats woven of split bamboo. The threshers beat the heads of the stalks against the mats to knock off the easily loosened grains, and then toss the straw to the side of the mats, where it is beaten with bamboo staves shaped like hockey sticks to remove any additional grains.

The first winnowing is carried out on the threshing mats, where the winnowers sift through the accumulated rice with their feet while waving woven bamboo winnowing fans to blow away the larger pieces of chaff. Then they load the rice into baskets and carry it to a mat-lined winnowing floor on which a short ladder has been erected. Young women pour the grain from the top of the ladder, while young men fan to blow away the remaining chaff. The cleaned rice is then loaded into a basket and carried to a mat-lined temporary barn. The baskets are counted as they are put in the temporary barn, before the rice is carried to the village. By mid-November all the rice has been harvested.

Year-End Ceremonies

In late November or early December, after the villagers carry all of their harvest back to the village, they return to the swiddens early one morning to call their souls and the soul of the rice back to the village. The village is then closed for an end-of-year celebration. Large animals, purchased with contributions from all households, are sacrificed; debts for village expenses (hosting visitors, fighting fires, etc.) are settled; and the male heads of all the households consume any remaining balance in the village coffers, in the form of home-brewed liquor purchased for the occasion.

Agricultural activities cease for a month or two after the year-end ceremony. Then the next year's swidden sites are selected and the round of activities begins again. The dry season is a time for weddings, for building and rebuilding houses, carding, spinning and weaving, and fishing. Some of the young men leave the village to seek dry-season jobs in order to accumulate enough money to get married. They might work in the mines or try to find agricultural work in the lowlands, where the harvest is a bit later. Also in the dry season a few of the men take the opportunity to enlarge their irrigated fields.

Irrigated Rice Farming

As noted earlier, in the 1930s Pa Pae villagers hired lowlanders to teach them how to construct and farm irrigated fields. Irrigated farming requires an initial investment in terracing and ditch and dam construction far greater than that of swidden cultivation. The vegetation must be cleared and the land leveled into terraces. The terraces must be diked to retain water, and a ditch must be dug to bring water to the field from a dammed stream. A heavy investment is also required for the water buffalo or oxen to plow or harrow the field, as well as for the plow, harrow, or hoe, which is used to build, clean, and repair dikes, ditches, and dams. Until recently, land costs were very low; if a potential irrigated field was within the area customarily "swiddened" by the household that was seeking to develop it, they would simply prepare it for irrigation, irrespective of the traditional community ownership of the land. If it was within someone else's customary swidden area, the prospective developer would ask permission (which would generally be granted unless the customary user was planning to develop it), and perhaps would make a token payment. In contrast to swidden fields, ownership of an irrigated field, once developed, can be registered, and the field can be mortgaged or sold. Also, unlike swiddeners, owners need not ask permission of religious leaders before planting. The only religious rituals connected with cutivation are those learned from the Northern Thai, who taught the Lua' how to make irrigated fields.

Irrigated Field Construction

Irrigated fields are often developed or expanded following clearing in the course of swidden cultivation. First terraces are leveled by hoeing to cut down the hillside, and then the loose soil is dragged to the low points. Dikes are made by piling clods of the clayey soil at the edge of the terrace, to a height of 15–20 cm. Conduits to carry the water from ditch to field and from one terrace to the next are made of large (10 cm wide) bamboo tubes, or sometimes by making rock-lined spillways in the dikes.

Annual Field Preparation

After the rains have begun, a small, easily watered section of the irrigated field is carefully fenced, hoed, irrigated, and leveled to serve as a rice nursery. Seed (usually soaked in advance to speed germination) is broadcast on top of the water. While waiting for the rains to soften the soil before plowing, the farmers clean and repair their dikes and ditches, and rebuild the dam. Individual households prepare their own fields, but farmers who share a ditch and dam cooperate in its maintenance. Dams are simple affairs built of stakes thrust into the streambed, and piled with mud, and rubble to raise the water high enough to force some of it into the ditch. No attempt is made to cut off the water flow completely. The unlined ditches generally follow contour lines and only rarely are channeled into wooden flumes to cross small creek beds.

When the soil in the fields is thoroughly soaked and soft, the farmers plow it with a simple buffalo- or ox-drawn hand-made wooden plow, tipped with a purchased metal blade. They break any large clods with a hoe. The fields are soaked again and then harrowed to a fine texture with a home-made animal-drawn wooden-toothed harrow. Both field construction and annual field preparation are usually men's work.

Transplanting

By the time the fields are prepared and the monsoon rains are well established, the rice in the nursery has grown to a height of 25 cm or more. The day before transplanting (or sometimes on the morning of transplanting) the seedlings are pulled up in the nursery, the tops are twisted off to an even length, and the seedlings are bundled and carried to the prepared field.

As with swidden planting, the farmer tries to organize transplanting so as to get the maximum amount of labor and to complete the job as quickly as possible. Transplanters (both male and female) take two or three seedlings from a bunch and press them with their thumb into the mud of the flooded field, trying to spread the plants evenly, about 25–30 cm apart. They do not use any measuring device to assure spacing, and sometimes, when informal teams race to complete planting a section, they may simply circle around and around, pushing the seedlings in as fast as they can.

Irrigating and Weeding

During the growing season the farmers regulate the flow of water onto and off of their fields as required by the growing plants, and in order to control the growth of weeds. Irrigated fields are weeded once or twice by combing with the hands through the growing rice plants. Because water is plentiful and the ditches serve no more than three or four farmers, allocation of water is rarely a problem and is handled informally. Irrigated rice is harvested using the same techniques as for swidden rice. Irrigation tasks are generally handled by men, but weeding may be done by either men or women.

Advantages of Irrigation

There are a number of advantages to irrigated farming despite the greater initial investment required for terracing, diking, and ditching. Once the irrigated fields have been developed, less labor is required for annual clearing or weeding than with swiddens. Also, the timing of operations is not nearly as critical as in swiddens, where burning and planting must be carefully coordinated with the onset of the monsoonal rains. Assuming that there is some rain, the quantity and distribution of rain over time is not as critical for irrigated farming, because irrigation controls water that comes from a fairly large watershed, as compared with swiddens, where the rain must actually fall directly on the field, and a prolonged local dry spell may kill the crops. Irrigated fields are generally prepared later in the year than swiddens, and the main irrigated rice crop is usually planted late in June or July, 2 months or so after swiddens are planted. Irrigated rice often ripens earlier than swidden rice, so there is little conflict in demands for labor for the two types of agriculture. Irrigated fields also have the potential to be cultivated more than once per year, and production per unit area per year appears to be higher on irrigated than on swidden fields. Given the apparent advantages of irrigated farming, one might ask why swiddening has persisted. The reasons seem to be the fact that swidden farming requires a much lower initial investment in labor and equipment; the lack of cash among subsistence swiddeners with which to hire labor and buy draft animals; and, until recently, the easy access to swidden land and the lack of access to market for irrigated cash crops.

Relations with the Market

Some villagers take a little of their produce (rice, pigs, grass mats, or sometimes forest products) to the lowlands to sell, and use the money to buy a few market goods. Traditionally these included cotton yarn, dye, metal blanks for tool blades, salt, salted fish, fermented tea, and matches. As transportation improved to Mae Sariang, the market there began to carry increasing numbers of manufactured goods, including flashlights and batteries, rubber thong sandals, and—after about 1960—canned fish and kerosene to fuel tiny lamps fashioned from condensed milk tins. In more recent years, as a result of the decline in cotton production, villagers have been buying clothing and blankets in town to replace their homespun; as *Imperata* roofing grass has become scarce and transportation to the village improved, they have begun to buy corrugated iron roofs, and two groups of investors in the village (together with some lowland relatives) pooled their money to buy diesel-powered rice mills.

Through the 1960s the village economic system was based on subsistence agriculture, with a minimum involvement in the market economy. There were no shops in the village, and only an occasional peddler passed through the village carrying a few goods from the lowland market. As early as the 1920s some young men, especially those from poor familes, left the village to seek wage work for several months or even several years in the mines or elsewhere, in order to accumulate enough money to get married. Poor people occasionally sought temporary work during the lowland harvest season, and in times of economic disaster whole families sometimes moved to the lowlands to live for a few years on odd jobs, until they could accumulate enough money to return to farming again in the mountains. For the most part, however, these villagers lived on what they were able to produce themselves.

In the 1970s this situation began to change in association with government development projects, which provided wage labor jobs within the village and pushed many of the villagers into growing soybeans (*Glycine max*) as a dry-season cash crop on their irrigated fields. A dry-season road was completed to the village in the late 1970s, and since this time the village has been exposed to increasing contacts with the market. Almost as soon as the road was completed, some lowland merchants opened a temporary shop where they sold matches, batteries, canned fish, sweets, candles, and other goods and bought (illegally) bark stripped from *Cinnamomum* sp. trees, which was trucked to the lowlands for use in joss sticks. By the end of the 1970s the balance had shifted from an emphasis on swidden to an emphasis on wet-rice production, and demands for cash were rapidly increasing.

Productivity of the Agricultural System

Swidden productivity is notoriously variable, with reported yields in Southeast Asia ranging from 500 to over 3,000 kg/ha (Kunstadter & Chapman, 1978: Table 1-2). In northern Thailand the reported range is from 814 kg/ha (northern Thai farmers on frequently used soils in Nan Province) to 1,849 kg/ha (Yao farmers on good soils in Nan Province). Lua' and Skaw Karen yields in the area reported on here averaged about 1,000 kg/ha in the late 1960s (Kunstadter & Chapman, 1978:13), a bit higher than yields reported for nonirrigated, rainfed fields in Thailand during the same period (Asian Development Bank, 1969: 160).

Variability in swidden productivity is the result of several factors: swidden is practiced on soils with a great range of fertility and water-holding capacity; swidden is commonly practiced on hill land where altitude, slope, and exposure vary; rainfall is often variable; and the onslaughts of pests are usually incompletely controlled. Annual averages for productivity even in a single village minimize the amount of variability experienced by individual farmers. At Pa Pae average yield, as measured by farmers' reports of yield per unit seed, has been relatively constant at a ratio of a little over 11 : 1 from 1967 through 1979. Swidden yields were about 1,000 kg/ha. In the same years, irrigated fields yielded at a ratio of about 29 : 1 or about 2,200 kg/ha (Table 5-2).

One major response of Pa Pae villagers to population growth in the 1960s and 1970s was to increase the number and yield of their irrigated fields. The amount of swidden rice planted increased and then declined between 1967 and 1980, the number of irrigated fields increased from 26 (about 0.5/household) to 55 (about 1.0/household) during the same period, while the amount of seed planted increased about 58%, and the yield per unit seed increased about 20% (Table 5-3). The productivity of Lua' farmers has been able to keep pace with the growth in population and the dependency factor because there is now a higher proportion of young children. Production of rice per person in the population increased only slightly between 1967 and 1979 (from 813.8 kg of unhusked rice/person to 842.3 kg/person), but rice produced per worker increased dramatically during this period, from 1,427.3 kg/worker in 1967 to 2,052.3 kg/worker in 1979 (Table 5-2C). Because this increase in productivity per worker occurred when the swidden system was under considerable stress and fallow times were being shortened, it is likely that the increase resulted primarily from the expansion of irrigated cultivation. Improved high yield irrigated rice varieties were not grown at Pa Pae until the early 1980s.

TABLE 5-2A. Rice yield per unit of seed (volume planted versus volume harvested), Pa Pae, 1967–1980.[a]

	1967	1968	1979	1980	Mean	S.D.
Swidden rice	11.23	10.31	11.02	11.84	11.1	0.631
Irrigated rice	27.34	25.50	31.34	32.10	29.07	3.166

[a]Source: Modified from Kunstadter (1985), Table II.

TABLE 5-2B. Rice yield per unit area (kg/ha), Pa Pae, 1967–1968.[b]

	1967	1968
Swidden rice	1,019	1,044
Irrigated rice	2,210	2,175

[b]Source: Kunstadter (1978), Table 6.6.

TABLE 5-2C. Rice production per person and per worker, Pa Pae, 1967–1979.[c]

	1967	1968	1979
Rice produced (liters/person)	813.8	823.1	842.3
Rice produced (liters/worker)	1,427.3	1,546.7	2,052.3

[c]Source: Kunstadter (1985), Table 3. Figures are in liters of unhusked rice.

TABLE 5-3. Volume of rice planted and rice harvested (liters unhusked), Pa Pae, 1967–1980.

	1967		1968		1979		1980	
	Seed	Yield	Seed	Yield	Seed	Yield	Seed	Yield
Swidden rice								
Early	1,205	11,660	1,470	13,220	786	8,400	500	3,860
Late	7,800	89,840	9,890	103,560	9,350	100,920	8,194	94,648
Glutinous	363	3,680	395	5,465	687	9,920	707	12,762
Total	9,368	105,180	11,855	122,245	10,823	119,240	9,401	111,270
Irrigated rice								
Nonglutinous	2,680	74,040	2,605	68,960	4,385	138,000	4,026	129,772
Glutinous	51	620	122	580	61	1,355	142	4,025
Total	2,731	74,660	2,727	69,540	4,446	139,355	4,168	133,797
Swidden + irrigated		179,840		191,785		258,595		245,067

Source: Kunstadter (1985), Table I.

Environmental Effects of the Traditional Swidden System

Local history, archaeological reconnaissance, aerial photographs, and evidence from soils and vegetation support the conclusion that swidden as practiced by the Lua' farmers of Pa Pae was a relatively stable system over a very long period. Pa Pae villagers say their ancestors lived in the immediate area of the current village site, and surface collections of Ming and Ching potsherds from the sites they identify confirm these statements. The antiquity of settlement in the area is also suggested by neolithic adze and axe blades, and even by paleolithic chopper tools, but Lua' villagers do not claim these as ancestral relics. Pa Pae villagers state that they have practiced swidden according to a regular 1-year cultivation, 9-year fallow cycle. Support for these statements is found in aerial photographs of the Pa Pae area from 1954 and 1965 and in observations on the ground in the 1960s and 1970s. Aerial photographs show a regular pattern of secondary forest regrowth, the location and density of which corresponds to the order in which the villagers say they cut field blocks (Zinke, Sabhasri, & Kunstadter, 1978:Figures 7.2, 7.3, 7.4).

Effects on Vegetation

Studies of vegetation in the fallow fields suggest that secondary forest regenerates to approximately the same species composition and biomass after 9 years of fallow as had been attained prior to the previous cycle of cutting and fallow (Kunstadter, Sabhasri, & Smitinand, 1978; Sabhasri, 1978). After harvest, biomass of the spontaneous vegetation regrowth on the swidden rises to about 9 mt/ha about 18 months after harvest, then to 26 mt/ha 4 years after harvest. There is little increase during the next 3 years of fallow, but by the 9th year of fallow (at the time the field is cut again) it reaches 63 mt/ha. This result compares with about 390 mt/ha in mature forest that is not being used for swidden.

Effects on Soils

Soil studies show a similar pattern. In the mature secondary forest most of the nutrients are bound up in above-ground vegetation. Cutting and burning deposits most of these nutrients on the soil surface. With rainfall during the growing season, they become available to the crop plants at the top soil levels. Nutrients are leached to lower soil strata during the early years of regrowth and then lifted above ground by the growing plants of the secondary vegetation. There appears to be relatively little change in the total amount of nutrients in the soil plus biomass over the cycle. Fire volatilizes some of the nitrogen, but nitrogen appears not to be a limiting element in this environment. Nitrogen is apparently fixed by the large number of leguminous plants in the secondary vegetation. The soil base materials are relatively poor in phosphorus, which may be a limiting element restricting productivity of this system (Zinke, Sabhasri, & Kunstadter, 1978).

Observations of the land forms and the soil during cultivation suggest that erosion is not a major problem. Sheet erosion is retarded by the practice of lining the contours of hill slopes with logs; gully erosion is retarded by maintaining unburned vegetation in the major gullies, and building log and brush frameworks for viny crop plants in the minor gullies. Speed of water flow over the exposed slopes is also reduced by maintaining uncut vegetation on the steeper ridges (Zinke, Sabhasri, & Kunstadter, 1978).

Ecological Consequences

The swidden fire appears to reduce temporarily the number of insect pests, but has relatively little detrimental effect on soil-building insects. As compared with mature forests in this area, the secondary vegetation in swidden areas is made up largely of species that are fire-resistant or fire-tolerant, and that tolerate high light intensity. Forest regeneration during the fallow period is rapid, so that bare soil is exposed to intense sunlight or heavy rainfall only following the swidden fire at the start of the growing season. There is vigorous regrowth of secondary vegetation after the final weeding and harvest, when there is a great deal of soil moisture. Trees quickly reestablish themselves from the stumps that remain in the fields after cutting and burning, and grasses and herbaceous plants are quickly shaded out.

The net result of this system of cultivation, as it was traditionally managed, is the preservation of small islands of mature forest and the maintenance of a diverse secondary forest made up of different stages of regrowth (Kunstadter, 1979). Along with the cultivation of a great variety of species in swiddens and gardens, this makes for much greater species diversity than would normally be found in a relatively small area. The villagers recognize and make use of this diversity by recognizing at least 322 species–use combinations from the fallow swiddens, as well as large numbers of species from other environmental types in the area (e.g., 132 from uncut forest, 87 from around the village, 42 from along streams, 14 from along trails, 13 growing wild in swiddens, 8 from the margins of fields) for diverse purposes (e.g., food, fuel, fiber, construction, medicines, and dyes) (Kunstadter 1979; Kunstadter, Sabhasri, & Smitinand, 1978:Table 2).[3]

This managed environment traditionally provided Pa Pae villagers, and Lua' and Karen living in similar villages, with almost all the materials they used in daily life; their involvement with the market was marginal. Until recently the inventory of goods purchased from outside their environment was quite limited and consisted primarily of luxury or prestige goods. Locally smelted iron tools were traditionally available from the Lua' village of Baw Luang; pottery was made in the nearby Lua' village of Chang Maw, but salt and silver had to come from the market, as did Chinese trade pottery. By the 1980s, however, Pa Pae villagers were depending on the market for increasing amounts of food, clothing, fuel, tools, and construction materials.

Landholdings and Change

From feudal times to the end of the 19th century, when the central government in Bangkok took over control of the north, Lua' and Karen hill villagers paid nominal tribute (orchids, cotton blankets, *Imperata* roofing grass) to the Northern Thai princes. Lua' villages held grants, in the form of inscribed metal sheets, from these princes. These grants recognized the Lua' villagers' rights to the land and to self-government, and exempted them from military service and corvée labor requirements (Nimmenahaeminda, 1965). The princes also advised the hill villagers of the proper use of swiddens in order to conserve soil resources.

Borders between lands belonging to neighboring villages were marked by landmarks (ridgelines, streams), but for the most part village territories were separated by uncut forest. Leaders of neighboring villages met to coordinate the date of swidden burning if there was any chance of fire escaping to the territory of the nearby village.

Households that "belonged" to a village had customary use rights to swidden plots, which they and their descendants used repeatedly from one field rotation system to the next. They might give, split, or exchange these rights with other village members, but they could not give or sell them to outsiders. In many villages, "belonging" implied payment of homage to the *samang* who derived his authority by descent from the ancient Lua' princely lineage. The situation in Pa Pae village is complicated because the village contains the descendants of people who came from three other villages as well as from Pa Pae itself, and each of these groups has slightly different traditional access to swidden land.

During a time of unrest in the early 19th century, the residents of two small neighboring Lua' villages moved to Pa Pae, where they helped to build fortifications. At that time land was plentiful, and, rather than walk all the way to the more distant of their old fields, these people swiddened only the closer of their original fields and used vacant land belonging to the original Pa Pae villagers. These people had their own *samang* lineages, until they died out, and so they did not pay tribute to the chief *samang* of Pa Pae. They rented their more distant unused fields for 10% of the annual rice crop, to Skaw Karen who were beginning to arrive in small numbers in this area. A little later, so the story goes, a water buffalo belonging to the chief *samang* of Pa Pae was accidentally killed by a man from another Lua' village. In recompense, the headman of that village sent several families to live temporarily at Pa Pae and pay tribute to the Pa Pae *samang*. In return for this tribute, these families were given access to Pa Pae swidden land. Several of the families returned home after the fine had been paid, but some families stayed in Pa Pae, where they formed a distinct lineage but had access to land the same as the descendants of other founding families.

When the Central Thai began to enforce their authority over the North, around 1900, all hill lands became part of the Royal Forest, and residents of the North became liable for a head tax, payable in cash. The head tax apparently forced Lua' and other hill people into greater participation in the market economy. For the most part this meant getting involved in wage labor, because there was little these people could sell for cash. The fact that their lands were now considered to be a part of the Royal Forest and that concessions could now be given to cut timber had little immediate effect on the relatively isolated hill villages like Pa Pae. Teak, the main commercial timber, grew mostly at lower elevations. It was not economically feasible to harvest logs so far away from large rivers, which were the main means of transporting them to market. Forestry regulations against the cutting of swiddens were not enforced. The major effect was the creation of a legal vacuum in which village leaders no longer had the authority to enforce the traditional land use rules.

The history of Karen villages in the area suggests that Karen population grew rapidly over the past 150 years, doubling every generation and forming new daughter villages. After the Central Thai took control over northern Thailand around the turn of the century, the Karen refused to continue paying rent to the Lua'. At first they respected the claims of Lua' villagers to land the Lua' included in their swidden cycle, but the Karen gradually expanded their own claims to take up most of the space between Lua' villages. By the late 1960s the Karen population (which had been only a handful a century earlier) vastly outnumbered the Lua' in what had once been exclusively Lua' territory. Karen were beginning to nibble away at the edges of land traditionally swiddened by Lua', and land shortages were beginning to be felt by all. A few hard-pressed Lua' villagers sold their swidden use rights to Karen (much to the distress of their fellow villagers, who were powerless to prevent these transactions). A few other impecunious Lua' lost their irrigated fields to Karen when they could no longer pay the interest on mortgages. The territorial integrity of the Lua' village was threatened.

Also by the 1960s, Hmong were establishing villages and cutting mature forests at the headwaters of the stream flowing through Pa Pae. Hmong used the threat of force to take this land. The Lua' had left this forest uncut to protect the watershed and because it was considered the sacred home of Lua' spirits. The lowland authorities, to whom Pa Pae villagers protested, declined to interfere because this was viewed as a dispute between two swidden groups, and all swidden cultivation was considered to be illegal.

Expansion of irrigated farming has prevented what otherwise would have been rapid impoverishment for the people of Pa Pae as a result of the loss or degradation of their forest enviroment. Some highland Lua' villages, such as Baw Luang, had irrigated fields as early as the 1890s. Irrigation was introduced into the Pa Pae area in the 1930s. A few farmers leveled and terraced fields in the larger, flatter valleys near the village, and groups of three or four men joined together to dig and maintain the ditches and small dams. Northern Thai ceremonials and land-use customs were adopted along with the techniques of building and farming the irrigated fields. Irrigated farming was free from the control of village religious leaders who governed swidden. These fields were considered to be owned by the individuals who had made them, and titles to them could be obtained if they were registered at the district office. Thus they could be mortgaged or purchased or otherwise acquired by outsiders, and disputes concerning them could be referred for settlement to district officials, not village leaders. This represented a major break with the landholding traditions associated with swidden. Northern Thai ceremonies do not require that the village elders confirm the choice of field, nor that the *samang* set the date for beginning cultivation or planning.

Pa Pae villagers increased their construction of terraced irrigated fields rapidly in the late 1960s and early 1970s. Terracing spread up the slopes and into the smaller, shorter, narrower, less well watered, and steeper tributary valleys of the major highland streams. By the late 1970s the balance in rice production at Pa Pae had swung away from swiddens in favor of irrigated fields.

Destabilization of the Traditional Socioeconomic System

The authority of traditional leaders was clearly weakened both by governmental action (declaring swiddening illegal and refusing to become involved in disputes over swidden land or to recognize the authority of village leaders to settle such disputes) and by the increased importance of irrigated farming. Their authority over land use was further diminished by religious conversion of large numbers of Pa Pae households to Christianity. Foreign protestant missionaries had passed through this area and made a few converts in a nearby Lua' village just before World War II. After the missionaries retreated during the war, the converts reverted to animism. In the late 1950s foreign missionaries came again and, by the late 1960s, had converted about half the Pa Pae families. When Christians were a small minority in the village, they continued to live up to the letter if not the spirit of their traditional obligations to village ceremonies, contributing their share to the purchase of sacrificial animals and receiving their share of the meat after animals had been sacrificed. By the late 1960s, however, as their numbers increased, they refused to continue these contributions, and it became increasingly difficult for the remaining animists to afford the necessary animals for the major village ceremonies.

By 1980 Pa Pae's population, which had been growing at about 2.7% per year, was more than 300. The old chief *samang* had died and had been replaced as government headman by his former assistant, whose mother was a *samang* but who was not himself a religious leader. The new headman was burdened with responsibilities for a series of government-sponsored development projects, but no longer had the authority within the village to control swidden burning or to settle disputes. The villagers no longer joined together to cut their swiddens in a single block, and they began cutting after a shorter fallow time. Swidden cultivation, which was still being conducted with the same techniques, could no longer be administered as in the old days, and the characteristic pattern of land use, which had been quite conservative and stable when population was low and there were no demands for cash, was beginning to disintegrate.

Shortages of local grazing land led lowland farmers in the late 1970s to contract with Pa Pae villagers to graze their water buffalo and cattle in the highlands. Many of the Pa Pae farmers were happy to take on this chore because it gave them a way of earning water buffalo, which they now needed in greater numbers to plow the expanded areas of irrigated fields. Traditionally livestock were turned loose to graze in fallow swiddens, where there was generally enough for them to eat. As the livestock population increased, the amount of fodder became inadequate, and cattle ate roofing straw and cotton plants. Villagers were forced to buy corrugated iron roofing and cotton yarn or ready-made clothing.

Forest concessionaires threatened to reforest the fallow swiddens and close them for further agricultural use, but were at least temporarily thwarted by the villagers' threats of violence. Development projects had mixed effects. Increasing amounts of land were being put into projects (e.g., a mulberry plantation for feeding silkworms) from which the villagers gained few benefits, but some of the projects provided wage-earning opportunities within the village. The road allowed easier transportation for villagers to market and to medical care, but also allowed easier access for motorized thieves. Villagers were adjusting to the growth of populations by increasing productivity (per field) through use of irrigation, but the opportunities for expanding irrigation were limited by the small size of the watersheds. Cash crops were introduced, but marketing opportunities were limited to the dry season when the road was passable.

It seems likely that the Lua' villagers of Pa Pae will continue to practice swidden as a supplement to their other occupations for many years to come. It is also likely that as population increases they will be forced to continue to reduce the length of fallow, and their swidden system will come to resemble that of the lowland Northern Thai. Species diversity will be lost as emphasis shifts to cash crops, and their forest will continue to be degraded.

Notes

1. For discussion of swidden elsewhere in Southeast Asia, see Conklin (1957), Freeman (1955), Izikowitz (1951), Judd (1964), Kunstadter, Chapman, & Sabhasri (1978), Pelzer (1945), and Spencer (1966). See Hanks (1972) for a description of the role of rice in Southeast Asian farming systems.

2. Research was supported in part by the National Geographic Society, the National Institutes of Health, the National Science Foundation, Princeton University, the University of Washington, and the East–West Center.

3. These figures for species–use combinations were derived from a collection of as many plant species in the area as possible. Plants were identified by the villagers as to local name and use, and the results of the collection were tabulated by type of location and use. Because a single species may have more than one use, there are usually more combinations of species and use than there are species.

References

Asian Development Bank. *Asian agricultural survey*. Seattle: University of Washington Press, 1969.

Conklin, H. C. *Hanunoo agriculture*. Food and Agricultural Organization, Forestry Development Paper No. 12. Rome: FAO, 1957.

Freeman, J. D. *Iban agriculture*. Colonial Office Research Studies No. 18. London: Her Majesty's Stationery Office, 1955.

Hanks, L. M. *Rice and man: Agricultural ecology in Southeast Asia*. Chicago: Aldine-Atherton, 1972.

Izikowitz, K. G. *Lamet: Hill peasants in French Indo-China*. Göteborg: Etnografiska Museet, Etnologiska Studier 17, 1951.

Judd, L. C. *Dry rice cultivation in northern Thailand*. Data Paper No. 52. Ithaca, New York: Cornell University, Southeast Asia Program, 1964.

Kamnuansilpa, P., Kunstadter, P., & Auamkul N. *Hilltribe health and family planning: Results of a survey of Hmong (Meo) and Karen households in northern Thailand*. Bangkok: Family Health Division, Department of Health, Ministry of Public Health, 1987.

Klankamsorn, B., and Charuppat, T. The forest situation in Thailand. In *Papers presented at National Forestry Conference, 1983*. Bangkhen, Bangkok: Royal Forestry Department, 1983.

Kunstadter, P. Subsistence agricultural economies of Lua' and Karen hill farmers, Mae Sariang District, northwestern Thailand. In Kunstadter, E. C. Chapman, & S. Sabhasri (Eds.), *Farmers in the forest*. Honolulu: The University Press of Hawaii, 1978.

Kunstadter, P. Ecological modification and adaptation: An ethnobotanical view of Lua' swiddeners in northwestern Thailand. In R. I. Ford (Ed.), *The nature and status of ethnobotany*. Anthropological Papers of the Museum of Anthropology. Ann Arbor: University of Michigan, 1979.

Kunstadter, P. Highland population in northern Thailand. In J. McKinnon & W. Bhruskasri (Eds.), *Highlanders of Thailand*. Kuala Lumpur: Oxford University Press, 1984a.

Kunstadter, P. *Demographic differentials in a rapidly changing mixed ethnic population in northwestern Thailand*. Nihon University Population Research Institute Research Paper Series No. 19, 1984b.

Kunstadter, P. Rice in a Lua' subsistence economy, northwestern Thailand. In D. Cattle and K. Schwerin (Eds.), *Food energy in tropical ecosystems*. New York: Gordon and Breach, 1985.

Kunstadter, P., & Chapman, E. C. Problems of shifting cultivation and economic development in northern Thailand. In P. Kunstadter, E. C. Chapman, & S. Sabhasri (Eds.), *Farmers in the forest*. Honolulu: The University Press of Hawaii, 1978.

Kunstadter, P., Chapman, E. C., & Sabhasri, S. (Eds.). *Farmers in the forest*. Honolulu: The University Press of Hawaii, 1978.

Kunstadter, P., Sabhasri, S., Aksornkoae, S., Chunkao, K., & Wacharakitti, S. Impacts of economic development and population change on Thailand's forests. In J. Furtado & K. Ruddle, (Eds.), *Tropical resource ecology and development*. Forthcoming.

Kunstadter, P., Sabhasri, S., & Smitinand, T. Flora of a forest fallow farming environment in northwestern Thailand. *Journal of the National Research Council of Thailand*, 1978, 10, 1-45.

Nimmenahaeminda, K. An inscribed silver-plate grant to the Lawa of Boh Luang. *Felicitation volumes in Southeast Asia studies presented to His Highness Prince Dhanivat Kromamum Budyalabh Bridyakorn*, 1965, 2, 233-238. Bangkok: The Siam Society.

Pelzer, K. J. *Pioneer settlements in the Asiatic tropics: Studies in land utilization and agricultural colonization*. New York: American Geographical Society, 1945.

Sabhasri, S. Effects of forest fallow cultivation on forest production and soil. In P. Kunstadter, E. C. Chapman, & S. Sabhasri (Eds.), *Farmers in the forest*. Honolulu: The University Press of Hawaii, 1978.

Spencer, J. E. *Shifting cultivation in southeastern Asia*. University of California Publications in Geography No. 19. Berkeley: University of California Press, 1966.

Wacharakitti, S., Boonnarm, P., Sanguantan, P., Boonsaner, A., Silapatong, C., and Songmai, A. The assessment of forest areas from Landsat imagery. *Forest Research Bulletin* 1979, 60. Bangkok: Faculty of Forestry, Kasetsart University.

Zinke, P. J., Sabhasri, S., & Kunstadter, P. Soil fertility aspects of the Lua' forest fallow system of shifting cultivation. In P. Kunstadter, E. C. Chapman, & S. Sabhasri (Eds.), *Farmers in the forest*. Honolulu: The University Press of Hawaii, 1978.

6

Upland and Swamp Rice Farming Systems in Sierra Leone: An Evolutionary Transition?

PAUL RICHARDS

Rice Farming Systems and Cultural Evolution

It has often been assumed that wet-rice cultivation represents an advance over rice cultivation techniques that depend solely on rainfall. The importance of upland rice (alternatively known as *hill* or *dry rice*) in parts of West Africa appears to be consistent with a more general picture of technological backwardness when tropical Africa is compared to Europe or Asia (Goody, 1971, 1976). Accordingly, agricultural development agencies in the region have neglected upland rice and focused their attention on the transfer of smallholder wet-rice technologies from Asia. Situating a West African rice-farming system in ecological and socioeconomic context, the present study questions whether it is very useful to discriminate between swamp and upland rice cultivation. The danger is that it diverts attention from the ways in which many West African rice farmers usefully *combine* upland and wetland cultivation methods.

The West African Rice Zone

African rice (*Oryza glaberrima*) is indigenous to West Africa. Today, the greater part of the West African rice crop derives from the Asian species (*O. sativa*). Asian rice was introduced perhaps four or five hundred years ago, either via Saharan trade routes or by the Portuguese (Carpenter, 1978). Rice cultivation is widely if patchily distributed throughout West Africa (Figure 6-1), from the coast as far north as the great river valleys of the Sahel (Buddenhagen, 1978; Chevalier, 1936; Dresch, 1949; Littlefield, 1981; Morgan & Pugh, 1969; Porteres, 1949). Much of this cultivation is concentrated in the West African rice zone, a roughly triangular region stretching from the Bandama River in central Côte d'Ivoire along the coast as far as the Senegal River and inland as far as the inland delta of the Niger in Mali (Figure 6-1). In the forest zone and in some of the wetter parts of the savanna zone, upland rice supplies the bulk of the

Paul Richards. Department of Anthropology, University College London, London, U.K.

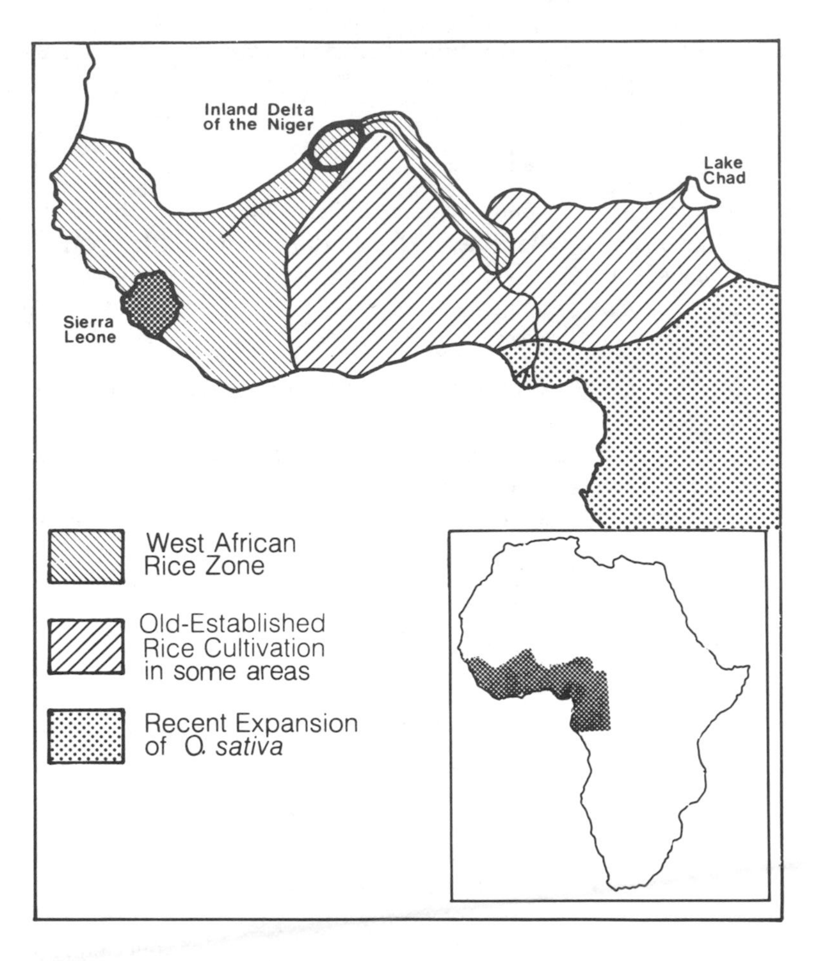

Source: After Carpenter (1978).

FIG. 6-1. Sierra Leone and the West African rice zone.

harvest. The emphasis in the drier parts of the savanna zone is on riverine cultivation making use of natural flooding, but with partial water control in a minority of cases (Buddenhagen, 1978; Harlan, 1975; Harlan & Pasquereau, 1969; Linares, 1970, 1981; Moorman & Veldkamp, 1978). Some modern schemes (e.g., the Office du Niger in Mali) employ full water control.

Sierra Leone, located in the heart of the West African rice zone, has the highest per capita rice output of any country in the region (about 125–150 kg/person/year). Population densities are typically 10–50 persons/km^2. These figures are low by Asian standards, although moderately high when compared to other parts of the West African rice zone. There is general agreement in the literature that in Sierra Leone labor shortage is a more serious constraint on production than land shortage (Karimu & Richards, 1981; Levi, 1974; Pearson, Stryker, & Humphreys, 1981).

Rice Farming Systems in Sierra Leone

Systematic studies of rice-farming systems in Sierra Leone were first undertaken in the colonial period (Glanville, 1938; Jordan, 1956; Little, 1951). More recent work includes studies by Johnny (1979), Njoku (1979), Njoku & Karr (1973), Richards, (1986), and Spencer (1974, 1975). The literature distinguishes five main rice farming ecologies:

1. Estuarine mangrove swamps (notably the estuaries of the Great and Little Scarcies)
2. Riverine grasslands (notably the lower reaches of the Sewa and Waanje, behind Turner's Peninsula)
3. Boliland swamps (seasonally flooded grassy plains developed over the ancient sediments of the Rokel River Series)
4. Inland valley swamps (seasonal or year-round swamps in minor valleys and deep-weathered depressions)
5. Uplands (interfluve areas suitable for rainfed rice cultivation).

It is common to refer to swamp and upland cultivation as if they were distinct farming systems. Over much of Sierra Leone, however, it is normal for farmers to combine cultivation of uplands and inland valley swamps. In northwestern Sierra Leone as much as one-third of all rice produced by farm households comes from swamps (Karimu & Richards, 1981). In central and eastern districts the proportion of swamp rice to total rice production is typically 10% to 20%. Perhaps 60–70% of national rice output comes from combined upland–swamp farms of this type. In the majority of cases both uplands and swamps are farmed by shifting cultivation, although swamps, once cleared, can often be used for several years continuously. Most uplands are abandoned for rice after one year, although the practice of planting a second crop of groundnuts (*Arachis hypogaea*); cassava (*Manihot esculenta*); or a mixture of cassava, groundnuts, and maize (*Zea mays*) is increasing.

Peasant rice farms in Sierra Leone fall mostly in the size range 1.0–2.5 hectares. Labor inputs and yields are lower on uplands than in swamps. Typical figures for uplands would be 200–240 workdays/ha and 800–1,200 kg/ha, and for inland valley swamps cultivated without water control, 200–300 workdays/ha and 1,000–1,500 kg/ha. The length of the average working day varies with the task in hand and the type of work group. Brushing and hoeing broadcast seed (known in Sierra Leone as *plowing*) are the two most demanding tasks in terms of energy required. A reciprocal labor cooperative of adult workers engaged in brushing or plowing will work from 9:00 A.M. to 4:00 P.M., with a break for a midday meal. Youth labor cooperatives (specializing in plowing) work from 7:00 A.M. to 7:00 P.M., but with many short breaks. The working day for weeding and harvesting tends to be as long as necessary, up to a maximum of 12 hours including breaks and diversions.

Upland and inland valley swamp rice cultivation in Sierra Leone may be compared with other West African food production systems in terms of energy-based input/output ratios (Table 6-1). The calculations—of necessity very rough and ready, given data of variable reliability and scope—purport to show how many days of subsistence (for a theoretical "reference adult") are earned by one day's worth of labor. Upland rice is at the high end, and swamp rice at the low end, of the spread of figures cited. Although rice yields on upland farms in Sierra Leone are not high, intercrops add considerably (perhaps 25–50% to the total food output of such farms. This is apparent in results for yield plots on 27 heavily intercropped farms in two villages in Moyamba District studied by Johnny (1979). In this case, intercrops accounted for well over half the total food output of the upland farms studied (Table 6-2).

A comparison of seasonal labor input profiles for rice in Sierra Leone with yam (*Dioscorea rotundata*) and Guinea corn (*Sorghum* spp.) in Nigeria brings out differences in the magnitude of seasonal labor constraints affecting these three major West African staples (Table 6-3). For each crop a bottleneck factor was calculated by aggregating the amount of labor required in any single month in excess of 10% of the total annual labor requirement. On this basis seasonal constraints in labor supply on Sierra Leone rice farms are greater than those experienced in yam cultivation in central Nigeria, but not as great as

TABLE 6-1. Energy-based input/output ratios for some West African farming systems.

Farming system	Input-output ratio[a]	Notes
Upland rice, Sierra Leone	1 : 9	Includes intercrops
Swamp rice, Sierra Leone	1 : 5	Excludes dry season crops
Sorghum, northern Nigeria	1 : 9	Ranges from 1 : 5 to 1 : 16
Sorghum/millet, northern Nigeria	1 : 7	Ranges from 1 : 6 to 1 : 9
Yam/cassava, eastern Nigeria (medium population density)	1 : 8	Compound farms 1 : 11, outer fields 1 : 6
Yam/cassava, eastern Nigeria (high population density)	1 : 5	Compound farms 1 : 7, outer fields 1 : 2
Yam, central Nigeria	1 : 7	Average of valley and upland farms

Source: After Diehl & Winch (1979); Johnny (1979); Lagemann (1977); Norman (1972); Richards (1983); Spencer (1974, 1975).

[a]Input/output ratios calculated on the following assumptions: working day about 5–7 hours; reference consumer's energy requirement about 11.500 kJ/diem; appropriate deductions for seed, milling, peeling losses, and so forth.

TABLE 6-2. Average yields per hectare, intercropped upland rice farms Moyamba area, southern Sierra Leone.

Crop		Yield (kg/ha)	Energy (kJ × 10^6)
Rice	*Oryza sativa*	1,063	8.94
Sorghum	*Sorghum margaritiferum*	200	1.73
Fonio	*Digitaria exilis*	?	?
Bulrush millet	*Pennisetum leonis*	?	?
Maize	*Zea mays*	104	1.49
Cassava	*Manihot esculenta*	747	2.76
Cassava leaves	*Manihot esculenta*	95	0.16
Sweet potatoes	*Ipomoea batatas*	178	0.62
Sweet potato leaves	*Ipomoea batatas*	64	0.11
Yams	*Dioscorea* spp.	87	0.22
Cocoyams	*Xanthosoma sagittifolium*	65	0.23
Cucumber	*Cucumis sativus*	261	0.18
Pumpkin	*Cucurbita* spp.	327	0.22
Tomato	*Lycopersicum esculentum*	149	0.13
Garden eggs	*Solanum melongena*	20	0.03
Beans	*Phaseolus lunatus*	297	3.40
Egusi	*Colocynthis citrullus*	68	0.11
Pepper	*Capsicum* spp.	9	0.01
Beniseed	*Sesamum* spp.	30	0.50
Okra	*Hibiscus esculentus*	57	0.10
Krain-krain	*Corchorus olitorius*	17	0.03
Sawa-sawa	*Hibiscus sabdariffa*	70	0.12
Hondii	*Amaranthus hybridus*	22	0.04
TOTALS			
Rice		1,063 (27%)	8.94 (42%)
Other crops		2,867 (73%)	12.19 (58%)
Total		3,930	21.13

Source: After Johnny (1979).

those experienced in grain cultivation in northern Nigeria. In the case of rice, the period of greatest labor shortage coincides with the preharvest "hungry season."

According to data compiled by Johnny (1979), work on upland rice farms is divided approximately 60 : 40 between men and women (Table 6-4). Men brush and clear farms, plow rice, fence farms against rodents, scare birds, and harvest. Women weed, scare birds, and harvest. Data for eastern Sierra Leone compiled by Spencer and Byerlee (1976) show that when both farm and non-farm tasks are considered, workloads are evenly divided between men and

TABLE 6-3. Labor input profiles and seasonal bottlenecks in West African farming systems.

Farm system	Percentage of annual labor input by month												Input/output data	Bottleneck factor (%)
	Jan	Feb	Mar	Apr	May	Jun	Jly	Aug	Sep	Oct	Nov	Dec		
Sorghum/millet (Zaria, northern Nigeria) (1)	2	1	1	12*	21*	13*	10	13	6	1	14*	7	(593 hr/ha; millet, 370 kg/ha sorghum, 767 kg/ha)	23
Yam (Tawari central Nigeria) (2)	6	15*	3	9	11*	10	9	5	6	16*	9	2	(1,680 hr/ha; yam 9,000 kg/ha)	12
Upland rice (Sierra Leone) (3)	1	3	4	5	7	10	14*	12*	16*	14*	10	4	(1,360 hr/ha; rice 1,100 kg/ha)	16
Rural households[b] in eastern Sierra Leone (farm and nonfarm work) (4)														
Men	7	8	6	5	7	14*	11*	14*	9	7	7	7	(1,027 hr/yr)	9
Women	7	4	3	2	4	14*	14*	16*	11*	10*	10*	5	(1,037 hr/yr)	15

Source: After Richards (1983); (1) Norman (1972); (2) Diehl & Winch (1979); (3) Spencer (1975); (4) Spencer & Byerlee (1976).

[a]Aggregate of monthly labor requirements in excess of 10% of annual total labor input.

[b]Main crop is rice.

*Labor bottleneck months.

TABLE 6-4. Male and female labor inputs on upland rice farms, central Sierra Leone.

	MALE		FEMALE	
Activity	Days/Ha	% Total Labor Inputs	Days/Ha	% Total Labor Inputs
Brushing	21.3	99	0.3	1
Felling	10.2	100	0.0	0
Burning/clearing	12.7	100	0.0	0
Plowing/sowing	23.8	60	16.1	40
Weeding	6.4	21	24.2	79
Pest control	24.6	86	3.0	14
Harvesting	24.3	36	43.2	64
Total	123.3	59	86.8	41

Source: After Johnny (1979).

women. One significant difference, however, is that the bottleneck factor is higher for women than for men (Table 6-3).

Since the 1930s (but more especially in the last 15 years) attempts have been made to persuade Sierra Leone farmers to adopt simple methods of water control for swamp rice cultivation. These measures are modeled on techniques employed in parts of Asia, e.g. India and Taiwan (Glanville, 1933; Richards, 1986). Initial development of a water-controlled swamp along these lines (clearing, stumping, leveling, and construction of channels and bunds) requires about 450–600 workdays/ha. Recurrent labor requirements (repair and maintenance of channels and bunds, and cultivation, including transplanting) are in the range 250–400 workdays/ha. With fertilizer and improved varieties, yields are high (especially if double cropping proves possible)—about 1,500–3,000 kg/ha—but often not high enough, in comparison with results obtained from local methods, to justify the initial capital costs and necessary additional investment in annual labor inputs (Airey, Binns, & Mitchell, 1979; Karimu & Richards, 1981; Lappia, 1980).

A major problem with water-controlled swamp cultivation is that its labor requirements tend to clash with the labor requirements of the upland farm. A smoother overall profile can be obtained by combining upland cultivation with local methods of swamp cultivation (Karimu & Richards, 1981).

Rice Farming in Social and Ecological Context

The following account of upland and wetland rice cultivation practices in Sierra Leone is based on fieldwork in Mogbuama, a village in central Sierra Leone at the foot of the escarpment separating the coastal and interior lowland plain from the granitic uplands of the northeast (Figure 6-2). A predominantly Mende-speaking village, Mogbuama is located a few kilometers to the south of the linguistic boundary between Mende and Temne country. Mogbuama farmers combine upland rice cultivation with two kinds of wetland management: cultivation of early ripening rices on moisture-retentive soils derived from colluvial and alluvial materials (old river terraces especially) and cultivation of long-duration rices in watercourses and inland valley swamps. The data refer to the 1983 farming season.

FIG. 6-2. Mogbuama and the Sierra Leone scarp zone.

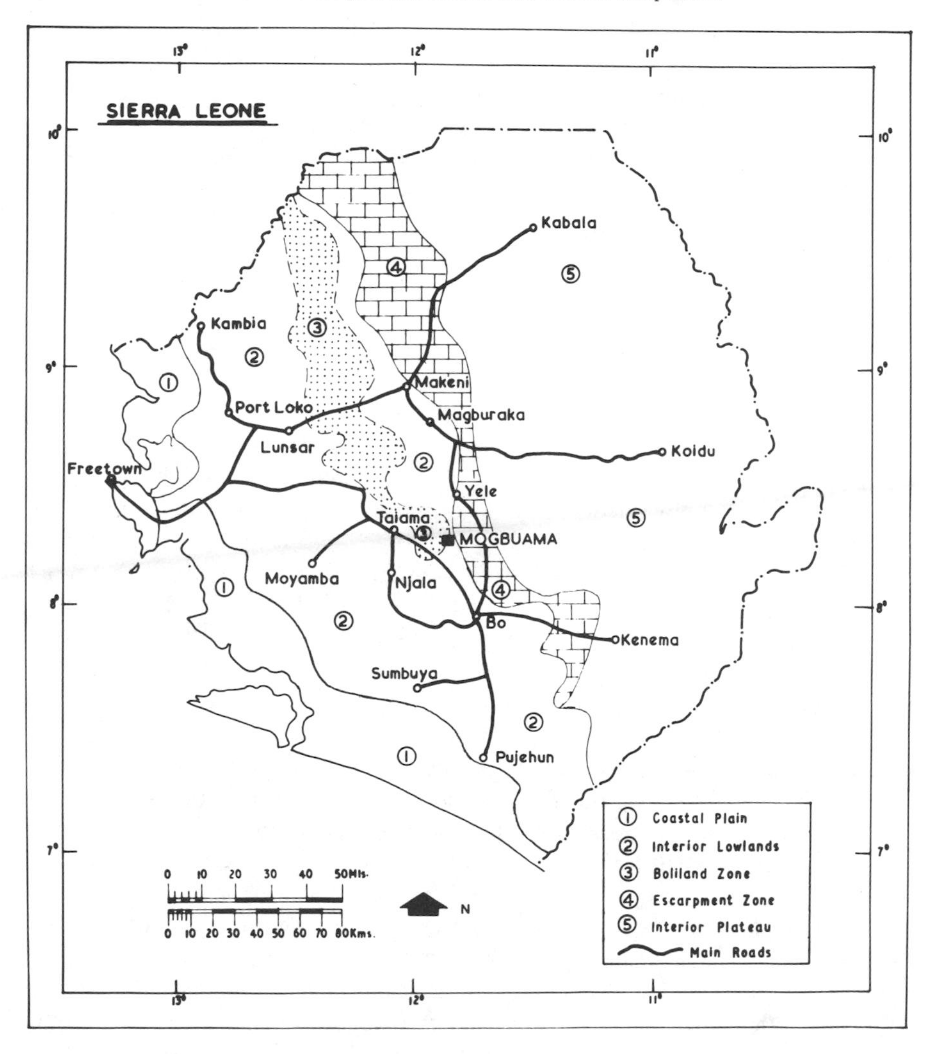

Social Organization

With a population in 1983 of about 575 persons, Mogbuama is the third-largest settlement in Kamajei Chiefdom, Moyamba District. In recent years any population increase due to a high birth rate appears to have been balanced by outmigration, principally to Freetown and the diamond-mining districts of eastern Sierra Leone. The 1963 and 1974 census totals for Kamajei Chiefdom were 6,555 and 6,557 persons, respectively. Kamajei Chiefdom was, until recently, a relatively isolated part of central Sierra Leone, connected to the railway at Mano by a single dirt road opened in 1945. During the early 1970s, however, the main road from Freetown to eastern Sierra Leone was rerouted across the chiefdom. Mogbuama is 7 km north of this new road but can only be reached on foot during the rainy season (June–December).

In 1983 there were 98 farming households in Mogbuama, four of them headed by women. In this context a *farming household* means a group supporting itself from an upland rice farm and associated wetland farming activities. Up to 25% of the rice harvest was sold, with the balance retained for subsistence. Mogbuama is noted for early rice, some of which is purchased by dealers from Bo, a large town 40 km to the east. Households with rice to spare often prefer to loan it locally (at the rate of two bushels to be repaid for every bushel loaned) rather than sell. In addition to cultivating rice, most adults pursue one or more of the following supplementary economic interests: hunting, tapping for palm wine, gin distillation, cultivation of groundnuts on land previously used for rice, cotton spinning (women), and weaving (men). Some wealthier heads of household own small tree-crop plantations. The main crop is coffee, much of which is planted in and around the settlement. Many of these plantations are pledged against debts and are consequently neglected and unproductive. A few adults support themselves by full-time pursuit of trade or a craft.

Household heads fall into one of two categories: citizens (*tălì*) or "strangers" (*hótá*). The implication of the first term is that the person concerned belongs to a land-owning group. Strangers are persons who have been attracted to settle in Mogbuama by abundant land or by the opportunity to ply a trade or craft. Each stranger gains access to land through attachment to a citizen who acts as sponsor. The sponsor is both landlord and legal guarantor for the stranger's conduct. A stranger is not entitled to plant tree crops. In Mogbuama about 20 heads of farming households have stranger status. Long-established strangers are sometimes treated as citizens. Incomers from other parts of Kamajei Chiefdom are not treated as strangers.

It is more difficult to assess the extent and impact of out-migration. Many young people have left Mogbuama as migrants to mining districts, but with the decline in the mineral economy a number have returned. Push factors of migration involve escaping debt, disputes, and litigation, or seeking relief from crop failure. Whether migrants return depends on the severity of their problems and their success elsewhere. Permanent out-migration is most likely for those who have passed through the school system. Mogbuama has had a small primary school for about ten years, and before that some young people were educated at schools in Senehun, Bo, and Njala. Currently about one-third to one-half of all children of primary school age are at school. I know of 20 or so Mogbuama citizens prominent at the regional and national level. The census figures cited here give a clue to the total number of out-migrants. The implication of a zero increase in total population for Kamajei Chiefdom between 1963 and 1974 is that the net rate of out-migration is equivalent to the national average rate of population increase, about 2% per annum.

Patron–client relationships are the foundation of the local political system. Wealthier household heads act as patrons or "big persons." Ten or fifteen male and three or four female patrons control among them the main village political and legal offices, and major positions in the Poro, Wunde, and Sande societies. The national political system is strongly clientelist in character, and village big persons are themselves clients of patrons at the regional and national level (Cartwright, 1978). Patrons also frequently lend money and negotiate crop sales. Some Mogbuama patrons have land, houses, and business interests in other settlements in Kamajei Chiefdom and further afield (Freetown, Bo, and Moyamba, for example). Wealth and patronage go hand in hand. In many respects, control of a patron–client network is as important a "capital" asset as investment in property, cattle, and tree-crop plantations, three of the main ways in which material wealth is accumulated in Mogbuama.

Although patron–client relationships are inegalitarian, it is possible to argue, nevertheless, that they are useful to both parties (Gellner, 1977; Scott, 1977). Mogbuama has land in relative abundance, but labor and capital are in short supply. In effect, patrons gain access to labor, and clients to capital, through such relationships. Patrons turn to clients to provide, say, political support during a chieftancy contest and timely labor at crucial moments in the farming cycle. Clients look to patrons for subsistence support (rice when the hungry season is severe), financial and other kinds of help with disputes and court cases, and brokerage services, for example assistance in coping with government departments, schools, and other external agencies (Murphy, 1981). The apparent absence of class consciousness in much of rural Sierra Leone is perhaps best explained by the argument that as long as clientelism continues to yield benefits to both rich and poor in an uncertain environment, its legitimacy is likely to remain unchallenged (Richards, 1986).

Mogbuama is only relatively lightly incorporated into the state system and national and international markets. The bulk of production is still for subsistence and local exchange, and direct interventions by government agencies are few and far between. In the past there have been a few agricultural extension efforts directed to the development of tree crops. More sustained contact between Mogbuama farmers and agricultural development agencies began as recently as 1982, when some villagers registered with the Moyamba Integrated Rural Development Project. This project offers credit, improved seed, fertilizer, and advice on water control in swamp rice cultivation; but understandably enough its impact had only been slight at the time of fieldwork. The farming systems to be described are largely if not entirely the result of indigenous initiative and informal diffusion processes (cf. Biggs & Clay, 1981; Chambers, 1983; Richards, 1985).

Land Tenure and Resources

Land in Mogbuama belongs to 12 descent groups. With a cultivable area of 34 km^2 and a population density of 17 persons/km^2, land shortage is not a general problem, but descent groups vary considerably in size, political influence, and amount of land controlled. Some descent groups own inadequate amounts of land for their size or are short of land close to Mogbuama. Farmers in such a position, or unattracted by the type of land available to their descent group, request the use of land owned by a more liberally endowed descent group. Such requests are generally granted freely.

Adult members of a descent group (male or female) claim land for rice farms by informing the head of the family at the beginning of each year where they intend to farm. Strangers, by contrast, require permission from the head of one or another of the established descent groups. A notional charge is levied for such permission. In 1983, this was about U.S.$1.50 and two bushels of rice. No stranger is able to secure land rights to plant tree crops. Of 98 rice farms in Mogbuama in 1983, 22 were made by strangers, mostly migrants from neighboring chiefdoms. Strangers' rice farms tend to be slightly smaller than the average (1.15 ha as opposed to 1.30 ha), accounting for about 19% of the total rice hectarage planted in 1983. In one or two cases rice farms made by strangers were undertaken on a sharecropping basis for an absentee patron.

It should be noted that Mende notions of descent group membership are flexible and respond to changes in local economic and political circumstances. This flexibility is most readily seen in the process whereby strangers are more fully incorporated into the community. Some strangers are short-term residents treated more or less like tenant farmers. If a stranger decides to remain, however, the bond between settler and landlord is liable to become a durable patron–client relationship. The landlord may help a male stranger find a wife, for example. In due course, an economically successful client might find himself offering his patron help in return. Some established clients (typically those who have speculated successfully in commodity deals) might eventually find themselves in a position to offer credit to a patron anxious, perhaps, to mobilize capital resources to campaign for political office. Where a number of patrons have become indebted in this manner to a successful stranger, it is feasible for the rules of descent to be revised. The stranger's household may then come to be recognized as a fully incorporated kin group, with land previously borrowed now its own property.

It is no surprise to find that several Mogbuama descent groups are of recent origin. Some were founded by migrants from the Guinea Highlands during the 19th century (Howard & Skinner, 1984). Several appear to have been established in return for the cancellation of trading debts incurred by local chiefs. One group established in this way is said to date from the trade recession during the 1930s. This degree of sociological and political flexibility appears to reflect the relative abundance of land in Mogbuama and a corresponding shortage of labor. Most communities in rural Sierra Leone appear to be anxious to build up their numbers by encouraging in-migration.

Mogbuama farmers have access to three major types of land. Land to the northwest of Mogbuama is seasonally flooded grassland, or *boliland*. The infertile boliland soils are developed from ancient metamorphosed sediments. To the west and southwest of Mogbuama there is an extensive area of relatively fertile, moisture-retentive soils, developed from river terrace materials where the Tibai and Mogboe rivers break slope at the foot of the granite escarpment (Figures 6-2, 6-3). These terraces, like the granite uplands to the east of Mogbuama, are naturally forested. In the granite zone soils occur in valley cross-sectional catenary sequences: gravelly soils on upper slopes, sandy soils (derived from colluvial materials) on lower slopes, and hydromorphic soils in valley bottoms (Figure 6-3).

When selecting land for rice cultivation, Mogbuama farmers choose between the two major types of forested land: the granite zone to the east of the village and the river terrace zone to the west (Figure 6-3). Only rarely are farms made in the grassy boliland proper. River terrace zone soils (*túmú* in Mende) are silty and more moisture retentive than soils in the granite zone, and are well suited for cultivation of early rice. Rices marketed from July to September fetch up to twice the price of rice in the main harvest season, October–December. The disadvantage of the river terrace zone is that it is more difficult to clear (the forest is thicker), and more vulnerable to the major meteorological hazard in central Sierra Leone, early rainfall. Heavy rain in February or early March makes it difficult to burn cleared vegetation. A poor burn deprives the farm of nutrients and imposes additional labor burdens on the farmer (unburnt branches have to be cleared by hand, and subsequent weed growth is much more rapid). The risks of a poor burn after early rainfall are much greater on moisture-retentive *túmú* soils than on the free-draining soils of the granite zone.

In the granite zone, soils on upper slopes with a high proportion of laterite gravel are known as *kòtú*. Lower on the catenary profile soils are sandy (*ŋà-nýa*). Rice planting in the granite zone commences a month to 6 weeks later than planting in the river terrace zone. The sequence is shorter-duration rices on colluvial lower slopes, medium-duration rices on the granite uplands, and long-duration swamp types in valley bottoms. Rice yields tend to be lower but more reliable than yields in the river terrace zone (1,000–1,200 kg/ha, compared to 1,200–1,500 kg/ha). The greater reliability stems in part from the fact that early rainfall is less damaging to prospects for a good burn.

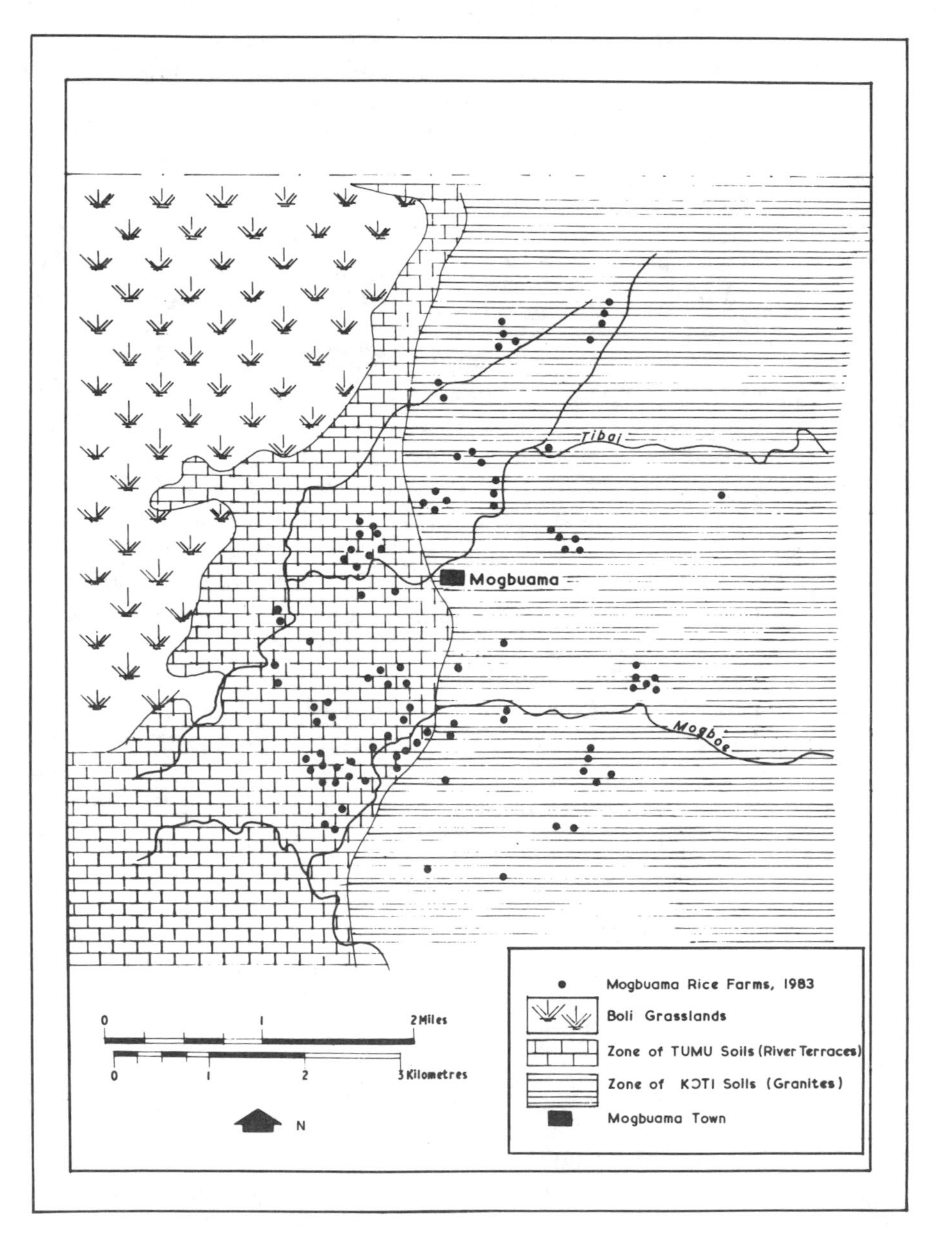

FIG. 6-3. Soil types and farms in Mogbuama.

In both the river terrace and granite zones, hydromorphic soils in valley bottoms are used for planting swamp rice. Seasonally flooded riverine grasslands along the two major rivers flowing through Mogbuama territory are used for cultivation of quick growing rices on the rising flood. These are known as *bàtì* farms. Work on such farms is risky (the rise of the flood is unpredictable), labor intensive, and in competition with other activities at the busiest time of the year. Yields are high, however, and rewards are great, since the harvest is ready when hungry-season shortages are at their most acute and food prices have peaked.

Inland valley swamps and smaller seasonal waterways are planted to long-duration swamp rices ready for harvest after the main upland rice harvest. This low-labor-intensity swamp farming complements labor requirements on the main upland farms. No water control is practiced. Some of the varieties planted are flood-tolerant (able to withstand submersion for short periods) and others are floating types (ones in which the first internode acts as a float chamber, elongating to match flood conditions). As a group, swamp varieties are known in Mende as *yàkà* rices. The term has something of a derogatory connotation, since all *yàkà* rices are considered poor in taste. Most of this rice is grown by dependents—wives and young men working under the authority of an elder—as a source of private income.

Fallow intervals in Mogbuama average 8–9 years on *túmú* soils in the river terrace zone, and about 12 years and upwards on the sandier soils of the granite zone (Figure 6-4). Farmers recognize two systems for cultivating upland rice: a short-fallow system with 1 year of rice, followed, perhaps, by a second year of groundnuts, or groundnuts intercropped with maize and cassava, and a long-fallow system in which rice is cultivated for 2 years in succession. Under the former system land is considered to be adequately fertile after 8–15 years of fallow (depending on soil type). The latter system is a method for dealing with mature secondary forest (land not cultivated for 30 years or more). In such circumstances the first crop of rice is generally less productive than the second crop. Farmers believe that very fertile soil in the first year causes rice to grow too tall, with a consequent reduction in panicle size. Grist (1975) confirms this characteristic response of a number of rice varieties to an oversupply of nitrates.

Waldock, Capstick, and Browning (1951 p. 19) note that the long-fallow system has posed problems in a number of districts in central and eastern Sierra Leone:

> Older bush is not in every case considered to give a return for effort involved . . . small sections of bush appear to be going back to forest because the farmers are unwilling to make the heavy effort involved in felling . . . either the return is not considered to be adequate for the effort involved or older farmers are unable to make the effort with so many of the younger men away from home.

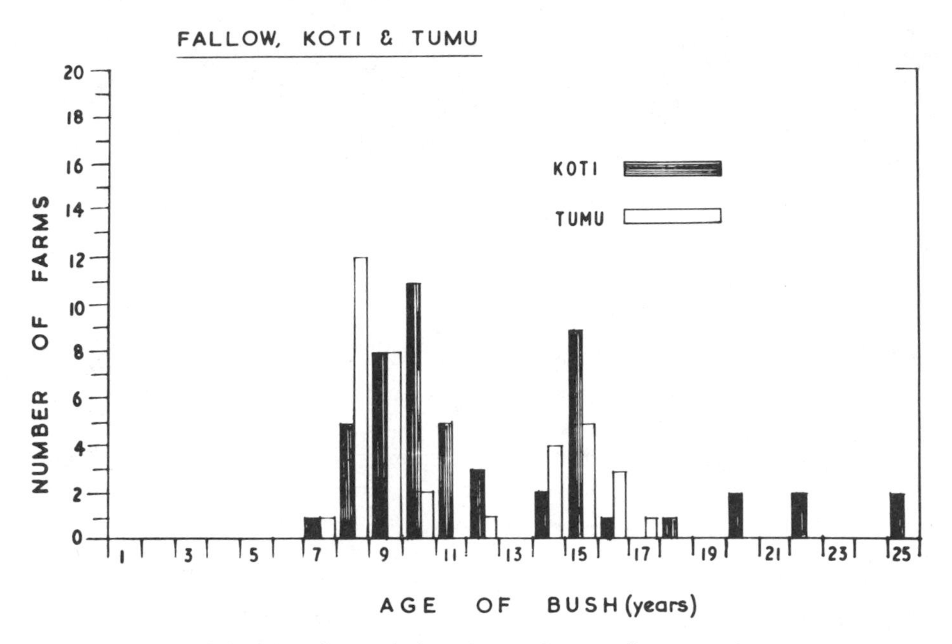

FIG. 6-4. Fallow periods on *kotu* and *tumu* soils compared.

In 1983 only one farm in Mogbuama was cultivated according to the long-fallow system. Although some descent groups have access to land not cultivated for a generation or more, few Mogbuama households can command the labor to clear thick forest.

Labor

Success in rice farming in Mogbuama is predicated on timely mobilization of labor. Thus control of labor is a key to understanding a number of features of village social and political life.

The rice farm is a "family" venture; that is, it is a combined project between one or more adult males, one or more adult females, and generally several children. The farm family is often, but not necessarily, linked by kinship: A man may farm with his father or uncle, a man may be married to one or more of the women, the children may be the children of the marriage. In some cases, however, a "farm family" may come together for no other purpose than to farm. I knew of cases where, for example, a man and a woman, both recently divorced, found it convenient to agree to share the responsibilities of a farm for a single season; where a woman trader, short of capital, proposed to join a former boyfriend and his wife in a farm; where a man and a female relative made a farm with the help of several foster children. At the end of the farming year temporary units of this sort might dissolve, never to form again. Members would be paid off (with a share of the harvest) like sailors at the end of a voyage.

It is important to note that the farm family is not necessarily a subunit of the descent group, nor is it necessarily a residential or consumption unit. It is best regarded as an informal, annually negotiated labor cooperative, bringing together two types of skilled labor. Adult males specialize in brushing, clearing, burning, plowing, and building farm structures (e.g., bird-scaring platforms and fences to exclude rodents). Adult females specialize in weeding, cooking, and cultivating intercrops such as sesame and cotton on the upland rice farm. Men and women divide harvesting between them. Children carry out a large number of supplementary tasks, such as fetching water and assisting in the weeding, and take a major part in bird scaring in the weeks before harvest.

The main features of the farm calendar (Figure 6-5) are as follows. In January men choose farm sites. From February to April they are busy brushing, burning, and clearing farms, and subsequently, from April to July, take the major responsibility for plowing (Figure 6-6). Women organize planting of intercrops. In July and August women undertake the bulk of the weeding, while men concentrate on pest control. Men and women share equally in harvesting, from August to November.

The head of household assumes control of the bulk of the rice crop. The rice crop from the main farm is first and foremost for feeding the family, but the senior man is free to sell any surplus or to divide this surplus among the other adults associated with the farm. A man who sells rice from the household farm will be expected to make certain expenditures on behalf of the women and young males who have assisted him, for example buying them cloth or clothing at Christmas. Women claim the majority of intercrops. Some, like cassava, lima beans (*Phaseolus lunatus*), maize, and vegetables, are grown mainly for subsistence. Others, like cotton (*Gossypium* spp.) and beniseed (*Sesamum indicum*), are grown mainly for cash.

The majority of farm households are headed by men. A farm will be female-headed where an older divorced or widowed woman chooses to be responsible for her own upkeep. Of 98 farms in Mogbuama in 1983, only 4 were female-headed. In these cases the women recruited younger relatives or strangers to carry out male work on the farm.

At times, the head of the farm household will demarcate a portion of the farm, or clear a separate plot, known as *gbɔ́lɔ́*, to meet the personal needs of one or more of those assisting him. At other times, members of a farm household will use any free time to farm plots of groundnuts (on last year's rice farm), swamp rice (generally flood-tolerant varieties in valley bottoms adjacent to the main farm), and/or sweet potatoes (*Ipomoea batatas*) in swamps during

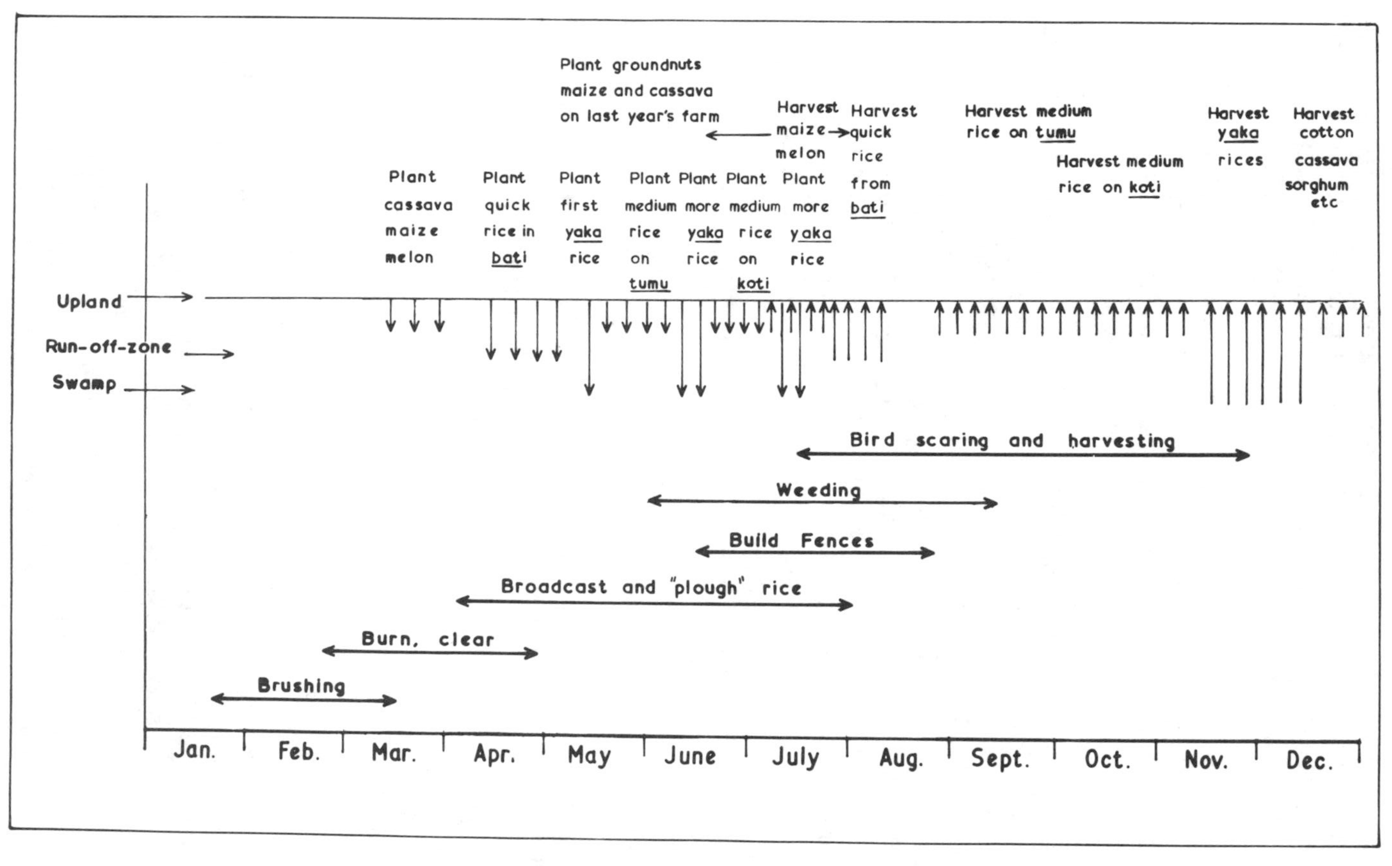

FIG. 6-5. The rice farming calendar in Mogbuama.

FIG. 6-6. *Bàtì* farm after clearing. The foreground ridge is *túmú* soils. The cleared area (center) is *bàtì* (riverine) swamp soils. The riverine galley forest is in the background.

the dry season. These additional plots are all private ventures. Any assistance provided by other members of the household will be paid for in cash or kind, and the harvest is the owner's sole property. Everyone in the farm household, from senior males to quite small children, aims to have an interest in one or more such plots.

Farm households vary in size from a man and his wife working alone up to units with 20–25 members. The average is a household of 7–9 persons: typically, one or two adult men, two or three adult women, and three or four children. Even the largest farm households, however, are unable to meet all their labor requirements from within the group. The major labor bottlenecks, in order of severity, are plowing, weeding, harvesting, and brushing.

In every case the difficulty stems from timing constraints imposed by ecological factors (Richards, 1985). If plowing is delayed too far into the rainy season, cleared farms become choked by weeds, and excessive cloud cover and rainfall inhibit the growth of young rice plants. The optimum period for weeding is 4–6 weeks after planting. Delays beyond this point sharply reduce yields (Nyoka, 1980). If the harvest is not gathered on time, the farm field is vulnerable to bird damage and theft. Tardy brushing leaves insufficient time for felled material to dry thoroughly before the first rainstorms. A badly burnt farm is weed-infested and infertile. Bottleneck problems are often cumulative. A poor burn, for example, necessitates extra work in clearing the field by hand, with the result that plowing may start late. Plowing and weeding pose the greatest difficulties because not only are they the most labor-intensive activities (in terms of workday/ha requirements) in the rice-farming cycle, but they also coincide with the period when household food and cash resources are at their lowest.

Labor to cope with these bottlenecks is recruited from outside the farm family. Some is hired on a daily basis. Younger and poorer villagers are prepared to work for U.S.$0.40/day plus meals. Only about 10–15% of nonhousehold labor is supplied in this way, however. The bulk is supplied by labor cooperatives.

Mogbuama has six different types of labor cooperatives:

Téê. A simple "by turns" arrangement, in which five or six friends or neighbors lighten the tedium of heavy work by laboring as a group according to a rota. Men form *téê* groups principally for rice planting and for fencing farms against rodents. Women sometimes form *téê* groups for weeding.

Kɔ́mbì. A formally constituted cooperative, founded for social or savings purposes, that also undertakes agricultural labor. Such groups typically have 15–30 members, mainly adults farming on their own initiative, and work is regulated by elaborate sets of rules and written constitutions. Where a *kómbì* has both male and female members, the men meet for brushing and planting, the women for weeding, and both men and women for harvesting. Mogbuama had two such groups in 1983, one founded as a savings club, the other as a dance society.

Bɛ̀mbɛ̀. A specialist group formed, by children, youths, or adults, for rice planting. The *bɛ̀mbɛ̀* varies in size from 5 to 15 members. Each member qualifies for a certain number of turns, which he may use on his own farm or sell for cash. The group may, at its discretion, decide to fill in any spare days during the planting season working for cash. On these days it hires itself to the highest bidder. A disciplined *bɛ̀mbɛ̀* achieves more work in a day than an equivalent number of hired workers. Consequently, cash rates of remuneration for an adult *bɛ̀mbɛ̀* are about 30–40% better than for ad hoc hired labor. Five *bɛ̀mbɛ̀* were formed in Mogbuama in 1983, two by adults, one by youths, and two by children.

Gbɔ̀tɔ̀. A specialist planting group for dependent youths. Where heads of farm households have several young men under their charge, one or more may be nominated to work in a *gbɔ̀tɔ̀*. The group operates like a *bɛ̀mbɛ̀*, except that the work (under the control of a managing elder) is tightly disciplined (various fines and punishments are meted out where individuals fail to keep up with the group) and accompanied by three drummers, and the hours are longer. Heads of household sometimes sell surplus turns for cash. No *gbɔ̀tɔ̀* was based in Mogbuama in 1983, but several Mogbuama households participated in groups of this kind organized elsewhere in Kamajei Chiefdom. In consequence, these groups were regular visitors to Mogbuama during the rice planting season.

Mbĕlé. A specialist rice harvesting group; the only one in which men and women work together. The operating rules are similar to those governing the *gbɔ̀tɔ̀*, but the atmosphere is more relaxed. Members frequently raise cash by selling spare turns.

In every case, the recipient of a labor turn, or the hirer, is responsible for feeding the group. Groups will fine the member concerned, or refuse to do the work, if the food is substandard. During the later part of the planting season (June–July), when food is scarce, the cost of this meal (together with incidentials such as palm wine, cigarettes, and kola nuts) is a more significant item of expenditure than the financial rate of remuneration required by the work group. In one case, the cash wage was $0.56 per capita, but the cost of the meal was about $0.80 per person. I suspect that a failure fully to take into account meals as payment in kind is responsible for the exceptionally low agricultural wage rates (by regional standards) reported for rural Sierra Leone in some economic surveys (Pearson, Stryker, & Humphreys, 1981).

Unless a farmer has chosen to cultivate a long-fallow farm, the early stages (brushing and clearing) only pose labor problems where early rain makes burning difficult (a hazard perhaps 1 year in 7). Much of the significance of early rain is that it exacerbates the labor problems experienced at planting time, because a poor burn fails to suppress weeds adequately.

The weeds in question are those that appear in the interval between clearing and planting. A thick carpet of herbaceous weeds such as *Sida stipulata* and *Ageratum conyzoides*, and convolvulaceous creepers such as *Ipomoea involucrata*, slows down the work force when hoeing the broadcast rice. On occasion, work rates drop to half those that might be expected on a farm without weeds. Much of the broadcast seed is lost as the hoers hack through and gather this mass of weed, and rebroadcasting may require up to half as much seed again to replace this loss.

The planting season, from the end of April to July, is an increasingly burdensome treadmill. Demand for labor for planting peaks during late May and June. Hire rates rise in response, adding to the labor recruitment difficulties of those households whose supplies of both food and cash are beginning to dwindle with the approach of the hungry season. Any delay at this stage worsens the weed problem, further reducing the efficacy of labor already priced beyond the means of a number of farm households. Some go into to debt while others rely on membership in a work group to complete their farms, though not all resist the temptation to sell outstanding turns for cash. A general end-of-planting panic fosters much wheeling and dealing, as farmers with incomplete farms outbid each other to buy up spare labor from labor group organizers. In June 1983 a young stranger, working in a *bὲmbὲ* but without a farm of his own, succeeded one night in selling his outstanding turn five times over, and promptly disappeared from the village with the proceeds.

Food and cash are in even shorter supply during the weeding season, but weeding is a somewhat less decisive labor bottleneck than planting, in the sense that an unweeded or partially weeded farm will produce something, but an unplanted farm can produce nothing. The harvest season is a period of high labor demand, but the harvest itself provides the resources with which to cope.

Indigenous Innovation

For most Mogbuama farmers, then, the central problem to be overcome each year is how to organize sufficient labor to complete rice planting without falling into a vicious cycle of debt and hunger. This is the key to understanding village farming strategies and the thrust of local ecological and technological innovation.

One way of coping with the problem of labor mobilization might be termed the sociological strategy—to build up a large household labor force and to take an active part in labor cooperatives. The latter may be by direct participation, by placing a dependent in a labor group, or by the exercise of political patronage. Patrons qualify for free turns when, for example, they act as "town chiefs" on behalf of a labor group. This might mean acting as a broker for the group in relation to outside agencies (development projects and local government agencies) or taking the group's part in disputes and court proceedings.

A second strategy involves the accumulation of sufficient resources of food and cash to be able to recruit a labor group at an opportune moment, or, failing this, to ride out the hungry season in good order (in this respect it is crucial to try to avoid debt, since heavy borrowing in one year exacerbates hunger in the next). Hunting, trapping, collecting honey, and distilling palm wine to make gin are all useful ways of acquiring assets that can be cashed for labor during the planting season. A few wealthier farmers have established small coffee plantations, and use cash from coffee sales to finance rice planting. A number of these plantations, however, are pledged to money lenders.

Intercropping is the main on-farm strategy of this sort. Labor groups insist on being fed rice. Guinea corn (*Sorghum margaritiferum*) and millet (*Pennisetum leonis*) are intercropped with rice largely to eke out household food supplies during the planting season, thus ensuring sufficient rice in hand to meet labor group feeding requirements. Cassava and maize have a similar role. Mention has already been made of the significance of local cotton varieties (*Gossypium barbadense, G. hirsutum*) intercropped on upland rice farms as a source of social security for women. Women will sometimes sell cotton cloth to raise cash to help a husband complete the household farm. Understandably, Mogbuama farmers are reluctant to contemplate abandoning upland rice without an alternative means of cultivating cotton. A nationwide program in the 1920s to introduce monocropped Allen cotton demonstrated conclusively the lack of feasible alternatives to existing varieties and cultivation practices (Division of Agriculture, 1926). Little if any research has been done on cotton since.

A third, ecological, set of strategies for coping with labor bottlenecks centers around soil management and choice of rice varieties. The moisture-retentive soils of the river terrace zone can be planted early, thus allowing some farmers to get ahead in the rush for planting labor. An early harvest when prices are high then provides windfall profits for hiring labor in the following season. The snag is that weed growth is always more intense than on the free-draining soils in the granite zone. Any delay or miscalculation in mobilizing labor for planting has serious consequences.

The additional weed growth associated with early rain and poor burn has devastating consequences for farms in the river terrace zone. The most recent occasion when early rain was a problem in Mogbuama was 1981. As a result, many farmers in the river terrace zone found it difficult to raise sufficient labor to complete planting their weed-choked farms. Some had to abandon more than half the area originally cleared. One man transferred his farming operations to the granite zone, vowing never to touch *túmú* soils again.

Farming in the river terrace zone is a high gain–high risk strategy (Figure 6-7). Farmers in the granite zone are in a safer short-term position, but are, perhaps, less likely to get on top of their problems in the long term. It is not surprising, therefore, that farmers in both zones are interested in any combination that gives some of the benefits of both worlds, provided the strategy in question eases, rather than complicates, labor supply problems.

The main approach adopted has been to select three sets of rice varieties: shorter-duration rices suited to early planting on *túmú* soils but that can also be grown on the gravel-free *ŋànyá* soils of the granite zone; medium-duration varieties giving reliable yields on both *túmú* soils and the gravelly upland soils of the granite zone; and long-duration, high-yielding, *yàká* varieties suitable for swamps and watercourses, but requiring a minimum of labor and attention.

In 1983, 98 Mogbuama farm households planted a total of 59 distinct rice varieties: 15 shorter-duration varieties, 34 medium-duration varieties, and 10 *yàká* rices. The majority of these rices are local selections. None of the recently released "improved" varieties has yet reached Mogbuama, except for three varieties I introduced during the course of my fieldwork (ROK 4, LAC 23 and CCA).

Each farm household experiments until it possesses, typically, a suite of one or two shorter-duration, two or three medium-duration, and one or two *yàká* rices, which the members feel they understand and which are suitable to their circumstances. These indigenous experiments (*hùngɔ́ɔ̀*) involve input-output trials, with careful note taken of the amount of seed planted and the amount harvested. When a farmer acquires some unfamiliar planting material, or selects seed with a particular property (e.g., long awns) from existing material, it is generally first tried out in the interzone between the upland and runoff zones, or between the runoff zone and swamp (Richards, 1986; Squire, 1943).

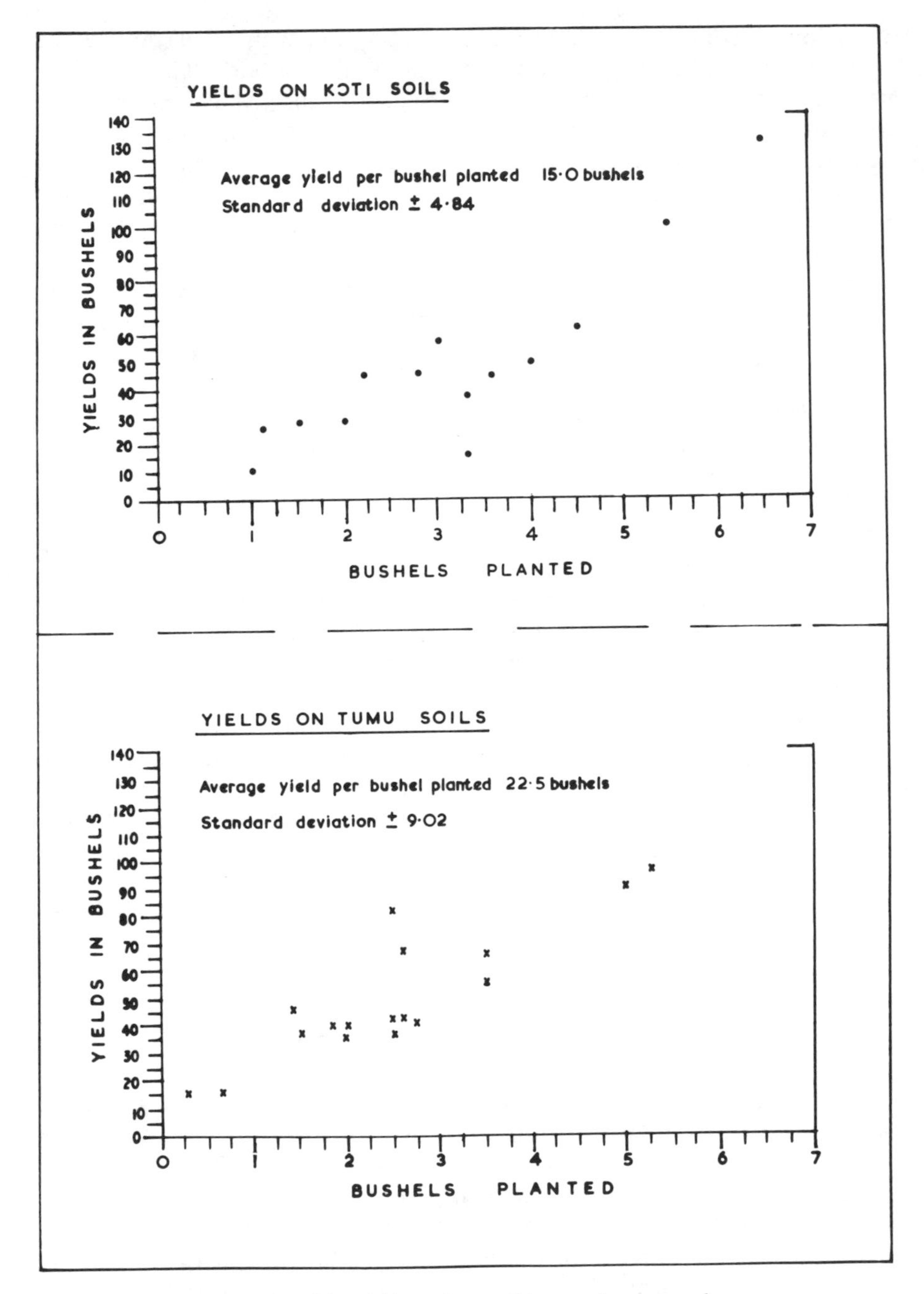

FIG. 6.7. Rice yields on *kotu* and *tumu* soils compared.

An ideal rice farm is one with the smoothest possible labor profile over the period February–December. For a typical medium-size household of 3 or 4 adults and 2 or 3 children, this will be about 1.2–1.5 ha, apportioned 15–20% to quick rice on moisture-retentive soils on lower slopes, 60–70% medium-duration upland varieties, and the balance *yàká* rices on private swamp plots. It should be added, perhaps, that the lack of emphasis on the *yàká* varieties, despite their high yields, is due to the fact that they ripen in November and December, when rice prices are at rock bottom.

Mogbuama farmers show more interest in experiments directed to improving the output of *túmú* or *bàtì* farms. Some farmers double-crop *túmú* soils subject to waterlogging in August–September, with a quick rice planted in April, followed by a crop of cowpeas planted in September. Other farmers intercrop *bàtì* farms with quick-ripening rice and a long-duration floating rice. The idea is to harvest the quick rice before the *bàtì* is completely drowned by the riverine flood in the later part of the rainy season, leaving behind a densely tillering, flood-tolerant rice that is sufficiently well established to cope with the deep water on the *bàtì* in August and September. Thus the *bàtì* is first harvested in June and July, and then again in September–October when the flood finally recedes. In one case examined in 1983, the first crop gave a yield of about 2,400 kg/ha, and the second crop about 600 kg/ha (Figure 6-8).

Although the basic idea of intercropping short- and long-duration varieties on *bàtì* land is well established, those making such farms show much experimental interest in finding the right mix of varieties and the right proportions in which to combine them. A quick variety that proves a shade too slow may be caught by the flood. A long-duration variety that tillers too profusely too early may compete with and thus depress yields of the quick variety.

Upland to Swamp: An Evolutionary Transition?

An official policy of discouraging upland rice and supporting swamp rice cultivation was first adopted by the colonial Department of Agriculture in Sierra Leone during the 1920s (Richards, 1986). At the time, shifting cultivation was thought to be ecologically unsound in all circumstances, a viewpoint modified by subsequent research (Ahn, 1970; Nye & Greenland, 1960). It was supposed that African cultivators were caught in a vicious circle in which shifting cultivation was both cause and consequence of their social and technological "backwardness." In Sierra Leone the Agriculture Department paid little if any attention to upland rice, because that would be tantamount to encouraging shifting cultivation. The route to permanent cultivation lay through swamp farming systems.

Nevertheless, an apparent problem had to be addressed. It was recognized that Sierra Leonian farmers were no strangers to swamp cultivation. Glanville (1938) notes 19th-century examples. If the superiority of these techniques was

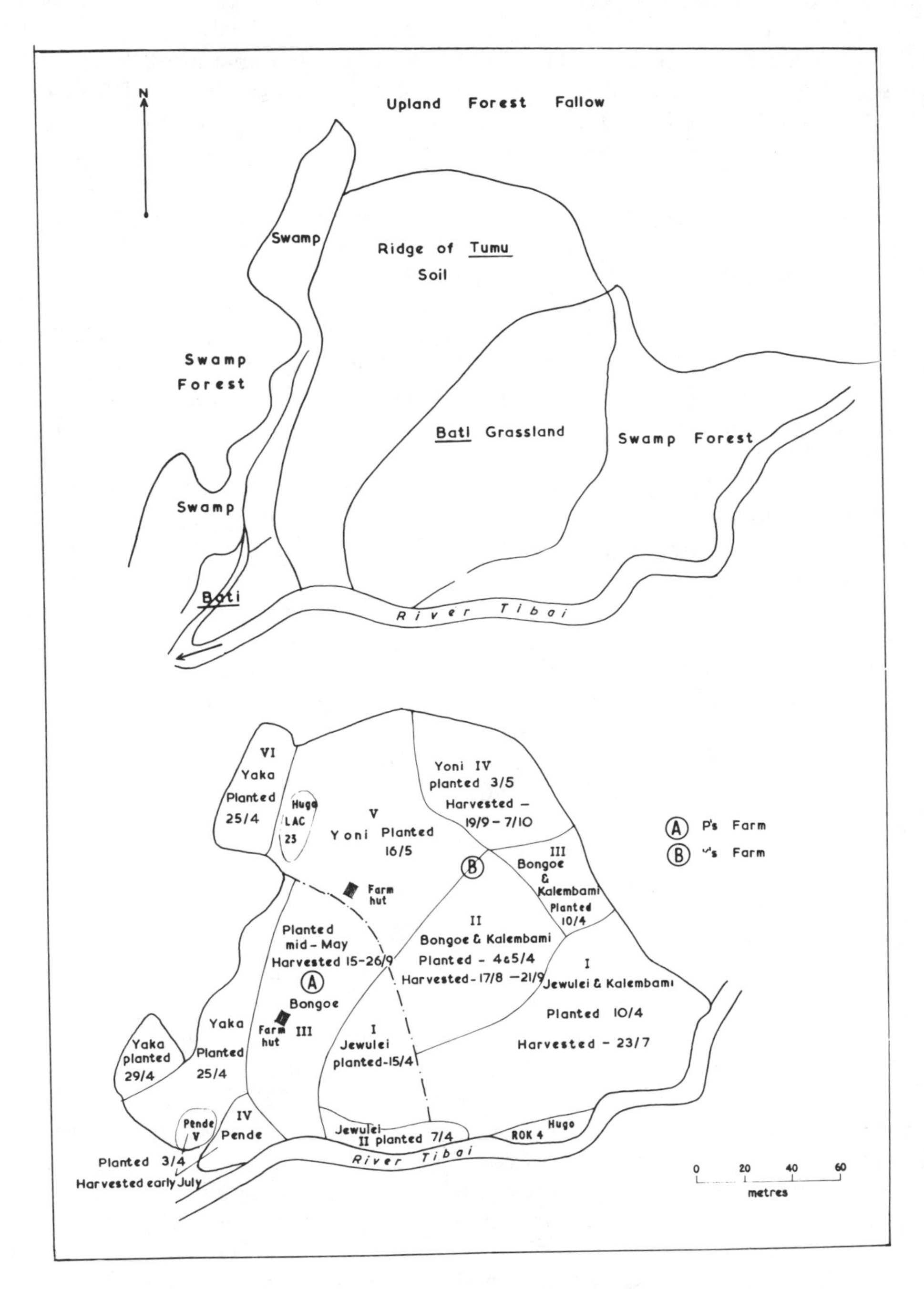

FIG. 6-8. Soils and rice-planting strategies: Two Mogbuama farms with *bàtì* land.

self-evident, as the colonial Department of Agriculture supposed, why had they not swept all else before them?

Glanville's answer is couched in terms of an evolutionary transition in which colonialism itself played a major part. Progress toward the more general use of swamps on a permanent basis was impeded, he argues, by "tribal" warfare and general social insecurity in precolonial times. The political obstacles to further spread of improved agricultural techniques had been removed by the subsequent imposition of the Pax Britannica. Under British protection, farmers had both the tranquillity and the incentives to adopt more intensive methods of production. These changes would be further hastened by increased population pressure and ecological deterioration on the uplands (cf. Bennett, 1976; Boserup, 1965; Wilkinson, 1973).

On Glanville's reckoning, between one-quarter and one-third of all rice in northern Sierra Leone was grown in swamps. Yet 40 years later, Karimu and Richards (1981) found that in the Makeni area of northern Sierra Leone the proportion of swamp to upland was still more or less what it had been in the 1930s. Even more significantly, perhaps, the majority of farmers in the Makeni area still preferred to use swamps on a periodic rather than a permanent basis. What is the explanation? Are Sierra Leone rice farmers caught in a time warp, or is there something wrong with the notion that a move from upland to swamp rice cultivation is a significant evolutionary transition?

There can be little doubt that the years immediately preceding the establishment of the British Protectorate in the Sierra Leone hinterland were turbulent, as indeed they were throughout much of Africa. Many historians, however, would contest the notion that this period was typical of the longer run of African history. Analyses of the kind undertaken by Ford (1971), Kjekshus (1977), and Suret-Canale (1971) argue that social and ecological instability in the late 19th century was more a consequence of the rapid growth of international trade and Great Power rivalry in African affairs than of the supposed "backwardness" of African rural societies.

The dislocative effects of the overseas slave trade, both direct and indirect, continued long into the 19th century (Inikori, 1982). Even where the slave trade had declined or been suppressed, the stability of many rural environments was threatened by ruthless exploitation of timber, ivory, and wild rubber. Peasant farmers and their households were at times "press-ganged" as carriers to head-load these commodities to the coast, to the detriment of food crop farming. The spread of epidemics of smallpox, veneral diseases, influenza, and the cattle disease rinderpest, apparently exacerbated by colonial penetration and the opening up of new long-distance trade routes, further undermined local economies. Many areas directly involved in overseas trade appear to have experienced demographic stagnation and even population losses in the late 19th and early 20th centuries (Hartwig & Patterson, 1978; Patterson, 1975, 1981).

Faced with environmental devastation and demographic reverses, it would not be surprising if the remedies adopted were sometimes makeshift from an ecological point of view. Could it be that the degree of emphasis on upland shifting cultivation in the repertoire of African land management techniques at the beginning of the colonial period was both temporary and unprecedented, as underpopulated communites once more set out to tame the wilderness that had sprung up during the preceding years of confusion and collapse? If so, the transition from upland to swamp cultivation witnessed by Glanville in the 1920s and 1930s might be better interpreted as the recovery of an older, more settled pattern of integrated land use, rather than a once-and-for-all change from "backwardness" to "modernity."

Olfert Dapper's account of rice-farming systems in the Cape Mount area of Liberia in the 17th century suggests that the combination of wetland and upland cultivation is of considerable antiquity rather than "transitional" as Glanville supposed. The historian Adam Jones (1983:165) summarizes Dapper as follows:

> [A first crop] grown in "low muddy places," was harvested as early as May [a *bàtì* farm?]. Probably the amount was small, but enough to feed the farmer and his dependents in the "hungry season." [Two other harvests] were similar in timing to those referred to by later writers: the intermediate crop was probably a short-duration variety, sown in "moistland" riverain areas [*túmú* soils?], with a fairly low yield; and the main crop was hardly different from the upland crop grown by most farmers today. . . .

The point at issue is not whether farming practices in places like Mogbuama are unchanged from the 17th century. The evidence of innovation and experimentation undertaken by Mogbuama farmers suggests that there would be great differences in detail. But Dapper's account, taken in conjunction with the preceding Mogbuama case study, suggests that such experimentation has taken place against a background of unchanging principle, namely, that given labor constraints and the range of ecological hazards faced by rice farmers in the humid sector of the West African rice zone, wetland and upland cultivation are best developed in a complementary manner.

For 50 years the main objective of rice development strategy in Sierra Leone has been to *replace* upland rice by swamp rice cultivation. In recent years this policy has attracted strong support from the International Agricultural Research Centers and development agencies such as the World Bank and the Food and Agricultural Organization of the United Nations (FAO). The replacement strategy is clearly at odds with the aims and initiatives of many peasant farmers. Its proponents defend it, however, on grounds of urgent ecological necessity. Their key assumption is that upland rice cultivation is on the point of collapse because upland fallow periods in Sierra Leone have undergone a rapid decline in the 20th century (FAO/LRD, 1980).

It is worth examining this assumption a little more closely. Reliable estimates of typical fallow intervals in earlier periods are hard to come by, but Dapper, in the 17th century, claimed fallow intervals on uplands as short as 3 years (Jones, 1983). A Freetown-based observer in the late 18th century (Winterbottom, 1803) reported typical figures in the range 4–7 years. Douglas Scotland, director of the colonial Department of Agriculture at the time of its move to up-country headquarters at Njala, cleared 6-year-old bush for his first set of rice experiments in 1912, on the grounds that this was then a typical figure for the area (Department of Agriculture, 1913). Around Njala today, fallow intervals range from 6 to 8 years. A nationwide survey of fallow intervals on uplands in 1978 returned an average fallow interval of 8.8 years (FAO/LRD, 1980).

These figures suggest that reports of remarkable reductions in fallow intervals over the last generation or so ought to be treated with caution. Where farmers themselves record major reductions in fallow period, it is important to determine whether they are referring to a deterioration in the fallow/cultivation ratio or to the apparent halving of the fallow interval that accompanies a shift from the long-fallow to short-fallow cultivation systems. It will be recalled, from the preceding case study, that old forest regrowth (land fallowed for 20–30 years or more) is sometimes planted to rice for 2 years in succession. Because thick regrowth of this sort is so difficult to clear, many farmers prefer shorter-fallow land (7–15 years) planted to a single crop of rice. There is, of course, no difference in the temporal intensity of land use between land cleared every 10 years and planted to rice once, and land cleared only every 20 years and cropped for two seasons. The change from long-fallow to short-fallow cultivation cycles may reflect shortage of labor, not land. It is ironic, therefore, that proponents of the replacement thesis should draw on this evidence to justify a program of labor-intensive swamp development that further exacerbates labor supply difficulties (Karimu & Richards, 1981).

As far as Sierra Leone is concerned, it is apparent that reports of the imminent demise of upland rice cultivation have been greatly exaggerated. There is both life in the old system and time for improvements to be developed. Any such improvements should be based on a thorough understanding of the advantages of combining upland and wetland cultivation. After half a century of research and development strategy shaped not by the realities of peasant rice-farming systems but by dogged adherence to a program rooted in evolutionist prescriptions, there is welcome evidence that attitudes are beginning to change. At the International Agricultural Research Centers there is renewed interest both in upland rice and in techniques of cultivating wet rice without water control (Greenland, 1984), and in Sierra Leone Rokupr Rice Research Station has begun to report results from a research program for improving upland rice yields that takes local intercropping strategies as its starting point (Rice Research Station, Rokupr, 1983).

Conclusion

When Sierra Leone farmers combine upland and swamp rice, they are not tentatively testing the water before deciding whether to move wholeheartedly to a more "advanced" system of production. As the case study of rice-farming practices in Mogbuama demonstrates, farmers combine upland and different types of wetland cultivation because this is advantageous to them in terms of efficient use of scarce labor and in terms of spreading risks. Failure to appreciate this point has undermined research and development initiatives in Sierra Leone for half a century or more, as a result of which nearly all available resources have been directed toward swamp innovations, under the assumption that swamp cultivation was a *replacement* for upland cultivation. What farmers themselves would like to see from development agencies are new inputs for upland cultivation and an integrated approach linking upland intercropping, runoff farming, and low-intensity swamp cultivation. Sadly, it would appear that cultural evolutionist dogma has, until recently, blinded policymakers and planners to the significance of experiments farmers have already undertaken along these lines, especially in the interzones between upland and swamp ecologies. Peasant ingenuity, for example double cropping in *bàtì* farms, counts for nothing as long as outsiders insist on seeing the interzone between swamp and upland as a boundary between different historical epochs.

Acknowledgment

I acknowledge, with respect to the fieldwork on which this study is based, a research grant from the Overseas Development Administration and the institutional support of the University of Sierra Leone. Responsibility for the opinions expressed is mine alone. Mende words cited in the text are given in the indefinite singular form and follow the orthographic conventions of Innes's *Mende–English Dictionary* (1969).

References

Ahn, P. M. *West African soils*. Oxford: Oxford University Press, 1970.

Airey, A., Binns, J. A. O., & Mitchell, P. K. To integrate or . . . ? Agricultural development in Sierra Leone. *IDS Bulletin*, 1979, 10(4), 20–27.

Bennett, J. W. *The ecological transition: Cultural anthropology and human adaptation*. Oxford: Pergamon Press, 1976.

Biggs, S. D., & Clay, E. J. Sources of innovation in agricultural technology. *World Development*, 1981, 9, 321–336.

Boserup, E. *The conditions of agricultural growth*. London: George Allen and Unwin, 1965.

Buddenhagen, I. W. Rice ecosystems in Africa. In I. W. Buddenhagen & G. J. Persley (Eds.), *Rice in Africa*. London: Academic Press, 1978.

Carpenter, A. J. The history of rice in Africa. In I. W. Buddenhagen & G. J. Persley (Eds.), *Rice in Africa*. London: Academic Press, 1978.

Cartwright, J. R. *Political leadership in Sierra Leone*. London: Croom Helm, 1978.

Chambers, R. *Rural development: Putting the last first*. Harlow: Longman, 1983.

Chevalier, A. L'importance de la riziculture dans le domaine colonial français et l'orientation a donner aux recherches rizicoles. *Revue de Botanique Appliquée et d'Agriculture Tropicale*, 1936, 16, 27–45.

Department of Agriculture (Government of Sierra Leone). *Annual report, 1912*. Njala, 1913. (Typescript).

Diehl, L., & Winch, F. *Yam-based farming systems in the southern guinea savanna of Nigeria*. International Institute of Tropical Agriculture, Agricultural Economics Section, Discussion Paper 1/79, 1979.

Division of Agriculture (Government of Sierra Leone). *Report on the recent attempts to establish the cultivation of cotton in Sierra Leone*. Freetown: Government Printer, 1926.

Dresch, J. La riziculture en Afrique occidentale. *Annales de géographie*, 1949, 58, 295–312.

FAO/LRD. *Bush fallow in Sierra Leone: An agricultural survey*. Technical Report no. 6. Freetown: Land Resources Survey, 1980.

Ford, J. *The role of the trypanosomiases in African ecology*. Oxford: Clarendon Press, 1971.

Gellner, E. Patrons and clients. In E. Gellner & J. Waterbury (Eds.), *Patrons and clients in Mediterranean societies*. London: Duckworth, 1977.

Glanville, R. R. *Sierra Leone: Rice cultivation. Report on a visit to Ceylon and South India with proposals for Sierra Leone*. Freetown: Government Printer, 1933.

Glanville, R. R. Rice production on swamps. *Sierra Leone Agricultural Notes*, 1938, 7.

Goody, J. *Technology, tradition and the state in Africa*, London: International African Institute, 1971.

Goody, J. *Production and reproduction: A comparative study of the domestic domain*. Cambridge: Cambridge University Press, 1976.

Greenland, D. J. Rice. *Biologist*, 1984, 31, 219–225.

Grist, D. H. *Rice* (5th Ed.). London: Longman, 1975.

Harlan, J. *Crops and man*. Madison: American Society of Agronomy and Crop Science Society of America, 1975.

Harlan, J., & Pasquereau, J. *Décrue* agriculture in Mali. *Economic Botany*, 1969, 23, 70–74.

Hartwig, G. W., & Patterson, K. D. (Eds.). *Disease in African history: An introductory survey and case studies*. Durham, NC: Duke University Press, 1978.

Howard, A., & Skinner, D. Network building and political power in northwestern Sierra Leone, 1800–65. *Africa*, 1984, 54, 2–28.

Inikori, J. H., (Ed.). *Forced migration: The impact of the export slave trade on African societies*. London: Hutchinson, 1982.

Innes, G. *A Mende–English dictionary*. Cambridge: Cambridge University Press, 1969.

Johnny, M. M. P. *Traditional farmers' perceptions of farming and farming problems in the Moyamba area*. Master's Thesis, University of Sierra Leone, 1979.

Jones, A. *From slaves to palm kernels: A history of the Galinhas country (West Africa), 1730–1890*. Wiesbaden: Steiner Verlag, 1983.

Jordan, H. D. Rice in the economy of Sierra Leone. *World Crops*, 1965, 17, 68–74.

Karimu, J. A., & Richards, P. *The Northern Area Integrated Agricultural Development Project: The social and economic impact of planning for rural change in northern Sierra Leone*. London: Department of Geography, School of Oriental & African Studies (Occasional Papers, New Series, 3), 1981.

Kjekshus, H. *Ecology, control and economic development in East African history*. London: Heinemann, 1977.

Lagemann, J. *Traditional African farming systems in eastern Nigeria*. Munich: Weltforum Verlag, 1977.

Lappia, J. N. L. *The economics of swamp rice cultivation in the Integrated Agricultural Development Project, Eastern Region, Sierra Leone*. Njala: Department of Agricultural Economics & Extension, Njala University College, 1980.

Levi, J. African agriculture misunderstood: policy in Sierra Leone. *Food Research Institute Studies*, 1974, 13.

Linares, O. F. Agriculture and Diola society. In P. F. McLoughlin (Ed.), *African food production systems*. Baltimore, MD: The Johns Hopkins University Press, 1970.

Linares, O. F. From tidal swamp to inland valley: on the social organization of wet rice cultivation among the Diola of Senegal. *Africa*, 1981, 51, 557–595.

Little, K. The Mende rice farm and its cost. *Zaire*, 1951, 5, 227–273, 371–380.

Littlefield, D. C. *Rice and slaves: Ethnicity and the slave trade in colonial South Carolina*. Baton Rouge: Louisiana State University Press, 1981.

Moorman, F. R., & Veldkamp, W. J. Land and rice in Africa: Constraints and potentials. In I. W. Buddenhagen & G. J. Persley (Eds.), *Rice in Africa*. London: Academic Press, 1978.

Morgan, W. B., & Pugh, J. C. *West Africa*. Methuen: London, 1969.

Murphy, W. The rhetorical management of dangerous knowledge in Kpelle brokerage. *American Ethnologist*, 1981, 8, 667–685.

Njoku, A. O., The economics of Mende upland rice farming. In V. Dorjahn & B. Isaac (Eds.), *Essays on the economic anthropology of Liberia and Sierra Leone*. Philadelphia: Institute of Liberian Studies, Liberian Studies Monograph, 6, 1979.

Njoku, A. O., & Karr, G. L. Labour and upland rice production. *Journal of Agricultural Economics*, 1973, 24, 289–299.

Norman, D. W. *An economic survey of three villages in Zaria Province: 2. Input–output study*: i. *Text*. Samaru, Zaria: Institute for Agricultural Research, 1972.

Nye, P. H., & Greenland, D. J. *The soil under shifting cultivation*. Farnham Royal, Bucks: Commonwealth Agricultural Bureaux, Commonwealth Bureau of Soils Technical Communication 52, 1960.

Nyoka, G. C. *Studies on the germination, growth and control of weeds in upland rice fields under different fallow periods in Sierra Leone*. PhD thesis, University of Sierra Leone, 1980.

Patterson, K. D. The vanishing Mpongwe: European contact and demographic change in the Gabon River. *Journal of African History*, 1975, 16, 217–238.

Patterson, K. D. The demographic impact of the 1918–19 influenza pandemic in sub-Saharan Africa: A preliminary assessment. In *African historical demography* (Vol. two). Edinburgh: Centre for African Studies, 1981.

Pearson, S. R., Stryker, J. D., & Humphreys, C. P. (Eds.). *Rice in West Africa: Policy and economics*. Stanford: Stanford University Press, 1981.

Portères, R. Le système de riziculture par franges univarietales et l'occupation des fonds par les riz flotants dans l'Ouest-africain. *Révue internationale de Botanique Appliquée et d'Agriculture Tropicale*, 1949, 29, 553–563.

Rice Research Station, Rokupr (Government of Sierra Leone). *Annual Report for 1982–83*. (Mimeographed, 1983).

Richards, P. Ecological change and the politics of African land use. *African Studies Review*, 1983, 26, 1–72.

Richards, P. *Indigenous agricultural revolution: Ecology and food production in West Africa*. Boulder, CO: Westview, 1985.

Richards, P. *Coping with hunger: Hazard and experiment in a West African rice farming system*. London: Allen and Unwin, 1986.

Scott, J. Patronage or exploitation? In E. Gellner & J. Waterbury (Eds.), *Patrons and clients in Mediterranean societies*. London: Duckworth, 1977.

Spencer, D. S. C. *The economics of traditional and semi-traditional systems of rice production in Sierra Leone*. WARDA Seminar on Socioeconomic Aspects of Rice Production in Sierra Leone, April 1974. (Mimeographed.)

Spencer, D. S. C. *The economics of rice production in Sierra Leone: 1. Upland rice*. Njala: Department of Agricultural Economics and Extension, Njala University College, 1975.

Spencer, D. S. C., & Byerlee, D. Technical change, labor use, and small farmer development: evidence from Sierra Leone. *American Journal of Agricultural Economics*, 1976, 58, 874–880.

Squire, F. A. Notes on Mende rice varieties. *Sierra Leone Agricultural Notes*, 1943, 10.

Suret-Canale, J. *French colonialism in tropical Africa, 1900–1945*. (trans. Till Gottheimer). London: Heinemann, 1971.

Waldock, E. A., Capstick, E. S., & Browning, A. J. *Soil conservation and land use in Sierra Leone*. Freetown: Government Printer, 1951.

Wilkinson, R. *Poverty and progress: An ecological perspective on economic development*. London: University Paperbacks, 1973.

Winterbottom, T. *An account of the native Africans in the neighborhood of Sierra Leone to which is added an account of the present state of medicine among them*. London: C. Whittingham, 1803.

PART III

MIXED TECHNIC AND PRODUCTION SYSTEMS

Introduction to Part III

The four farming systems discussed in this section have long traditions of intensive cultivation in densely settled conditions. In each case environmental constraints affect the expansion and intensification of the system. Both traditional and modern technologies are employed to help alleviate these constraints and to sustain rather high levels of output. Despite the use of fossil fuel inputs, labor remains a critical component to cultivation. Finally, each system is engaged in some level of commodity production, although, with the possible exception of the Punjab case, consumption production remains important. For these reasons, mixed systems typically display substantial variety in terms of the attributes examined here (Figure III-1). Not only do output intensities vary by as much as 4,000 kg/ha/yr, but technologies range from mainly paleotechnic inputs to largely neotechnic inputs, and production ranges from almost wholly consumption to wholly commodity.

The fifth case study involves a comparison of three Himalayan (Nepalese) villages. Located within 15 km of one another, each is situated in dramatically different environmental and economic circumstances. Environmental diversity, of course, is a hallmark of mountain agroecosystems. Economic diversity has been created by differential proximity to a highway and by the recent eradication of malaria, which has promoted the settlement of a subtropical valley at lower elevation. The two older settlements at higher elevations average over 100 people/km^2, while the newer village has only about 57 people/km^2. In each case, annual cultivation or multiple cropping takes place. Cropping practices vary from HYV rice and wheat on irrigated, terraced plots to maize and cereal/fallow on rainfed plots. Corresponding outputs are about 6,100–4,300 kg/ha/yr, respectively. Taken as a whole, the three villages average about 5,250 kg/ha/yr. The degree of commodity production, which ranges from 10 to 37% of total output, is directly related to the size of individual farms and their access to the highway.

Musha, in the Nile Valley, is an example of a traditional system in rapid transformation, through government inducement, into a fully mechanized market system. Population densities in the village are about 900 people/km^2, but farmers respond to both governmental and market demands as well as local population demands. Land is irrigated and multicropped (1.5–2.0 harvests/yr) in a rotation that involves cotton (*Gossypium barbadense*), HYV wheat, maize or sorghum (*Sorghum durra*), and vegetables. Annual yields per plot reach

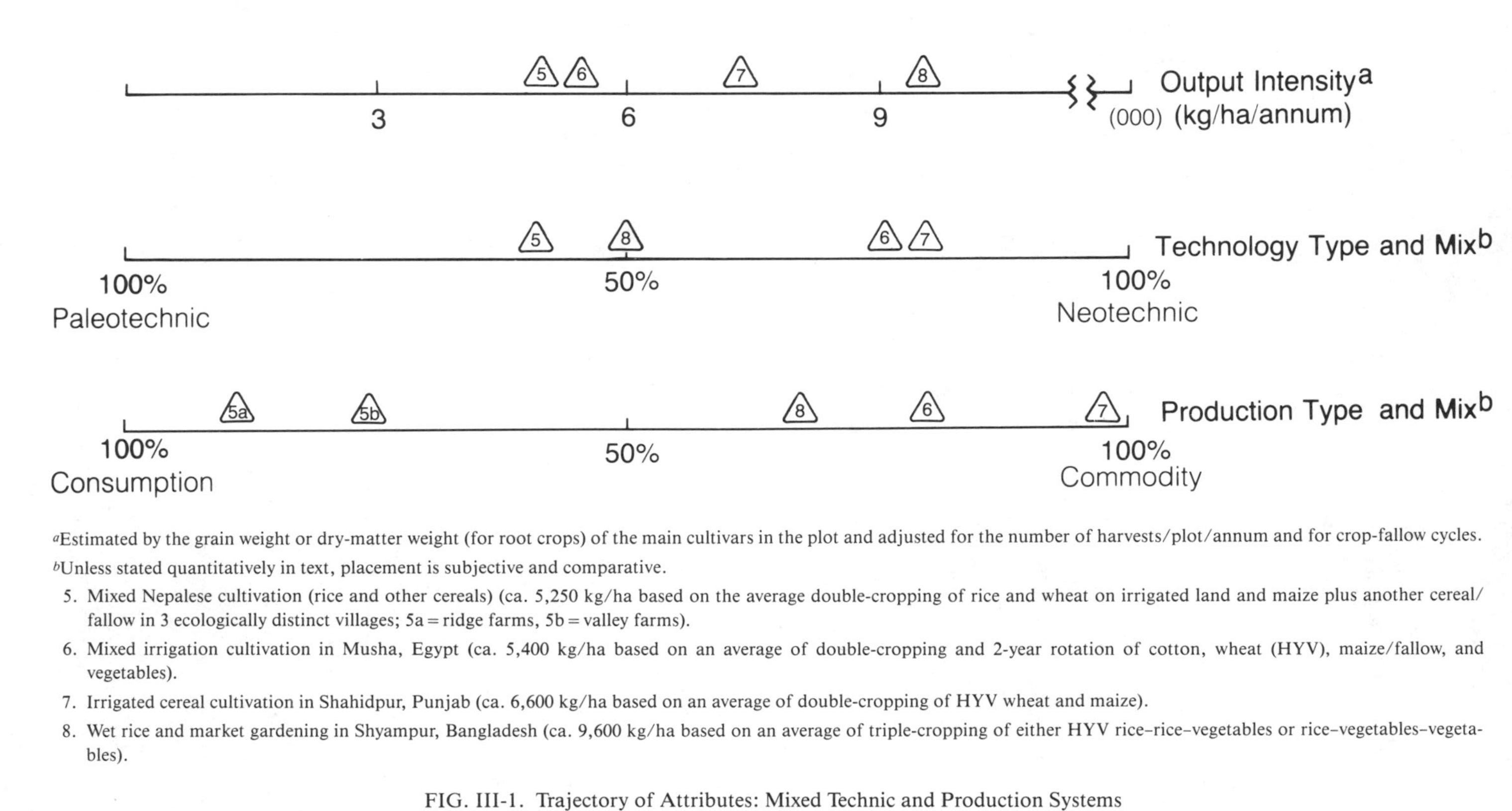

[a]Estimated by the grain weight or dry-matter weight (for root crops) of the main cultivars in the plot and adjusted for the number of harvests/plot/annum and for crop-fallow cycles.

[b]Unless stated quantitatively in text, placement is subjective and comparative.

5. Mixed Nepalese cultivation (rice and other cereals) (ca. 5,250 kg/ha based on the average double-cropping of rice and wheat on irrigated land and maize plus another cereal/fallow in 3 ecologically distinct villages; 5a = ridge farms, 5b = valley farms).
6. Mixed irrigation cultivation in Musha, Egypt (ca. 5,400 kg/ha based on an average of double-cropping and 2-year rotation of cotton, wheat (HYV), maize/fallow, and vegetables).
7. Irrigated cereal cultivation in Shahidpur, Punjab (ca. 6,600 kg/ha based on an average of double-cropping of HYV wheat and maize).
8. Wet rice and market gardening in Shyampur, Bangladesh (ca. 9,600 kg/ha based on an average of triple-cropping of either HYV rice–rice–vegetables or rice–vegetables–vegetables).

FIG. III-1. Trajectory of Attributes: Mixed Technic and Production Systems

5,400 kg when averaged for each cultivar in the four-cycle rotation. Although diesel pumps and tractors are used, a majority of farms rely on human or animal labor. The state has had a direct and significant hand in creating the current "high-tech" system in Musha, but may be depressing output and returns to farmers through pricing policies.

The seventh case study is that of Shahidpur village in Punjab, India, with a population density of 548 people/km^2. As in the village of Musha, change in Shahidpur has been toward a mechanized, market-dominated farming system based on irrigation and HYV cultivars, resulting in major, if not variable, increases in output intensity. Land is double-cropped to various cultivars, with HYV wheat, maize, and rice among the most important ones. Depending on the particular crop rotation, average yields can reach about 6,600 kg/ha, a much heralded result of the so-called green revolution. To accomplish this feat, fossil fuel technologies constitute at least 75% of all inputs to the system. Shahidpur has apparently moved to that threshold stage in which consumption production may soon vanish. One cost of this development has been increased disparities among farm units.

Shyampur, a small village on the urban fringe of Dhaka in Bangladesh, supports one of the densest populations imaginable for an agricultural setting, over 2,000 people/km^2. In this low-lying, flood-prone village, the traditional system of paddy cultivation has rapidly given way to the multicropping of rice (for direct consumption) and vegetables and jute (for the market). Because rice production has not kept up with demand, farmers have shifted to market gardening as a partial alternative. Some rice must be purchased with the profits. Triple-cropping of rice and vegetables has led to an average production of 9,600 kg/ha/yr. Principal neotechnic inputs are hybrid cultivars, fertilizers, and pesticides, but high labor inputs remain critical to the system. Flooding, which is beyond the ability of the village to control, may be a barrier to further improvements in production.

The variability evident in these four systems is expected because of the range in population, technological and production types, and mixes thereof. The considerable differences in output intensities among them are exacerbated by the choice of cultivars and by environmental and socioeconomic conditions. Population pressures seem to be broadly related to each system's output intensity, but other factors may be equally important, such as the degree of involvement with market production, access to neotechnic inputs, and major environmental constraints (e.g., Brookfield, 1962; Turner, Hanham, & Portararo, 1977). Other interesting facets of these mixed systems include: the increasingly significant role that off-farm employment plays as either mechanization and/or the ratio of people to farm size increases, given accessibility to such employment; and the economic disparities that are developing among farmers of the same locale as societal norms transform from communal to individual foci (see Bayliss-Smith & Wanmali, 1984).

References

Bayliss-Smith, T., & Wanmali, S. (Eds.). *Agrarian change and development planning in South Asia*. Cambridge: Cambridge University Press, 1984.

Brookfield, H. C. Local study and comparative method: An example from central New Guinea. *Annals of the Association of American Geographers*, 1962, 52, 242–254.

Turner, B. L., II, Hanham, R. Q., & Portararo, A. V. Population pressure and agricultural intensity. *Annals of the Association of American Geographers*, 1977, 67, 384–396.

7

Agricultural Ecology of the Mid-Hills of Nepal

JACQUELINE A. ASHBY AND DOUGLAS PACHICO

Nepal faces two of the major problems of development—the need to increase food agricultural productivity and the need to relieve population pressure on the land. In the Nepalese hills, the traditional resource system of the small subsistence farm is reaching the limits of its viability (Ashby, 1982). At the same time, new resources are being created by modernization policies pursued since 1951. New agricultural inputs are being introduced, roads are being constructed, new markets for agricultural produce and for labor are opening up, access to schooling is expanding, and government administration increasingly penetrates rural society (Calkins, 1976; Hagan, 1976).

This process of change has important implications for the analysis of hill farming systems in Nepal. The traditional practice of agriculture involves complex adaptations to the Himalayan mountain environment, with its steep gradient and its striking contrasts in altitude, microclimates, soils, water resources, and vegetation within small areas (Guillet, 1983). The ancient technology of terraced rice production on soils suitable for paddy; the cultivation of maize and other cereals suited to upland soils; and the integration of livestock, pasture, and forest in the farming system are all components of adaptive strategies, yet these do not represent a traditional equilibrium relationship between humans and the environment.

Historically, the expansion of area cultivated through the clearing of forests and construction of new terraces has been a key factor in sustaining a growing population (MacFarlane, 1976). This long-term process is reaching the limits of its viability as the supply of uncleared land suitable for agriculture approaches exhaustion. The clearing of steep slopes for grazing, firewood, and cultivation has led to increasing problems of severe erosion.

The growing imbalance between resources and population may be corrected by modernization policies to provide farmers with new technologies and alternatives to past methods of subsistence livelihood (Banskota, 1979; Mathema & Van Der Veen, 1978). While some farmers face a shrinking resource base

Jacqueline A. Ashby. International Fertilizer Development Center and Centro Internacional de Agricultura Tropical, Cali, Colombia.

Douglas Pachico. Centro Internacional de Agricultura Tropical, Cali, Colombia.

for subsistence agriculture, others can increase productivity with new inputs (Blaikie, Cameron, & Seddon, 1977; Feldman & Fournier, 1976). The ways in which Nepalese farmers have responded to the challenge of mountain ecology in the context of social and economic change is the theme of this comparative analysis of three ecologically and socioeconomically distinct farming systems.

An understanding of Himalayan hill agriculture requires a general model of the hill farm system as a framework for interpretation (Harwood, 1979). This chapter outlines such a model for farms in the Himalayan foothills of Nepal and discusses farmer decision-making objectives. These examples show how variations in the overall farming system model appear in different combinations of ecological and socioeconomic resources. Important features of three farming systems are discussed in detail to illustrate how farmers have adapted ecological, economic, and social-organizational resources to meet their objectives.

Characteristics of the Himalayan Hill Farming System

One of the most important characteristics of Himalayan hill farming in Nepal is their subsistence component. Diverse crop enterprises provide fuller employment throughout the year with less severe labor peaks than would result from monocrop specialization. These crop enterprises include irrigated and rainfed rice as well as wheat, maize, millet, and numerous minor crops and vegetables. They are interrelated in several ways. First, they must share a more or less fixed amount of labor, land, cash inputs, and power for traction, and the hill farmer must allocate resources among various crop enterprises. Second, timing the tasks of growing various crops can affect the amount of resources available to each of them. Third, crops are interrelated through intercropping and relay cropping.

There is also interdependence between crop and livestock enterprises on the hill farm. Livestock provide power for traction and manure to maintain soil fertility. Crops and their residues are important components of the livestock ration. Livestock products supplement the food available to the family from crops. Cash may be generated from livestock and used to purchase inputs such as fertilizer to increase crop productivity.

The objective of ensuring a necessary level of subsistence is culturally defined and varies among different social groups. Meeting this objective often entails strategies in which long-run consequences are discounted in favor of meeting immediate needs (Lipton, 1968; Scott, 1976). Individual farmers' decision making in the Himalayan hill farm system can be interpreted as rational or adaptive in terms of the short-run requirements of subsistence livelihood, even though environmental consequences, such as deforestation and soil erosion, may imply the long-run disintegration of the system.

The objectives that farmers bring to the management of their farm system are explicable by theories of profit maximization, satisficing, maximization of returns to the scarce factor, or decision making under uncertainty. These objectives differ according to the size and location of farms. Especially important is the distinction between small farms engaged in subsistence production and large farms engaged in commercial production. In the analysis that follows, farmers' decision-making rules are interpreted as adaptive strategies for coping with immediate problems of subsistence livelihood. In subsistence production, profit maximization models of farmer behavior are often inappropriate because, once a family has produced enough food to meet its needs, there is little incentive to produce a surplus (Sahlins, 1972). This type of decision making has been termed *satisficing* (Simon, 1959), because once a desired level of production is attained, a farmer who is a satisficer will not expend additional resources to obtain further gains. The actual situation of many small farmers in the hills of Nepal, however, is that they cannot obtain even minimal levels of satisfaction, as indicated by the overall food deficit in the hill region (Ministry of Food and Agriculture, 1971). In such a case, farmers respond by trying to derive the most returns to the resources they havc at hand, which often means maximizing returns to their scarcest or limiting factor (Doyle, 1974; Norman & Palmer Jones, 1976). Since different groups of farmers face different limiting factors, their strategies will vary. Smallholder farmers tend to have abundant labor relative to land, their scarce factor, and hence maximize output per unit of land. Largeholder farmers, conversely, tend to maximize output per unit of labor (Yotopolous, 1977).

A key sociological feature of hill farming systems in Nepal is a very unequal distribution of land (e.g., Messerschmidt, 1974). Differential access to land structures the flow of resources between the components of the farm. Smaller farm units experience different pressures in food provision, availability of cash, use of family labor or hired labor, and many other aspects of their farm operation compared with farm units owning more land. Shortage of land drives hill farm families to self-exploitation in the form of extremely land-intensive activities for which the return may be very low but which are indispensable for achieving an adequate subsistence. A pattern of "agricultural involution" (Geertz, 1963) relying on the use of increasingly labor-intensive techniques to cope with rising population density, is characteristic of some hill farming systems with fertile soils. Other systems adapt by exporting labor that cannot be absorbed locally. Thus, variation in access to land fundamentally affects the choices that farmers make and the patterns of social change. The following analysis looks at how farms of different sizes manage their agricultural resources in several key enterprises of the hill farming system.

Although the maximization of returns to the most limiting factor is a basic principle, the intricate relationships between diverse resources for hill farming mean that a wide variety of adaptive strategies exist for the management of these resources. The three hill farming systems described in the next section of this chapter typify this intricate set of relationships between land resources, cropping combinations, population density, caste composition, and the historical pattern of human settlement associated with differences in elevation in the Himalayan foothills of Nepal.

Environment, Social Structure, and Agriculture in the Three Hill Farming Systems

Hill farming systems in Nepal are differentiated ecologically by elevation. The three farming systems in this study are located in the eastern foothills of the Himalayas and are situated in a range of microclimates. Although they lie within a 15-km radius of each other, they vary in elevation (above sea level) from ridge communities at 1,900 m to a midaltitude valley community at 1,500 m and a subtropical valley community at 1,000 m (Figure 7-1). The data presented here were collected from random samples of farmers in each system who were interviewed on several occasions about their agricultural practices and other aspects of household organization (Ashby, 1980; Pachico, 1980).

The Mid-Valley System

Many of the relationships between the ecology and the social organizational features of Himalayan hill farming systems are only comprehensible with reference to historical changes in local settlement patterns that reflect growing population pressure on the land. Probably the most ancient settlement of these three farming systems is Sanga, the midaltitude valley community, a dense cluster of dwellings overlooking terraced fields that slope gently down to the Punimata River (Figure 7-1). As is typical of hill settlements, houses are plastered with red ocher and roofs are thatched with straw. Half a dozen houses of the more affluent are brick constructions, occasionally with corrugated tin roofs. Each house has a small front courtyard where livestock are tethered and fed during the day. The more prosperous households have buildings attached for livestock, and sometimes a small fenced patch for vegetables and fruit trees. Between clusters of houses are upland terraces, interspersed with the *occubanyan* or *peple* tree and ponds where buffalo are taken for washing. At intervals along the main footpath that parallels the ridge are piped water taps, the primary source of water for household use.

The village is bordered on the south by the Arniko Highway; an important local market town is about 1 mile away; and the capital, Kathmandu, is about 2 hours away by bus. Proximity to market centers contributes to the ability of farmers to participate in a high-cash-flow agriculture associated with extensive use of fertilizer and newly introduced crop varieties. The current scarcity of land is reflected in a six- to tenfold inflation in land prices in the past 30–40 years. Pasture has been brought into cultivation, and virtually all the ancient forest has been cleared, although since 1964 the government has protected the few remaining forest lands from cutting. Villagers in Sanga must therefore buy their firewood from nearby villages where there remains forest that can be cut.

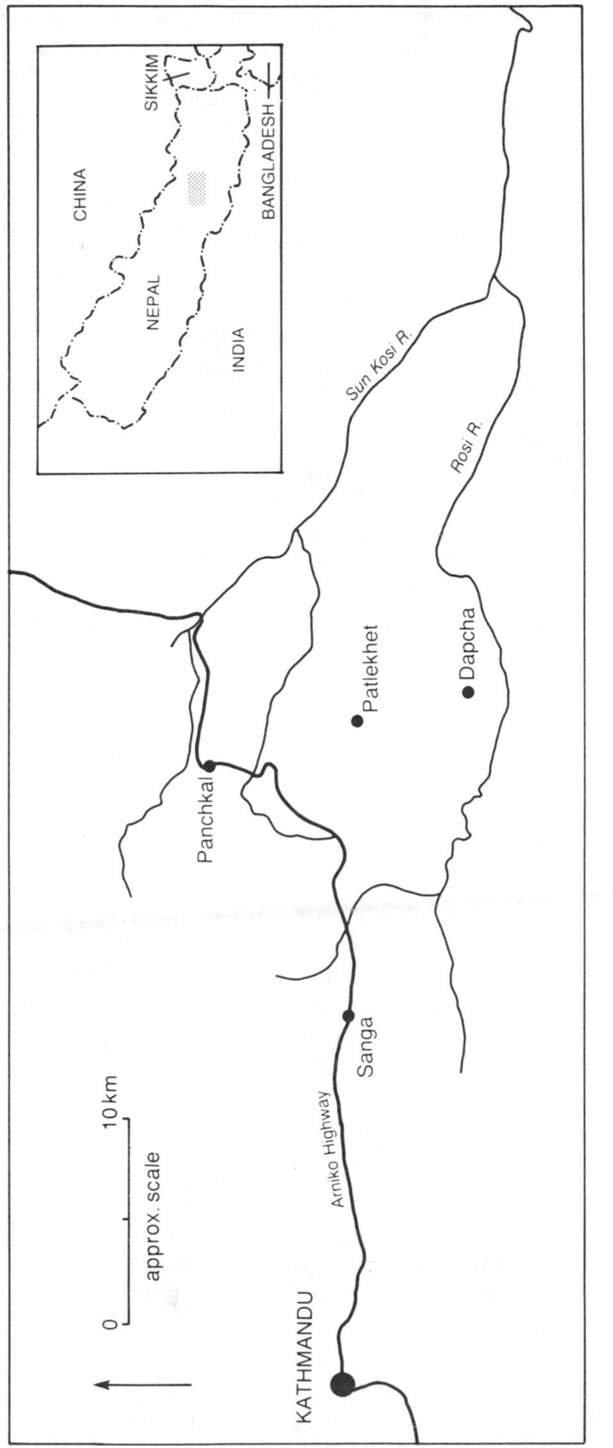

FIG. 7-1. Location of Sanga, Patlekhet, Dapcha, and Panchkal.

The lack of forest and public pasture limits livestock to that which can be supported on the farm. This limitation, coupled with the small farm size, has led to the almost complete disappearance of draft animals in Sanga, and most tillage operations are done by hoe. Farmers consistently report that their fathers plowed with oxen, but they themselves cannot afford to do so.

This village is something of an exception in Nepal today, with good access to roads and markets, high cash flow, extensive use of new technologies, no public forests or pasture, and very little draft power. In these respects it portrays some of the opportunities for development that could occur on small farms in the hills with improved access to markets and new technologies.

The Ridge System

The ridge farming system comprises two settlements, Patlekhet and Dapcha, about two and a half hours walk south from the Arniko Highway (Figure 7-1). The hillsides are steep and rocky, and supplies or products for market must be portered back and forth. Dwellings are less densely clustered than in the mid-altitude valley and are generally less prosperous in appearance, as evidenced by the absence of large brick houses or corrugated iron roofs. Houses are scattered among small upland terraced fields, which slope steeply down the hillside to the valley floor 1,000 m below (Figure 7-2). In Patlekhet village center there is the one tea shop and one small general trading store, where maize is bought and sold and cigarettes, matches, biscuits, and other items are available. The other village, Dapcha, is much larger and was formerly an important entrepot for goods exchanged between Kathmandu and the eastern hills. The advent of the Arniko Highway, which has so benefited Sanga, has disrupted the traditional trade routes, and the Dapcha market has suffered a considerable commercial decline and consequent emigration of merchants. Several cloth shops, tea shops, and general merchandise stores remain, but the presence of an agricultural extension agent, a paramedical health worker, and a post office do not compensate the village for its loss of trade.

These ridge villages share some of the characteristics that typify the severe conditions prevailing in the hills of Nepal. Use of purchased inputs for agricultural production and sale of surplus products are limited by the villages' remoteness. Emigration to seek outside employment, seasonal or semipermanent, is an important economic activity and a major source of cash income. In contrast to Sanga, in these villages oxen are used for most tillage. Public pastures and forests provide firewood and fodder for livestock. The pasture is always heavily grazed, and the forests are cut back for firewood and animal fodder or to clear land for cultivation. After an antimalaria spraying program began in 1957, extensive terraced fields were created on the valley floor. The increase in rice production enabled farmers to pay taxes on ridge land that had previously been left uncultivated because of the tax burden. Thus the greater availability of land in the valleys for planting rice was a stimulus to forest

FIG. 7-2. Patlekhet. Lower green terraces are in rice, brown terraces are ripening maize. Village settlement is near the border of these two land types, and upland maize terraces extend above up the slope.

clearing and creation of new upland fields for maize. Streams in the valley floor are used to irrigate lowland terraces, but on the ridges water is scarce and must be fetched and carried for a distance of up to half an hour's walk.

The Subtropical Valley System

North of the midaltitude valley, the Arniko Highway ascends a ridge and then winds steeply down into the subtropical valley of Panchkal, the third farming system analyzed in this study (Figure 7-1). Until the mid-1950s this valley was malarial. The principal inhabitants were the Danwars, a hunting and gathering people (Bista, 1967). After malaria eradication, there was immigration to the valley primarily by the upper-caste Brahmins and Chetris from the nearby ridge villages. Since the subtropical valley has been inhabited by settled agriculturists for only a short time, population pressure has not yet led to farm fragmentation. Houses are scattered among sizable fields, often unterraced, across the valley floor. On the eastern side, a river supplies year-round irrigation to low-lying fields. The long frost-free growing season, coupled with irrigation, facilitates growing three crops annually. Panchkal's surplus production can be readily shipped to the Kathmandu market on the Arniko Highway. This favorable resource position is as yet only slightly altered by emerging problems of firewood scarcity and soil erosion on the upland fields caused by rapid deforestation. This farming system represents some of the characteristics of the agricultural frontier, where inputs are utilized less intensely than in the midaltitude valley and the ridges and where there is major potential for increasing agricultural productivity.

Important differences exist among the three hill farming systems in the pattern of land distribution and in population densities (Table 7-1). The majority of farms are owner-operated; only in the midaltitude valley is there renting or sharecropping, involving 9% of the cultivated land. In the midaltitude farming system, the average farm size is 0.65 ha, and the person/land ratio is 11.6 persons/ha. The lower 55% of the households own 47% of the land. The average farm size and the distribution of land in the ridges is almost identical, with a person/land ratio of 10.2 persons/ha. In the subtropical system, however, the average farm is 1.2 ha, perhaps because the recently settled farms have not yet been fragmented through partiple inheritance. The person/land ratio is only 5.7 persons/ha. The larger average family size in the midaltitude valley and ridges reflects the greater population density in these two farming systems, which support approximately twice as many persons per unit of land as does the subtropical valley (Table 7-1).

Land Classification

Land is classified into three types: irrigated paddy land (*khet kulo lagne*), rainfed paddy land (*khet kulo nalagne*), and upland (*bari*) which is unsuited for paddy cultivation. *Khet* comprises 73% of the land in the midaltitude valley and 64% of the land in the subtropical valley. The ridge farming system is based primarily on *bari* (66% of land), suitable only for upland crops, but ridge farmers also have *khet* land in the low-lying valley bottoms. The majority of farms in the subtropical valley cultivate rice with gravity-fed irrigation (79% of *khet*), whereas in the midaltitude valley most low land is rainfed (only 41% of *khet* is irrigated).

The term *khet* is used by farmers to classify soils that are clay-based and on which they can impound water, usually the collected runoff from upper slopes. Irrigated *khet* refers to these types of soils with access to water from small streams. Upland or *bari* soils are generally light, sandy, and well drained, and are located on the upper slopes where there is less runoff. Both *khet* and *bari* are terraced, but in the ancient midaltitude settlement of Sanga and in the ridge villages, no new terraces have been created in the past 15–20 years. This situation is typical of Himalayan farming systems experiencing population pressure (MacFarlane, 1976). Existing terraces on *khet* land are maintained to

TABLE 7-1. Distribution of land resources in three hill farming systems, Nepal.

Farming system	Farm size category	Land per capita (ha)	Family size	Percentage of households	Percentage of land
Midaltitude valley	Small (0.5 ha)	0.04	5.7	55	19
	Medium (0.5–1 ha)	0.09	8.5	29	34
	Large (1 ha)	0.17	11.8	16	47
Ridges	Small (0.5 ha)	0.05	5.7	51	22
	Medium (0.5–1 ha)	0.10	7.2	30	34
	Large (1 ha)	0.16	9.0	19	44
Subtropical	Small (0.5 ha)	0.06	5.6	23	7
	Medium (0.5–1 ha)	0.13	5.8	33	22
	Large (1 ha)	0.25	7.1	44	71

Source: Pachico (1980).

impound runoff, whereas *bari* has been terraced differently to prevent soil loss, slope erosion, and land slides. In the subtropical valley, Panchkal, *khet* fields are flat and bunded to impound water, whereas gently sloping upland fields where soils will not permit puddling are seldom bunded.

Social Stratification and Land Distribution

The distribution of land resources is a fundamental feature of the social stratification system. Within each village the larger farms tend to own a higher proportion of the better-quality land. For example, 75% of the land owned by the small farmers in the subtropical valley is upland; the corresponding figure for the larger farmers is only 45%. Likewise, in the midaltitude valley 35% of the land of the large farmers is irrigated *khet*, the best-quality land, whereas only 23% of the land of the small farms is irrigated *khet*.

Social stratification associated with land ownership reflects caste. The majority of households in each location consist of Hindu Brahmin or Chetri castes, also the predominant caste group in the midhill region. Nonetheless, caste and ethnic group differences are important. The midaltitude valley has no non-Hindu groups. In the other areas, non-Hindu groups are present, occupying lower status and less privileged positions. Tamangs constitute 29% of households in the ridge area, and 24% in the subtropical valley are Danwar. Hindu Brahmins and Chetris are conventionally the highest-status groups, and the official national language, Nepali, is their native tongue. The Hindu occupational castes in these areas are local artisans—Danli (tailor), Sunar (goldsmith), Sarki (cobbler)—who also engage in farming. The Tamangs in the ridge farming system speak their own Tibeto-Burman language, with Nepali as a second language, and their culture and religion is Buddhist. They represent the indigenous hill population, which was displaced politically and economically in the influx of Indian Hindu castes into Nepal after the 12th-century Moslem invasions of India (Caplan, 1970). The Danwar in the subtropical valley also speak their own language.

Control over larger landholdings is predominantly in the hands of members of the Brahmin and Chetri castes. The predominance of Brahmin and Chetri castes in the area, however, means that a substantial proportion of the small farmers are also Brahmin or Chetri. There is, then, considerable variation in the caste status associated with size of landholding.

Crop Production and Income

Differences in elevation, land types, population density, and distribution of control over land resources among the three farming systems are related to the importance of different agricultural production activities in each one. Crop combinations vary with elevation and the availability of soils suitable for rice cultivation (Table 7-2). In terms of area planted, rice (*Oryza sativa* L.) followed by wheat (*Triticum aestivum* L.) are the major crops in the middle elevations. In the ridges, maize (*Zea mays* L.) is the major crop, followed by winter wheat. In the subtropical valley, rice, maize, and potatoes (*Solanum tuberosum* L.) are important crops, grown principally in rotations.

Although subsistence production is the most important goal of these farms, each hill farming system is involved in the cash economy. The majority of households in the midaltitude and subtropical valleys sell some farm produce, whereas fewer than half in the ridge farming system report any sales. The two valley systems market on the average of one-third to two-fifths of the annual rice harvest and about one-quarter of wheat production. Very little maize is sold. A substantially smaller surplus is produced for sale in the ridges: less than 10% of all crop production is marketed in the ridges, compared with 27% in the midaltitude valley and 38% in the subtropical valley. In the past, grain was bartered to local middlemen, who would then have it portered to Kathmandu or sold to shopkeepers in the local market towns. Since the Arniko Highway was constructed in the late 1960s, it has become more common for farmers in the valleys to negotiate directly with shopkeepers from the Kathmandu Valley, and payment is now made in cash. In the ridge farming system an important source of cash used to be the sale of milk and *ghee* (clarified butter) in the local market town, but this source of cash income has become less important with the decline in numbers of livestock. Households in the ridges have to meet their cash needs principally from nonfarm employment or from agricultural laboring. This pattern of cash income sources for difference farm sizes (Table 7-3) illustrates that on the small farms, where land is a scarce factor, crops are primarily for consumption. Labor, whether involved in agricultural or nonfarm employment, is the small farms' principal source of cash.

The transition to a cash labor economy has been a major impact of

TABLE 7-2. Agricultural production characteristics of three hill farming systems in Nepal.

Agricultural production characteristics	Farming system		
	Midaltitude valley	Ridges	Subtropical valley
Percentage of total area planted[a]			
Rice	74	45	59
Maize	13	57	39
Millet	17	10	0
Wheat	63	57	45
Percentage of annual production[b]			
Rice	42	27	40
Maize	7	25	15
Millet	3	1	0
Wheat	15	8	6
Other crops	7	11	19
All crops	74	72	80
Animal products	26	28	20

[a]Areas sum to more than 100% because of intercropping, relay cropping, and rotations.

[b]Production valued at average farm gate price for each product.

TABLE 7-3. Cash income of farms of different sizes in three hill farming systems, Nepal (cash income by source in U.S.$).

Farming system and farm size[a]	Total cash income per capita (U.S.$)	Percentage of cash income from crop sales	Percentage of cash income from livestock sales	Percentage of cash income from wages	Percentage of total crop value sold
Midaltitude valley					
Small farms	26	19	22	58	12
Medium farms	37	34	37	29	22
Large farms	60	62	18	20	40
Ridges					
Small farms	19	5	20	75	4
Medium farms	17	15	20	65	8
Large farms	33	27	27	45	17
Subtropical valley					
Small farms	13	23	18	59	10
Medium farms	29	65	25	10	32
Large farms	56	72	6	22	40

Source: Pachico (1980).

[a]See Table 7-1 for definition of farm size classes.

outside market forces. Twenty years ago farmers used almost exclusively *parma* or exchange labor. However, increased labor requirements for new high-yielding varieties could not be met by the family and exchange labor system. It is easier than in the past to find labor for hire in the valley systems. Ridge community inhabitants migrate to the valleys for seasonal agricultural labor because food shortages compel them to purchase food. This is the case particularly at the end of winter, when valley farmers hire labor to harvest their winter wheat crop. But the nonagricultural labor market is also an especially important source of cash earnings for small farms in the midaltitude valley and ridge systems, where over 40% of households report nonfarm earnings. In contrast, less than one-quarter of farms in the subtropical valley engage in nonfarm employment.

These differences in food availability and sources of cash income reflect varying degrees of market involvement. Hill farms can be categorized into four types, reflecting their degree of market integration: (1) commercial rice farms, (2) commercial livestock farms, (3) subsistence farms, and (4) agricultural laboring farms. These farm types were distributed across the three farming systems by selecting those that market at least one-third of the total value of crop and livestock production, and earn above the median cash income for all farms in the sample survey. At least half of all cash earnings on commercial farms were from sales of farm produce. This type was divided into large commercial rice farms (owning one hectare or more of land) and small commercial livestock farms that derive most cash from sales of milk. Large commercial farms have average annual cash earnings of U.S.$532; average farm size is 1.8 ha with 2.0 family workers (aged 15–65 years) per hectare cultivated. Small commercial farms' cash income averages U.S.$371, and they cultivate only 0.67 ha with a ratio of 4.4 family workers/ha.

Farmers deriving one-third or more of gross annual income from nonfarm employment of family members who work in the market towns, were defined as *part-time* farms. These farms own less than 1 ha of land, averaging 0.41 ha in size, with a high person/land ratio of 6.2 family workers/ha. Cash income averages U.S.$231 per annum. The remaining farms in the sample are subsistence operations, which average only U.S.$32 cash income annually and cultivate an average of 0.57 ha with 4.4 family workers/ha. Part-time and subsistence farmers hire little or no labor and market very little of their farm produce.

The distribution of four different types of farms among the three farming systems (Table 7-4) reflects the differences between the prosperous rice cultivating valleys and the poorer, subsistence-oriented maize cultivation system in the ridges. Where maize provides little marketable surplus, there is a majority of subsistence farms. Where access to markets is most advantaged, in the midaltitude valley, fewer small farms are primarily subsistence operations, and many have gone into sales of milk or of family labor to generate cash incomes. The subtropical valley is distinguished by the highest proportion of large commercial farms.

TABLE 7-4. Farm type composition of three hill farming systems in Nepal: Percentage distribution of farms in each system.

	Farming system		
Farm type	Midaltitude valley	Ridges	Subtropical valley
Subsistence farms	33	57	42
Part-time farms	27	24	5
Small commercial farms	24	7	18
Large commercial farms	14	12	35
Total	100	100	100

Crop production is increasingly subsistence-oriented as farm size decreases. Some small farms, however, have successfully diversified into milk production for the market, which provides cash for purchased inputs (such as fertilizer) needed to grow new crop varieties. Others have gone into nonfarm employment. Large farms, on the other hand, produce crops for the market and sell a substantial surplus over their own subsistence needs.

Cropping Patterns

The annual cropping cycle in Nepalese hill farming systems revolves around the monsoon rains, which usually come in full force around June or July. The monsoon sets the time for rice planting on *khet* land. On land that is not suitable for rice cultivation (*bari*), maize is planted in the premonsoon showers in April–May. The two different land types and two different monsoon cultivars enable farmers to spread land preparation and planting over a few months, since planting occurs at different times. Similarly, harvest times are staggered, with maize typically harvested around September–October and rice in November–December. This promotes more even labor scheduling on the farm during the cropping year.

Although many crop rotations are found in all three systems, significant differences can be observed between the systems in the relative importance of various rotations (Table 7-5). In the midaltitude valley, a rice–wheat rotation occupies almost all of the *khet* as well as the majority of land in the area. Maize relay-cropped with millet is the main rotation on *bari*. In the ridge system, a summer *bari* crop of maize, intercropped with soybeans, followed by winter wheat, is the most important cropping pattern. The rice–wheat rotation on *khet* is second in importance. In the subtropical valley, rice–wheat again predominates on *khet*, but maize intercropped with soybeans (*Glycine max*), followed by a winter fallow, is the main *bari* rotation. Unique to the warmer, well-irrigated subtropical valley is triple cropping on *khet*, with two rice crop

TABLE 7-5. Crop rotations in three Nepalese hill farming systems (percentage of total crop area by rotation).

Crop rotations	Midaltitude valley	Ridges	Subtropical valley
Lowland (*khet*)			
Rice–wheat	69.9	31.2	38.0
Rice–potatoes	2.9	[a]	6.5
Rice–fallow	[a]	9.6	2.6
Rice–rice	0	0	2.2
Rice–potatoes–rice	0	0	2.1
Rice–potatoes–maize	0	0	1.4
Rice–maize	0	0	0.8
Upland (*bari*)			
Maize/soy–wheat[b]	0	40.2	5.5
Maize–millet	19.0	8.5	0.9
Maize/soy–fallow	6.1	10.1	34.6
Maize/soy–mustard	1.9	[a]	4.6

[a]Less than 0.5%.

[b]Includes barley.

harvests. Most of the cropping patterns can be found in each system, though in differing degrees of importance. Farmers' choice of cropping patterns is influenced by a combination of factors: soils, climate, water availability, labor supply, access to markets, capital endowments, subsistence requirements, and integration into the cash economy. How farmers evolve adaptive strategies to these conditions in their choice of crop combinations and technology will now be discussed for some of their major crop decisions.

Rice Cultivation

Rice plays a central role in each of the three farming systems. It is the single most important cultivar in terms of area planted and value of sales, and as a food staple in the two valley systems. In the ridge system, however, only a small proportion of the land is suitable for terracing, so that rice is grown principally for food and less as a cash crop. Rice is a focus of innovation because of the green revolution's introduction of new seed–fertilizer technology, making rice variety choice a process of weighing traditional cultivation objectives against new options.

Traditional rice cultivation techniques are broadly similar across the three hill farming systems. Rice is sown in seedbeds in early spring. Seedlings are transplanted by women once the monsoon rains have flooded the terraces. These terraces are bunded to impound rainwater, and those with access to irrigation are served by gravity-fed channels from the rivers. Through informal consultations, farmers seek a community consensus over planting dates and variety for contiguous blocks of fields. This is done both to coordinate water from upper to lower terraces, and to ensure that harvest occurs simultaneously because livestock graze the rice stubble as soon as the crop is harvested.

Land is prepared by repeated plowings with bullocks, except in the mid-altitude valley where men cultivate by hand with the *kodali*, a heavy digging hoe. In the midaltitude valley chemical fertilizer is often broadcast at transplanting, but elsewhere farmers apply chemical fertilizer less frequently, though the use of compost is universal. Rice is hand weeded, usually twice, and chemical control of diseases or insects is nil in all systems. After draining the fields at the end of the monsoon rains, the crop is harvested with sickles, threshed by flailing on stones, winnowed in the wind, and dried in the sun.

Despite these very broad similarities in the techniques of rice production, substantial variation occurs with respect to the rice variety farmers plant. Analysis of choice of rice variety illustrates how farmer decisions are the result of an interaction between environmental conditions, farmer resources, and farmer objectives. There are six principal rice varieties that are grown in the three systems. The importance of these varieties differs both by systems and by resource endowment. *Taichin*, a newly introduced fertilizer-responsive dwarf variety, is most important in the midaltitude and ridge systems, especially among small farmers. *Pokhareli* is a long-grained, highly preferred and high-priced variety, popular among commercial farmers throughout the three farming systems. *Thapachinia*, *kathe*, *mansura*, and *chote marsi* are traditional varieties in the region, and are grown principally in the ridge and subtropical valley systems. Finally, *bhaidya* is a short-season variety grown only in double cropping in the subtropical valley.

Farmers' choice between the two recently introduced varieties—*taichin* and *pokhareli*—shows how wide a variety of factors enters into this decision. Resource requirements of the different varieties are particularly important (Pachico & Ashby, 1983). *Taichin* is much more difficult to thresh than other varieties, requiring 60% more total harvest labor than *pokhareli*. The rice harvest is a peak period of labor activity, and rapid turnaround time to prepare rice plots for the subsequent wheat crop is a critical factor in wheat yields. As a result, farmers who cannot readily supply the extra labor needed for hand threshing of *taichin* prefer to plant *pokhareli*. Furthermore, because *taichin* is more responsive than other varieties, average fertilizer dosages and cash costs are greater for *taichin*.

Total cash costs for growing *taichin* are higher than for *pokhareli*, and returns to variable costs are lower. Returns to variable capital are 226% for *pokhareli*, considerably above the 172% return obtained with *taichin*. As a result, *pokhareli* is more attractive than *taichin* for largeholder farmers who are primarily concerned with obtaining high returns to their more limiting factor of labor.

In contrast, *taichin* is more attractive to smallholder farmers who have plentiful family labor. *Taichin* actually yields higher returns per unit of land than *pokhareli*, a crucial consideration on small farms where land is the principle limiting factor. This is particularly compelling for small rice farms where a large proportion of the crop is for consumption. For smallholders it is appropriate to compare the two varieties in a subsistence budget with the assumptions that all labor is provided by the farm family, that fertilizer is the only cash cost, and that enough of the output is sold to meet fertilizer costs while the rest is milled and reserved for home consumption (Table 7-6).

Although *pokhareli* is superior as a commercial variety because of its higher price and lower fertilizer costs, *taichin* is far superior as a food staple. *Taichin* produces 50% more food because of its higher yield and higher conversion rate from paddy to milled rice. Consequently, for smallholders producing for their own consumption and for whom maximizing returns to land is crucial, *taichin* is a preferable variety. Therefore, the pattern of resource endowments and farmer objectives make *taichin* attractive for smallholder, subsistence-oriented farmers, whereas *pokhareli* is more suited for larger and more commercial farms. This pattern is reflected in the importance of *taichin* in the midaltitude and ridge systems, while in the subtropical valley, *pokhareli* is preferred.

The traditional rice varieties also present farmers with a range of different characteristics, which are selected according to agronomic, social, cultural, and economic needs. Of the traditional varieties, *kathe mansura* and *chote marsi* are better adapted to lower altitudes where cold is less of a problem. *Chote*, the highest yielding of the traditional varieties, is popular among all farmers with low-lying rice terraces in the ridges and subtropical valley. Smallholder farmers grow it because it yields more food per hectare than other traditional varieties, while largeholders grow it to feed their workers. *Kathe mansura* is lower yielding but has a better grain quality, so it is a traditionally preferred eating variety among largeholders who can afford to sacrifice yields in order to have a higher-quality, higher-status variety. *Kathe*, however, is being displaced by *pokhareli*, which has better yields and a shorter growing season.

TABLE 7-6. Subsistence budget including costs and receipts for two rice varieties, midaltitude valley.

Output	Taichin	Pokhareli
Unmilled rice, kg/ha	4,707	3,628
Sales to meet cash costs of production, kg/ha	1,120	579
Net unmilled rice, kg/ha	3,587	3,049
Net milled rice, kg/ha	2,045	1,433
Net milled rice, kg/dollar cash cost	0.15	0.20
Net milled rice, kg/day family labor	5.10	3.66

Source: Pachico (1980).

Another traditional variety, *thapachinia*, has suffered a major decline in popularity and has been largely displaced by *taichin* and *pokhareli*, since these newer varieties are superior in both subsistence and commercial terms. It has, however, maintained a role in rice production because it matures 3 weeks earlier than other varieties, in time for the traditional fall harvest festival, at which rice must be served. Most largeholders plant some *thapachinia* for ceremonial purposes, but smallholders cannot afford to dedicate an entire plot to it because of low yields. Instead, they grow a band of *thapachinia* around the perimeter of a field sown to another variety. The *thapachinia* is harvested early for the festival, and a trench is dug where the *thapachinia* was in order to drain the fields to facilitate harvest of the major variety. This practice illustrates how farmers of different resource bases exploit the particular characteristics of different crops or cultivars in order to meet a complex set of objectives that are both economic and social in nature.

Wheat

After the rice harvest, wheat is the main winter crop planted on *khet* or lowland rice plots, accounting for nearly all winter *khet* in the midaltitude valley and about three-quarters of *khet* in the other systems. In the ridge system, about one-third of *khet* is left fallow in winter after rice is harvested because of insufficient water and distance from the farmstead. Ridge fields are low yielding, and theft from the unsupervised plots is a risk. In the subtropical valley, winter wheat competes with highly profitable potatoes, but it is restricted by incompatibility with triple cropping.

As with rice, wheat production has been heavily influenced by technical change. The high yielding dwarf Mexican wheat varieties have had an important impact. Average wheat yields in the midaltitude valley have been raised to nearly 2 t/ha, almost twice the national average. Achievement of these high yields is dependent on high levels of chemical fertilizer application (averaging 70 kg of nitrogen/ha in the midaltitude valley) and also on irrigation, which is especially crucial. Adoption of the Mexican wheats has been practically universal where there is good winter irrigation. In the valley systems, wheat is grown only on winter *khet*.

In contrast, wheat is grown on both winter *khet* and *bari* in the ridge system. While some two-thirds of the area of *khet* wheat in the ridge system is planted to the new varieties, less than half is planted to the new varieties on *bari*. The same farmers are planting traditional varieties on *bari* and new varieties on *khet*. Insufficient awareness of the new varieties does not limit their use in the ridge system. Rather, farmers have learned through experience that the new varieties express their yield advantage only in conditions of good moisture and high fertility. On the poor, unirrigated, and upland *bari* soils, traditional varieties remain superior. In the ridge system, the use of the new varieties of both wheat and rice is relatively restricted to *khet* because of insufficient irrigation and capital constraints. The high yields of the new wheat varieties depend on high fertility conditions, but chemical fertilizer use is severely constrained in the ridge system because of the high cost of transportation.

Another decision that wheat farmers face is whether or not to intercrop mustard. Some 20% of farmers in Sanga intercrop mustard (rape, *Brassica* spp.) as an oilseed crop in wheat. Mustard seed is broadcast along with wheat seed, and the mustard is harvested by uprooting it about a month before the wheat is harvested. Most farmers indicate that they believe the mustard adversely affects wheat yields. The main reported cause appears to be damage to wheat during the harvesting of mustard. Since care is required to minimize this damage, mustard harvest is best done by family labor. Farms with intercropped mustard tend to be smaller than average. Some farmers believe that mustard competes with wheat, especially for fertilizer, and also shades it.

Mixed cropping of mustard and wheat also occurs in the ridge villages, where almost all *bari* wheat is cultivated with mustard. Very few ridge farmers grow mustard with wheat on *khet*. On the more fertile and moister *khet* soils, wheat is more densely planted and high yields can be obtained. Lower yields occur from intercropping mustard on *khet* than on *bari*. Ridge farmers can meet their subsistence needs for oilseeds on their *bari* without suffering as great a loss in wheat as they would if they intercropped mustard with their lowland wheat.

Farmer decisions with respect to their wheat crop illustrate that the interaction of physical and socioeconomic factors influences how farmers adapt their agricultural practices to their particular situation. In the valley systems, where irrigation is adequate and where farmers participate in a high cash flow agriculture that facilitates fertilizer purchases, the new wheat varieties dominate. In the ridges, where irrigation is poorer, capital scarcer, and fertilizer more expensive, the traditional varieties persist. Similarly, the farm resource base is important in determining whether mustard will be intercropped with wheat. Subsistence-based, small farms are more likely to use this practice.

Lowland Cropping Rotations in the Subtropical Valley

Multiple cropping through the winter season is a possibility in the subtropical valley because winters are frost-free and irrigation is available. There are several alternative multiple cropping rotations, but rice–wheat is the main one for farms of all sizes, followed in importance by rice–potatoes.

An analysis of farmers' decision to plant potatoes illustrates relationships between the physical resources available for agriculture, the social system affecting the availability of labor, and the crop's economic characteristics. Potatoes are integrated into different rotations on farms of different sizes. On small farms, three-quarters of potatoes are grown in a three-crop rotation, the most important of which is rice–potatoes–maize. On large farms, potatoes are generally grown in the double crop of rice–potatoes.

Soil, water, and labor each play a role in the selection of crop rotations with potatoes. In the three crop rotations, turnaround time between crops is tight and may be a limiting factor. Land preparation is the major constraint on a rapid turnaround from one crop to the next. Heavy soils require more time to dry and more effort to till; hence, triple cropping is not possible, and farmers leave the land in a winter fallow. Only where irrigation is adequate and soils light enough to dry rapidly and be tilled easily can farmers use three-crop rotation. Smallholder farmers have more labor available and are therefore more likely to achieve the rapid turnaround essential for successful multiple cropping. Although largeholder farmers may be able to hire labor, the amount they need is not always available at the right time.

High labor inputs for potato cultivation commonly exceed the capacity of the farm family, unless potatoes are grown on only part of the lowland. To some degree, hired labor can supplement family labor; but even without the cost of hiring labor, potato production requires relatively high cash inputs. High cash requirements for seed, fertilizer, and labor not only may surpass the financial resources of small farms but also may involve considerable exposure to risk. Potato production is risky from several standpoints. Seed and fertilizer must be purchased, and over 50% of the high labor requirement for potatoes occurs at the land preparation and planting stage. Farmers, therefore, have a substantial initial investment that can be jeopardized by crop failure due to insects, disease, pests, or failure of irrigation. Potatoes are also risky because they are a market rather than a subsistence crop. If farmers face low prices or are unable to sell their potato output, they can neither consume it nor store it for a favorable market. With available technology, potatoes will not keep in the hot, humid climate that prevails in the subtropical valley in the months following potato harvest. Thus farmers must bring their potatoes to market immediately after harvest and sell them for what they can obtain.

The severe problems involved in marketing potatoes are a major reason that potato cultivation remains only a minor commercial enterprise. The subtropical valley has an advantage in its warm climate and irrigation; farmers there harvest a crop of potatoes earlier than in the midaltitude valley. The valley is also closer to the Kathmandu market than are other early-season potato producers in the Terai. Consequently, subtropical valley farmers can bring potatoes to market when they are scarce, receiving a price 40% higher than that received by midaltitude farmers who come to market a month later. Lower potato prices for farmers in the midaltitude valley diminish commercial production there, and high transportation costs from ridge villages virtually eliminate commercial potato production in those villages.

Farmers' decisions whether to plant potatoes in the subtropical valley must be considered in the context of their winter crop alternatives of wheat, early paddy, and lowland maize. Potatoes earn the highest net returns/ha but require high cash outlays. Potatoes also have the highest labor requirement of all crops. Wheat remains a very attractive alternative because it has the lowest capital and labor requirements, faces less market risk, and can be used for home consumption.

Early paddy is similar in terms of high input and high returns/ha. Its use is restricted by its high demand for irrigation. It also requires a high labor peak at harvest, which coincides with the time of transplanting the main monsoon rice crop. It is, however, a more preferred food and a more profitable enterprise than winter maize. Winter maize is grown principally on small farms in a three-crop rotation of rice–potato–maize. Maize, which does not require as much capital or labor as early rice, readily fits into a triple crop rotation. Farmers with ample labor and irrigation may take a spring crop of rice. This is a new practice in the area, which, though currently relatively unimportant, may become increasingly popular. Spring maize is grown mainly by smallholder farmers in triple cropping systems. Wheat remains the most important winter crop and the least resource-intensive alternative.

Maize-Based Upland Cropping in Midaltitude and Ridge Systems

Whereas the lowland cropping systems are integrated into the market with sales of rice and wheat, as well as potatoes in Panchkal, upland cropping systems tend to be subsistence-oriented. Very little of the most important upland crop, maize, is marketed. Its role in providing household food sets some distinctive parameters for farmers' decision making about the management of upland cropping systems, especially when the weight given to subsistence versus cash crops is compared among farms of different sizes.

Farmers choose from an array of maize varieties, maize–millet and maize–soybean associations, and maize–wheat rotations with soybean–mustard intercrops or maize–legume intercrops. The choice of the maize variety hinges on the interaction between maize and the main associated upland crops, soybeans and finger millet. There are three major types of maize grown: *Kumaltar, seto*, and *pahelo chepto*. *Kumaltar* is a new variety that has been introduced by the government agricultural extension service. Farmers know it is high yielding but requires high amounts of chemical fertilizer to realize its yield potential. *Kumaltar* plants are larger but require a longer growing season than local maize varieties. *Pahelo chepto*, yellow corn, is said by farmers to yield well, though less so than *kumaltar*. *Pahelo*, preferred for the quality of its cornmeal, is grown on the more fertile upland soils. *Seto* matures earlier than the other varieties and is used for popcorn, an important snack for field workers. Yields average 3,089 kg/ha for *kumaltar*, 2,250 kg/ha for *pahelo*, and 1,701 kg/ha for *seto*.

Different patterns of new-variety adoption appear in different farming systems and reflect needs and preferences. In the midaltitude village of Sanga, as farm size increases, the portion of area planted to *kumaltar* increases. Conversely, as the size of farm decreases, the percentage of area planted to *seto* rises. *Seto* is cultivated on poorer soils and is the most popular variety among smallholder farmers.

This preference is related to the importance of millet to smallholder farmers. In Sanga, as in widespread areas in the hills of Nepal, finger millet is commonly relay-cropped with maize. Millet is sowed in seedbeds and then transplanted, like rice, at the time of the second maize weeding. Men turn the soil with a *kodali*, while women transplant the seedlings. Around the end of November, millet is harvested by women cutting the seedheads off of the plant. Later the straw is harvested as an animal feed.

A large percentage of small subsistence-oriented and upland farms grow millet; the crop is less important on large commercial farms. This is the inverse of the pattern of use of the maize *kumaltar*. Millet cannot be grown in relay with *kumaltar*, which stunts the millet, both because of incompatible growing seasons and because it is taller with broader leaves than the traditional variety, *seto*. Finger millet and *kumaltar* are mutually exclusive alternatives.

Largeholder farmers do not bother to plant millet because of the advantages of cultivating soybeans intercropped with maize. Soybeans are usually planted a day or so after the maize. Women dibble the soybean seed into the soil, two or three seeds together, about 50 cm apart. Soybeans typically ripen 1–2 weeks after maize. They are harvested by uprooting the entire plant and threshed by beating with sticks. Larger farms, and those on which labor is a limiting factor, prefer the maize–soybean system, which gives higher returns to labor even if labor must be hired. But for subsistence farms with small landholdings and ample family labor, the maize–millet relay crop is preferable to the maize–soybean intercrop since it produces more food/ha. In the ridge farming system, however, smallholder farmers choose to grow maize in association with soybeans in preference to millet because this combination maximizes food production.

This explains why the relationship between the maize *kumaltar* and small farms in the valley does not occur in the ridges. Farmers in the ridges choose from the same group of maize varieties as in the midaltitude valley. The improved variety, *kumaltar*, is more common because maize is more important to food supply and millet and soybeans play a different economic role in the ridges than in the valleys.

Soybeans make up about 12% of the total value of crop output from small farms in the ridges, making it their third most important crop after rice and maize. Soybeans are also the most important cash crop since most rice and maize are consumed at home. Most important, soybean cultivation in association with maize permits the ridge farmers to procede with a subsequent crop of winter wheat intercropped with mustard on upland fields. They may be related to soybeans' nitrogen fixing ability. If, however, farmers grow millet relay cropped with maize, as in the midaltitude valley, they cannot plant wheat.

After the millet crop is harvested there is insufficient soil moisture to support a subsequent crop. Thus millet is usually cultivated only where soils are too light and moisture retention too low to permit the cultivation of a crop of wheat.

In the subtropical valley system, as in the other systems, maize is the principal monsoon *bari* crop, and *kumaltar* is grown on two-thirds of these fields. This preference reflects the commercial orientation of farmers. A wide array of alternative intercropping combinations accompany upland maize which are compatible with growing *kumaltar*. Some 80% of farmers reported intercropping soybeans and chickpeas (*Aicer arietinum*) with maize. A minority of farmers also grow black gram (*Phaseolus mungo*), horse gram (*Dolichos uniflores*), or cowpeas (*Vigna unguiculata*). The complexity of the intercropping systems resulting from these combinations is illustrated in Table 7-7, which depicts the practices used by a sample of 12 farmers. Each line in the table represents the combinations of practices used by an individual farmer. No two are identical.

Farmers face a number of decisions by intercropping with maize. First, since almost all farmers intercrop soybeans, either a local or a new "Chinese" variety, must be chosen. Soybeans are planted in one of three ways. The most common is to plant the seeds two or three per hill by hand, a day or so after maize planting. Broadcasting prior to maize planting is also common. One farmer planted his soybeans in a furrow behind an ox plow team, the same method used in maize planting. Finally, the farmers also choose among other crops to intercrop with maize.

TABLE 7-7. Intercropping practices with maize in subtropical valley system.

Local soy variety	New soy variety	Furrow plant soy	Hill plant soy	Broadcast soy	Black gram	Chickpeas	Horse gram	Cowpeas	Millet
1[a]	0	0	1	0	0	1	1	0	0
1	1	0	1	1	1	1	1	0	0
0	1	0	1	0	0	1	0	0	0
0	1	0	0	1	0	1	0	0	0
0	1	0	1	0	0	1	1	1	0
0	1	1	0	0	0	1	0	0	0
0	1	0	0	1	0	0	0	1	0
0	1	0	1	0	0	0	0	0	0
0	1	0	1	0	0	1	0	0	1
0	0	0	0	0	0	1	0	0	1
0	0	0	0	0	0	1	0	0	1
1	1	0	0	1	0	0	0	0	0

Source: Pachico (1980).

[a]Each line represents the practices used by one farmer. 1 = farmer uses practice; 0 = farmer does not use practice.

This array of choices leads to a diversity that is only partly explicable in economic terms. For example, soy is broadcast to conserve labor. The need for additional food leads a few farmers to relay crop millet with maize. Some of the differences in the growing of minor crops, however, are related to personal tastes and preferences. There is no consensus about exactly which combinations are superior, and there is too much diversity in these systems to identify "best practices." Each farmer makes his own artful judgment as to which practice is best.

Livestock

Livestock forms an integral part of hill farming systems, although its role differs substantially depending on the resource base of individual farmers and the relative importance of subsistence and market oriented farmers and the relative importance of subsistence and market oriented activities (Shrestha & Evans, 1984). The role of livestock in the ridge system is typical of the Himalayan farming system in Nepal. Bullocks provide the traction for soil preparation and the recycling of nutrients through the animals is the principal means of maintaining soil fertility. Crop residues are returned to the fields as manure, and nutrients are transferred from grazing lands to crop land. The harvesting of tree foliage as fodder for livestock is common practice, and thereby nutrients are cycled from forests to cultivated land. While this is a sound strategy for an individual farmer, the overgrazing of communal pastures and the over exploitation of the remaining forests are now leading to severe problems of erosion. Major land slips sometimes destroy large areas of rice terraces and upland fields. For the individual livestock owner, the optimal strategy is to maximize off take from the common lands to feed his livestock and manure his fields. In the aggregate, however, this individually rational strategy results in ever more severe problems of environmental degradation. Planners concerned with preserving the Himalayan watershed view the overstocking of livestock as a major problem.

Besides providing traction and compost, livestock also make an important contribution to the diet, especially through milk and yogurt. Animals also provide an occasional source of cash for many farmers through local sales of both stock and milk products. In the ridges cattle are important for traction and milk. Many ridge farmers, however, are only able to keep small ruminants, especially goats, as they represent less capital and are easier to feed.

The livestock situation in the midaltitude valley is quite different from that in the ridges. Due to higher population density, small farms, and the disappearance of communal pastures and forests, many farmers are unable to keep bullocks for traction. Buffalo are not used for traction. Since the early 1970s this village has been selling fresh buffalo milk for the urban Kathmandu market, and for many farms this has become a major source of cash.

Raising buffalo is labor and cash intensive in this midaltitude valley. Because of insufficient pasture or forest, all buffalo are stall fed. Rice straw is the biggest component of the diet, supplemented by millet straw, small quantities of green corn stover, and grass from bunds or path verges. In the winter, green fodder is scarce, and weeds are cut (so they can regenerate) in irrigated wheat fields. Grass cutting is laborious, requiring 4 to 6 hours per day. Especially during lactation, buffalo rations are supplemented with maize, millet, or soybean flour, or with oilseed cake. These items are often purchased. Smallholder farmers (less than 0.05 ha), do not produce sufficient rice straw to keep buffalo, so 90% of them with buffalo buy rice straw from larger farms.

The high labor intensity of raising buffalo is unattractive to largeholder farms. Smallholders face capital constraints that limit their ownership of buffalo; the average cost of a she buffalo is roughly equal to their average annual total income. Farmers with medium-sized holdings are the principal entrepreneurs in milk buffalo, and their farms represent the small commercial type of farm whose principal source of cash is its livestock. They tend to have both sufficient family labor and enough land to care for and feed buffalo.

The midaltitude valley is exceptional in its access to markets, which makes livestock profitable. It provides an example of how large numbers of livestock can be maintained by moving to stall feeding. This generates cash and manure for cropland and alleviates the problem of erosion. Although this system provides one means of impeding environmental degradation in the ridge system, the difficulties of achieving this solution are numerous. First, the existence of good transportation and access to markets, which is essential, cannot be easily or rapidly achieved for many remote villages. Second, farmers in the midaltitude valley made the transition to stall feeding only when no alternative existed. Insufficient pasture and forest forced them to move to the more intensive system. It is by no means clear that they would have made the transition without this push.

Productivity Comparisons

One approach to assessing productivity differences across farming systems is through production function analysis. This technique, derived from neoclassical economic theory, measures the statistical relation between inputs to the production process (e.g., land and labor) and the output. The marginal productivity of inputs can be calculated from production functions, which in turn can be compared both across farm systems to see which are the most productive. Marginal productivities can be compared with input costs as a test of farm efficiency, since the equivalency of marginal productivity with cost is a condition of efficiency.

In this study, a Cobb–Douglas production function is used (i.e., linear in the logarithms of variables), and it is estimated by means of ordinary least squares regression. Because important differences in technologies exist between the three farming systems, separate functions are estimated for each farm system. In all cases the total value of crop production is determined by the area cultivated, the quantity of family labor active in agriculture, the quantity of hired labor, and inputs of nutrients. In all three systems nutrients are measured by kilograms of chemical nitrogen; in the ridge system, a separate variable for quantity of manure is also included. When the composite variable was included in the equations for the two valley systems where chemical fertilizer use is more important, manure did not make a significant contribution to output, so it was deleted from the equations in the valleys (Table 7-8).

The marginal productivity of resources is computed for each of the farming systems (Table 7-9). Land has the highest return in the midaltitude valley, U.S.$453/ha, while the lowest returns to land prevail in the ridge valleys, U.S.$199/ha. These results are reasonable since the valley land is generally more fertile, and the levels of other complementary inputs are higher in the midaltitude valley than in the ridges. Comparing the midaltitude valley to the subtropical valley, it can be seen that in the former the productivity of land is higher and the productivity of labor lower. This result seems logical because the productivity of any resource depends in part on the levels of the other resources used. Holding one resource constant, such as land, as the usage of other resources increases, the marginal productivity of land will be increased. Thus the high productivity of land in the midaltitude valley is due to the fertility of the soil and also the high levels of labor and fertilizer that are used in conjunction with land.

Similarly, the productivity of hired and family labor are highest in the subtropical valley. This is consistent with the relatively more extensive labor that accompanies the larger average farm sizes in this valley. Differences in

TABLE 7-8. Whole farm production functions for three farm systems in Nepal (all variables in logs; standard errors in parentheses).

Variables	All farms	Midaltitude valley	Ridges	Subtropical valley
Land	.63	.69	.52	.83
	(.04)	(.08)	(.05)	(.09)
Family labor	.13	.08	.10	.08
	(.05)	(.06)	(.07)	(.12)
Hired labor	.04	.03	.04	.08
	(.01)	(.01)	(.02)	(.02)
Nitrogen	.12	.19	.06	.03
	(.01)	(.05)	(.02)	(.02)
Compost			.19	
			(.04)	
Constant	5.92	5.82	5.20	5.56
R^2	.72	.83	.72	.76

TABLE 7-9. Marginal productivities calculated from whole farm production functions, three farming systems, Nepal (U.S.$).

Variables	Midaltitude valley	Ridges	Subtropical valley
Land ($/ha)	453	199	296
Family labor ($/yr)	19.7	7.5	10.2
Hired labor ($/day)	0.66	0.58	1.0
Nitrogen ($/kg)	1.0	2.0	0.8

technology and the use of other inputs aside, the comparative returns to land and labor between the midaltitude and the subtropical valley should not seem surprising given the differences in person/land ratios that exist between the two systems. In the midaltitude valley, where more labor is used per hectare of land, the productivity of land is relatively high and the productivity of labor low when compared with the subtropical valley. On the other hand, in the subtropical valley, where there is less labor per hectare, the productivity of labor is greater and that of land less than for the midaltitude valley.

The productivity of land and labor for the ridge villages is in all cases lower than in either of the two valleys. These low estimates of the returns to labor as well as land in the ridge villages reflect the relative poverty of the farmers on the ridges, where the land is less fertile and the farm sizes small. There the person/land ratio is as high as in the midaltitude valley, but the land is generally poorer and there is less use of complementary inputs such as fertilizer and new crop varieties.

Small-Farm Development in Hill Farming Systems

This analysis of Himalayan farming systems in Nepal has focused on the primary concern of the majority of farmers: how to ensure subsistence given the physical, economic, and cultural resources at hand. The comparison of ridge and valley systems demonstrates how the penetration of market forces into the subsistence economy interacts with the traditional modes of resource allocation in the cropping system and with regard to livestock. Increasingly, ridge farmers find it more difficult to sustain traditional practices in the face of changes that have depressed local trade, depleting grazing and forest resources, and diminished their capacity to produce enough food for a growing population. Their solution is to seek employment outside their community. By contrast, in the midaltitude valley system, the internal pressures of population increase on scarce land resources have led to intensification of traditional production techniques. These were fortuitously compatible with technical innovations in rice and wheat varieties. The development of markets for milk was another response to population pressure. Yet a critical component of small farms' economic survival has been access to nonfarm employment, which enables smallholders to participate in a high-cash-flow system of agriculture. The subtropical valley system, by comparison, apparently can continue to intensify through technological innovation in agriculture.

As integration into the market economy increasingly affects their lives, hill farmers bring new objectives to management of their farms that go beyond provision of subsistence. Such objectives include educating their children; obtaining employment in the towns; and gaining access to clean piped water, Western-style medical care, and consumption of manufactured foods. Each of the three farming systems discussed in this chapter represents a facet of the future of the hill farm and of the ways in which these farmers will approach such objectives.

References

Ashby, J. A. *Small farms in transition: Changes in agriculture, schooling and employment in the hills of Nepal*. Ph.D. dissertation. Ithaca, NY: Cornell University, 1980.

Ashby, J. A. Technology and ecology: Implications for innovation research in peasant agriculture. *Rural Sociology*, 1982, 47, 234–250.

Banskota, M. *The Nepalese hill agro-system: A simulation analysis of alternative policies for food production and environmental change*. PhD dissertation. Ithaca, NY: Cornell University, 1979.

Blaikie, P., Cameron, J., & Seddon, D. *The effects of roads in west central Nepal: Part I: Summary*. Norfolk, England: University of East Anglia, 1977.

Bista, B. P. *The people of Nepal*. Kathmandu: Ministry of Information, 1967.

Calkins, P. H. *The impact on income, employment and nutrition of developing horticulture in the trisuli watershed, Nepal*. PhD thesis, Cornell University, 1976.

Caplan, L. *Land and social change in east Nepal: A study of Hindu tribal relations*. London: Routledge and Kegan Paul, 1970.

Doyle, C. J. Productivity, technical change and the peasant producer. *Food Research Institute Studies*, 1974, 13(1).

Feldman, D., & Fournier, A. Social relations and agricultural production in Nepal's Tarai. *Journal of Peasant Studies*, 1976, 3, 447–464.

Geertz, C. *Agricultural involution: The process of ecological change in Indonesia*. Berkeley: University of California Press, 1963.

Guillet, D. Toward a cultural ecology of mountains: The central Andes and the Himalayas compared. *Current Anthropology*, 1983, 24, 561–567.

Hagan, A. R. *The agricultural development of Nepal*. Columbia, MO: Agricultural Experimental Station, University of Missouri, 1976.

Harwood, R. *Small farm development: Understanding and improving farming systems in the humid tropics*. Boulder: Westview Press, 1979.

Lipton, M. The theory of the optimizing peasant. *Journal of Development Studies*, 1968, 4, 327–351.

MacFarlane, A. *Resources and population: A study of the Gurungs of Nepal*. Cambridge: Cambridge University Press, 1976.

Mathema, S. R., & VanderVeen, M. G. *Socioeconomic research on farming systems in Nepal*. Kathmandu: Ministry of Food and Agriculture, 1978.

Messerschmidt, D. A. *Social status, conflict and change in a Gurung community of Nepal*. Ph.D. dissertation, Oregon University, 1974.

Ministry of Food and Agriculture, H. M. G., *Nepal. Farm management study in the selected regions of Nepal 1968–9*. Kathmandu: His Majesty's Government, 1971.

Norman, D. W., & Palmer-Jones, R. W. *Economic methodology for assessing cropping systems* Paper presented to the Symposium on Cropping Systems Research and Development for the Asian Rice Farmer, International Rice Research Institute, Los Banos, Phillipines, 1976.

Pachico, D. H. *Small farmer decision-making: An economic analysis of three farming systems in the hills of Nepal*. PhD dissertation, Cornell University, 1980.

Pachico, D., & Ashby, J. A. Stages in technology diffusion among small farmers: Biological and management screening of a new rice variety in Nepal. *Agricultural Administration*, 1983, 13, 23–37.

Sahlins, M. *Stone age economics*. Chicago: Aldine-Atherton, 1972.

Scott, J. C. *The moral economy of the peasant*. New Haven: Yale University Press, 1976.

Shrestha, R. L. J., & Evans, D. B. The private profitability of livestock in a Nepalese hill farming community. *Agricultural Administration*, 1984, 16, 145–158.

Simon, H. A. Theories of decision making in economic and behavioral sciences. *American Economic Review*, 1959, 49, 253–283.

Yotopoulos, P. A. The population problem and the development solution. *Food Research Institute Studies*. 1977, 16(1), 1–13.

8

Mechanized Irrigation in Upper Egypt: The Role of Technology and the State in Agriculture

NICHOLAS S. HOPKINS

Egyptian agriculture combines one of the most carefully man-made landscapes in the world: a pattern of large, stratified villages, a long history of government intervention in the practice of agriculture, and a sense of dynamism and rapid change. In this chapter the farming system of Upper Egypt is examined primarily through the case of one large village, Musha, in 1980–1981.[1] The emphasis is on the contemporary farming system, although some attention is also given to trends in irrigation and mechanization.

This analytic description of the farming system is placed in the geographical and ecological setting of one village. Particular emphasis is given to the social organization of work in agriculture, both within the household and, more important, outside the household, in the form of hired labor, rented machinery, and sale of produce. This labor process takes place within a particular technological context, here marked by a distinctive irrigation system and a pattern of widespread but not universal mechanization.

One of the most striking features of Egyptian agriculture throughout its history is the role of the state. Since the 1952 revolution this role has grown, and the state is now the silent partner of each farmer. Not only has the state created the infrastructure for agriculture, from irrigation canals to land tenure rules, but, through a cooperative system, it advances operating credit in the form of inputs to the individual farmer and attempts to specify crop area, seed type, and the destination of the harvest. Hence, any analysis of Egyptian agriculture must consider the role of the state and its ability to extract surplus and direct the processes of change.

Egyptian Agriculture and Irrigation

The Nile Valley and Delta have been farmed intensively since Pharaonic times, using the basin system of irrigation (Butzer, 1976; Hamdan, 1961). Upper Egypt consists of a narrow strip of land on one or both sides of the Nile from

Nicholas S. Hopkins. Department of Sociology, Anthropology, and Psychology, American University in Cairo, Cairo, Egypt.

Aswan to Cairo, which was irrigated according to the basin system. Lower Egypt is equivalent to the Nile Delta. Here the river spreads out into a series of channels, whose floodwaters were also controlled using the basin system until the early to mid–19th century.

The current farming system is of recent origin, having been created in the late 19th and early 20th centuries by a combination of Egyptian, French, and British engineers (Barois, 1889; Willcocks, 1889). The state—precolonial, colonial, and postindependence—played the principal role in the transformation of the irrigation system (Brunhes, 1902). The original system was based on control of the annual flood of the Nile through a system of basins combining dikes and sluices designed to maximize the effect of inundation and silt deposits. Today, water is completely supplied through a network of canals. Under the former system, water was supplied only once a year, and only one crop was possible. Under the present system, water is supplied throughout the year, and double cropping is the norm.

The first person to envision this change was Mohammed Ali Pasha, ruler of Egypt from 1805 to 1848, who wanted to transform Egyptian agriculture for large-scale cotton production. A new irrigation regime was required because cotton demands much water during the summer months when the river is normally low, and because it cannot be harvested before the annual flood. The solution was to regulate the flow of water by barrages at the head of the delta, just north of Cairo. The barrages were not completed until the 1860s, and then imperfectly. Nevertheless, various makeshift devices made it possible to expand cotton cultivation in the delta. The technical changes during this period involved the construction of a network of canals that would contain water most of the year simply because they were deeper than the level of the river water as it declined, and of barrages to raise the head. This system, combined with various lifting devices such as the steam pumps that were common by the 1870s, made it possible to introduce double cropping and cotton cultivation in the delta (Rivlin, 1961). The British were able to perfect the delta barrages after taking over responsibility for the Egyptian irrigation system in 1882. Later they built other barrages in Aswan, Asyut, and elsewhere, that contributed toward sustaining the water at a sufficiently high level to fill the canal network. This system remained essentially stable until the completion of the High Dam at Aswan in 1964.

The Upper Egyptian irrigation system began to change with the construction of the Ibrahimiya Canal in 1873. This canal took off from the Nile at Asyut and made possible the summer cultivation of cotton and sugarcane in large areas of Minya and Beni Suef provinces. For 30 years this canal simply took water from the Nile without any kind of barrage, and consequently its usefulness was fully dependent on the level of water in the Nile itself. The completion of the Asyut barrage in 1902 improved the Ibrahimiya by channeling more water into it. This change allowed the full width of the valley in Minya Province to be brought under canal irrigation.

In the remainder of Upper Egypt, south of Asyut, and including the village of Musha, the basin system of irrigation remained in force until the construction of the Aswan High Dam. There was no effort to modify the layout of the basins themselves, but there were projects to improve the sluice gates (e.g., by constructing them of masonry with metal gates), to improve the dikes themselves, and to improve the network of feeder canals (Clot Bey, 1840: II, 473–475; Ross, 1892).

Despite these improvements, the system still relied on the annual flood (Crary, 1949; Hurst, 1952: 38–46). The river rose in late summer, and by August or September the water level was high enough to justify opening the sluice gates in the dikes to allow it to enter the basins. The basins then filled to a depth of 1–1.5 m, variable in part because the floor of the basins was not level. After the water had been on the ground for 5–6 weeks, sluice gates at the lower end of the basin were opened, and the basin emptied. As the water receded, the farmers planted in the moist "ooze and slime." It was important not to let the water drain out faster than the farmers could plant. Since the land was too muddy for draft animals, planting was largely a matter of broadcast sowing. Only the winter crops could be grown, principally wheat and bersim. Summer crops were grown on limited areas that were raised above the level of the flood, such as the banks of the Nile itself, and were watered using mostly animal-powered lifting devices. Although some features of this basin system of irrigation lasted until the construction of the Aswan High Dam in 1964, the introduction of pumping of groundwater in the 1930s led to double cropping and other changes, as in the case of Musha.

Musha

The village of Musha is atypical of Nile Valley villages because it is relatively large, has many large landowners, and is relatively close to a big city (Figure 8-1). Also, education and white-collar employment are relatively common. Yet its agricultural problems are representative of Egypt as a whole. It shares with other Egyptian villages its basic household structure, its problems of adaptation to the introduction of mechanization and changing government policy, its alternatives for agricultural labor, and its incipient class structure (Bösl, 1984).

Musha is located within a broad ecological zone that stretches from Sohag to Asyut and corresponds to the former command basin of the Sohagiya Canal. The Nile Valley consists of three parallel ecological strips. The first is a raised bank along the river, which held back the flood. The villages on the bank's relatively sandy soil specialize in bananas, palms, and certain vegetables; residents also fish and trade their catch to villages away from the river. Along the desert edge are cemeteries, quarries, and monasteries. Many of the villages are built in the desert and were halting places for caravans from Kharga Oasis and beyond. Between the river and the desert are the basin villages, built on mounds raised above the old level of the floodwater. The topsoil is thick,

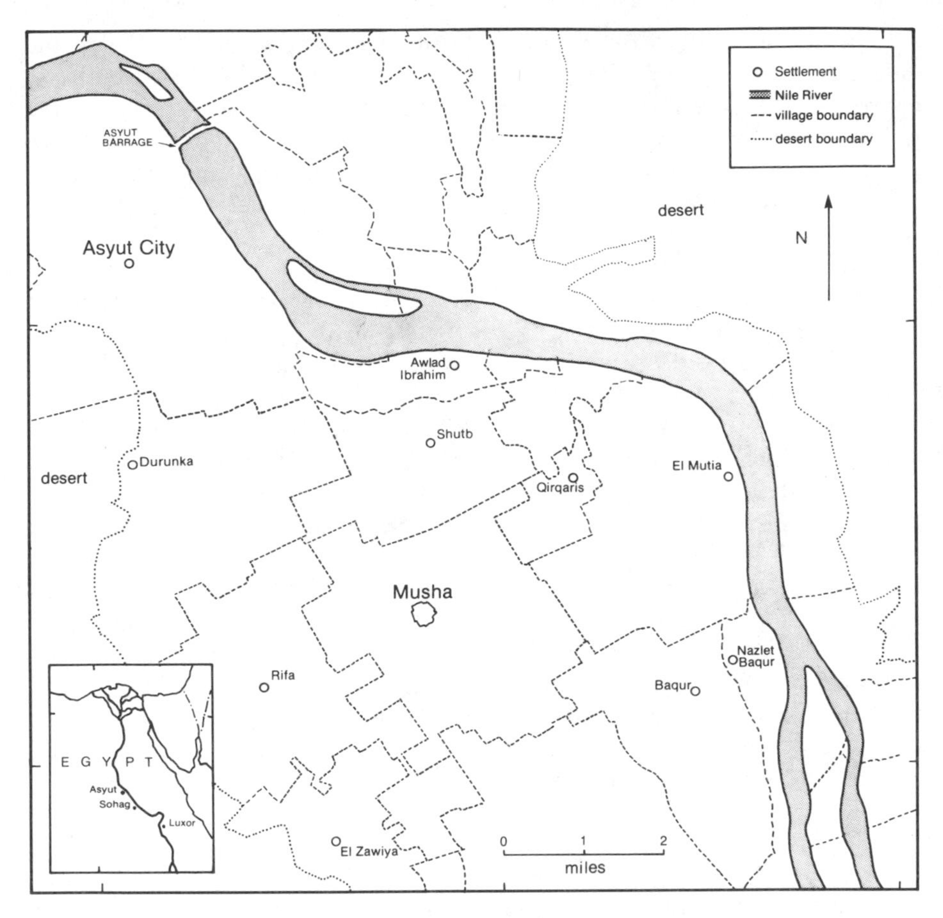

Source: After Bosl (1984:249). Reprinted by permission.

FIG. 8-1. The environs of Musha, showing village boundaries in the Zinnar Basin.

heavy, and dark, and agriculture predominates. Environmental differences between these three strips give rise to trade in agricultural products, and to labor flows, generally from the river and desert villages into the basin villages.

The village is just south of the city of Asyut, about 400 km south of Cairo. Its population is around 18,000, and the village land area is a little less than 2,000 ha (5,000 feddans). The bulk of the population live in the nucleated village, where the density is at least 60,000 persons/km². Nominally, each household possesses and resides in its own house. The houses are important for storage, for stabling animals and raising poultry, for certain work purposes, and for eating and sleeping.

The agricultural landscape is highly organized. The land is flat and, to an outsider, featureless. The village land is divided into 70 agricultural tax sections. The sections in Musha are usually oblong (Figure 8-2). The fields are laid out in strips, following the long dimension of the oblong. Both the sections and the strips are oriented either northwest–southeast or northeast–southwest. Some fields are as narrow as 15 m and as long as 1,000 m. The preference for strip fields may reflect the ecological conditions of basin agriculture where, if the land were not completely even, the most productive sectors might not be in the same place each year because of differential flooding. Long strips best

FIG. 8-2. Musha: Field layout.

Source: After 1905 cadastral survey.

equalize the chances of each farmer getting usable land for his crop (see Lyons, 1908, for details on field and village layout).

The general changes in Egyptian irrigation and land tenure systems during the 19th century also affected Musha. A number of events contributed to ecological and political change in the village. They included the improvement of the basin system of agriculture, the partial regularization of the water supply after the construction of the first Aswan Dam in 1902, the cadastral survey of 1905 confirming the private ownership of land, the introduction of steam and diesel pumping engines, the construction of the Aswan High Dam in 1964, and the switch from basin to perennial irrigation after the introduction of the mandatory cooperative system in the period around 1964–1965.

People, Land, and Agriculture

Demography

Between 1907 and 1927, there was a decline in Musha's population linked to male out-migration (Baer, 1982:76; Willcocks, 1889:247). After 1927, this demographic trend was reversed, and population has continued to rise since then. More men than women were recorded after 1927 (98:100 men in 1937, and 93:100 men in 1976). Although causality cannot be determined, the reversal of the population trend coincided with a marked increase in the productive forces of the village. In the early 1930s, the introduction of diesel pumps made double cropping possible and also allowed the introduction of cotton into the crop rotation. These two changes considerably increased the need for labor in the village and cut back on the seasonal migration of men to the north. On the other hand, very poor families, who could no longer survive in the village and sought a subsidized urban existence, continued to migrate. Education also contributed to migration.

More recently migration outside Egypt has become an issue. In Musha, 10% of the households sampled had a history of migration abroad. Extended to the population of Musha as a whole, this suggests that there were 250–300 Mushans working abroad in early 1981. Migrants were typically married middle-aged men from the lower-middle stratum of the population. It cost about U.S.$500 to migrate abroad in 1981. This limits the involvement of the poor because it constitutes about one year's income.

Sociologically, migration abroad resembles the old migration to the port towns to work as stevedores and porters in that the migrant goes alone and expects eventually to return enriched to his home community, where his wife and children remain as hostages. Migrants are gambling that a few years of effort and privation will raise them to a higher standard of living, to some kind of take-off point. Migration by men alone assures the receiving country that most will eventually return home, and it ensures that most money will be sent home to care for the family.

Land Tenure

Along with water and people, land is a basic resource for Egyptian agriculture. Land tenure in rural Egypt is primarily freehold tenure. Each parcel has an owner who retains the ultimate title. The owner has sometimes ceded use rights to another, via the mechanism of legal rental, which protects the tenant and his heirs against expulsion. More temporary use rights (the right to cultivate the land for one season, for instance) are also sometimes ceded to other individuals. Owners may trade land-use rights without affecting the underlying ownership, in order to create larger plots for more efficient cultivation. Within this pattern, there is enormous complexity in practice. One of the basic implications of the system is that many people have no access to land (roughly 45% of all households in Musha are "landless").

Three sets of records regulate access to land. A nationwide cadastral survey was carried out in Egypt under English auspices from 1898 to 1907, covering Musha in 1905. The records, based on tax sections, from this cadastral survey are still used to help settle land quarrels. These are the only records that link a certain owner with a given plot of land.

At the time the village cooperative credit society was set up in 1965, a second register was compiled, useful for inheritance and tax purposes. This register contains the names of the landowners in the village with an indication of the amount owned. It is based on a list of owners instead of an area of land and is keyed to the 1905 cadaster. To match it with actual parcels on the ground requires additional information.

A register of landholders is also kept at the agricultural cooperative. The holding units are a good proxy for farm enterprise size. There is only one holder for each 3.5 owners, evidence of considerable consolidation in control of land. The average holding is about 1.4 ha (nearly twice the national average). A *holding* is defined as land owned, plus land legally rented-in, minus land legally rented-out. In most cases, the holding includes both land owned and land rented. This list makes no effort to localize the land. It is in effect a record of shares held in the common village patrimony. The holding is the basis for determining both the responsibility for growing government-required crops and the amount of government-subsidized inputs a farmer is entitled to receive.

The government collects tax on all landownings larger than 1.3 ha. One local tax collector estimated that about 60% of Musha's land is taxed, yielding a tax revenue on the order of U.S.$60,000. The average tax is around $50/ha/yr. In theory the tax is collected from the owners, not the holders, but the holders may pay directly if the owner is no longer present in the village.

Musha is a village of largeholders. Some 15.3% of the landholders officially farm 51.1% of the land in holdings of 2.0 ha or more, and the 10 largest holdings cover 8.9% of the land. The concentration of land in Musha would be even more striking if actual farm enterprises could be counted instead of holdings, particularly since land held in adjacent villages is not covered in these figures. Musha has more largeholders and fewer smallholders than either the national averages or the results of our Ten Village Survey in 1982 (Table 8-1).[2] Despite this concentration of control over access to land, Musha has been relatively unaffected by land reform, and so typifies the village where a stratum of largeholders has been dominant throughout the socialist period.

Cultivars

The principal cultivars in Musha are divided into summer crops and winter crops. The main summer crops are cotton (*Gossypium barbadense*), maize (*Zea mays*), and sorghum (*Sorghum durra*). The main winter crops are wheat (*Triticum aestivum*), beans (*Vicia fava*), lentils (*Lens culinaris*), chickpeas (*Cicer arietinum*), and bersim (*Trifolium alexandrinum*). For more details, see El-Tobgy (1976). There are two fairly obvious changeover points in the year—in November for the switch from summer to winter crops, and in April for the reverse. The rotation cycle covers 2 years instead of the more common 3-year rotation (El Shagi, 1969; Richards, 1982). A cycle starting with cotton in the first summer is followed by wheat in the next winter and by maize, sorghum, or fallow in the second summer. The cycle is completed in the second winter with

TABLE 8-1. Landholding in Musha compared with the Ten Villages and with national figures.

A. Numbers of holders in each category

	Musha		Ten Villages		National
Holding size (ha)	*N*	%*N*	*N*	%*N*	%*N*
<0.4	201	14.2	2,805	36.5	39.4
0.4 to <1.2	718	50.8	3,886	50.5	40.7
1.2 to <2.0	277	19.6	694	9.0	12.4
2.0 and up	216	15.3	308	4.0	7.5

B. Area covered by holdings of a given size

	Musha	National
<0.4	2.5%	12.35%
0.4 to <1.2	25.7%	33.82%
1.2 to <2.0	20.8%	19.81%
2.0 and up	51.1%	34.04%

Sources: Musha: Cooperative list of landholders, 1979; Ten Villages: Cooperative lists of landholders, 1981; national: Harik (1979:39), from Ministry of Agriculture, Egypt.

bersim, lentils, and chickpeas. These crops are harvested relatively early to allow for timely planting of the cotton, to begin the new cycle. This rotation is to some extent imposed by the government (though less now than formerly), and in theory entire blocks of fields are given over to single crops. In fact, in 1981 only cotton was cultivated in imposed blocks.

In addition to the cycle crops, certain other cultivars are grown in Musha. Orchards are excluded from the cycle. Over 100 ha are given over to grapes, and some of the large farmers have orchards of pomegranates (*Punica granatum*) and other fruits. In addition, onions, peppers, watermelons, and small patches of vegetables such as *mulukhia* (*Corchorus olitorius*) are raised for home consumption. Like many villages in Egypt, however, Musha remains dominated by the open-field crops. The 1982 Ten Village Survey showed that the four most common crops were bersim (grown by 89.5% of the farmers), wheat (86.4%), cotton (80.9%), and maize (78.4%). Those who grew some combination of these four crops were 41% of the total sample surveyed.

Animal husbandry is extremely important for small farmers. The most common animal in Musha is the water buffalo, kept principally for its dairy products (fresh milk, cheese, *ghee*), followed by the cow, also used primarily for its dairy products. A survey of 236 smallholder households in Musha in 1979 showed the presence of 237 buffalo and 95 cows; altogether 150 of the households (64%) had at least one of these animals.[3] If landless households are included, those keeping at least one buffalo or cow are between one-third and one-half, and even more than that keep sheep, goats, or poultry.

The animals produce goods that can be consumed and sold, and are themselves the subject of speculation. Along with bread, cheese is the most common food in Musha; water buffalo is the only red meat commonly sold and eaten. In contrast to some areas of Egypt, no use of these animals for traction or other work is made in Musha. Some individuals deal in animals, trading for prices that are generally higher to the north than to the south, or bringing young animals home to fatten for resale. Animals are thus important in the economy of some households because they permit some degree of capitalization and profit making that is independent of the rigorous agricultural cycle.

Technology and Social Organization

Equipment

The traditional tool kit was very simple and reflected a reliance either on animal power or on human effort (Winkler, 1936). In the past, the basic all-purpose tool was the short-handled hoe, used to weed and to irrigate or to design field layout for irrigation. In addition, a small sickle was used for harvesting, and a digging stick was used for planting crops such as cotton. The traditional shallow plow was pulled by a yoke of cows. Flood irrigation required no special tools. Threshing was carried out with the help of a threshing sledge, pulled around in a circle by a pair of cows. Winnowers relied on the wind and used a winnowing fork and sets of sieves for the final cleaning. Transport was either by donkey (for small amounts or short distances) or by camel (for larger amounts and longer distances, frequently used to bring the crop in from the fields). Winnowing, camel transport, and sometimes plowing or threshing were carried out by specialists rather than by each household on its own behalf. Most tools were wooden and made by carpenters. Blacksmiths were rare and may not have been located in a village such as Musha; the metal pieces needed could have been brought in ready-made.

By 1978 the number of tractors in Egypt was estimated at 25,000 and rising rapidly. Mechanization has almost replaced animal power in agriculture. In Musha some camels and donkeys are kept for transport. Many tasks requiring human energy, however, are still done by hand. All 57 farmers surveyed in 1981 said they used machines at least some of the time, whereas 60% of this group used animals for some operations.

Although first introduced in the 1930s, tractors did not become widespread in Musha until after World War II, and they did not dominate field preparation until around 1965–1970. In June 1981 there were 48 functioning tractors in Musha, with a total of 3,100 horsepower. The average size tractor is 65 hp, and there is roughly one tractor for each 40 ha.

The equipment accompanying the tractor is fairly standard. All the tractors, apart from a few antiques, are rubber-wheeled with hydraulic lifts. Plowing is done with an adjustable chisel plow. Most tractor owners have a drum thresher, which is powered by a belt from the tractor's takeoff wheel. These drum threshers are mostly manufactured in the Egyptian delta; they were developed for wheat but can also be used for sorghum and beans. The final addition to the tractor is a four-wheeled wagon, which can be used to transport fertilizer and insecticide and to bring crops in from the field. Many of these are also manufactured in the delta towns.

The largest farmers are more likely to own tractors and pumps. Those holding 8 ha or more own two-thirds of the tractors. The seven largest enterprises in the village, which farm about 20% of the land, own 27% of the tractors and have shares in 46% of the pumps. Nine families owned two tractors, and one owned three; of these, the three who were not large landowners were making their living by renting tractor services.

Some tractors belong to a family working enough land (40 ha or more) to make essentially full-time use of it (10 or 11 cases). Others belong to a family with substantial land, but the tractor is rented out regularly (16 cases). Still others belong to one or more people who operate it as a principal source of income since they have relatively little land, less than 6 ha (15 cases). About one case in six involves a partnership between people of different households. Many others are owned by a family partnership (pairs of brothers or cousins, for instance).

The rental market in Musha is dominated by a few active custom operators. The 53 farmers who identified the last tractor they used cited 25 different tractors, but the four most frequently cited tractor owners were mentioned from 6 to 9 times, and they totaled 57% of all such mentions. These four tractor owners correspond roughly to the four points of the compass. Each has two tractors and so is definitely committed to the tractor rental market.

Working with a tractor requires a driver and a manager, who is usually the owner (Figure 8-3). Many of the drivers are teenagers; when they are older men, they sometimes become partners in the enterprise and have more managerial responsibility. The manager schedules the use of the machine, keeps track of payments in a large notebook, and arranges for maintenance. During the peak season in May–June, the tractor may plow in the morning when the soil is moist, haul during the middle part of the day, and run threshers in the late afternoon and evening when the crop is as dry as possible. The manager must also take into account the location of the work—trying to plow adjacent fields simultaneously, for instance. Most tractor owners find it a profitable business, and many seek to expand their operations. Thanks to government subsidies, costs are relatively low, and owners can charge as much as the market will bear. Farmers, on the other hand, are more concerned about labor costs than machinery costs, and so tolerate relatively high rental charges.

Egyptian agriculture depends on chemical fertilizers (such as urea or su-

FIG. 8-3. Tractor plowing in the vicinity of Musha. Note escarpment in the background (N. Hopkins).

perphosphates), home-produced animal manure, and pesticides. In Musha, 92% of 59 farmers surveyed in 1981 said they used chemical fertilizers, and 66% said they used manure. These products are generally applied by hand. Back sprayers and spraying wagons for pesticides are present but are not widely used. Aerial spraying of the cotton crop, a recent innovation, provides an additional reason for the areas planted in cotton to be contiguous. Except for cotton, seed is kept from one harvest to the next sowing in the village, if not by the farmer himself.

Irrigation

In Musha the most significant shift in irrigation was stimulated by the introduction of pumping machines by private farmers in the basins (Figure 8-4). The first steam pumps in Musha were installed just before World War I. By the 1930s they had been replaced by diesel pumps, and this more flexible system dominated the village by the mid-1930s. The pumps lifted groundwater to provide a supplement to the flood during the period from January to the end of the summer. Not only doublecropping but also the cultivation of cotton became possible under this system. The result was increased land values, increased demand for labor in the rural areas, and the creation of relatively large landholdings. These landholdings were controlled by a few traditionally rich families and some others that had made their money by labor recruitment. Cotton brokerage is another, more recent, source of rural wealth. The roots of Musha's social stratification go back to the 1930s.

In the 1960s the completion of the Aswan High Dam changed the picture again (Abul-Ata, 1977). When the annual flood ceased, farmers drew their water from new government-constructed feeder canals covering the whole territory. However, at least in Musha, the feeder canals were constructed to link up the existing diesel pumps. The task of these pumps then changed. Instead of drawing groundwater from 7–10 m, they simply raised the water about 50 cm from the government canals to the level of the fields.

Year round, the government supplies water to the feeder canals one week out of two. The diesel motor powers a pump that raises the water from the level of the canal to the level of the field. The water flows through a network of ditches until it reaches the farmer's field. A farmer who wants to irrigate must first arrange with the pump guard to provide water for the ditch that serves his field (if the pump guard has not already informed him) (Figure 8-5). Then he must provide the necessary labor, and open up a break in the ditch bank so that the water will flow into his field. He is required to pay the owner of the pump a set fee per watering (U.S.$2.00 per feddan, or U.S.$5.00 per hectare, per watering in 1981), and he must pay the pump guard an annual fee according to field size. Very few pumps have only one owner. The average is probably around three, and the maximum around eight or ten. One owner usually manages the pump, and takes charge of the accounts. The actual work, including purchase of fuel and maintenance, is done by the mechanic and the guard.

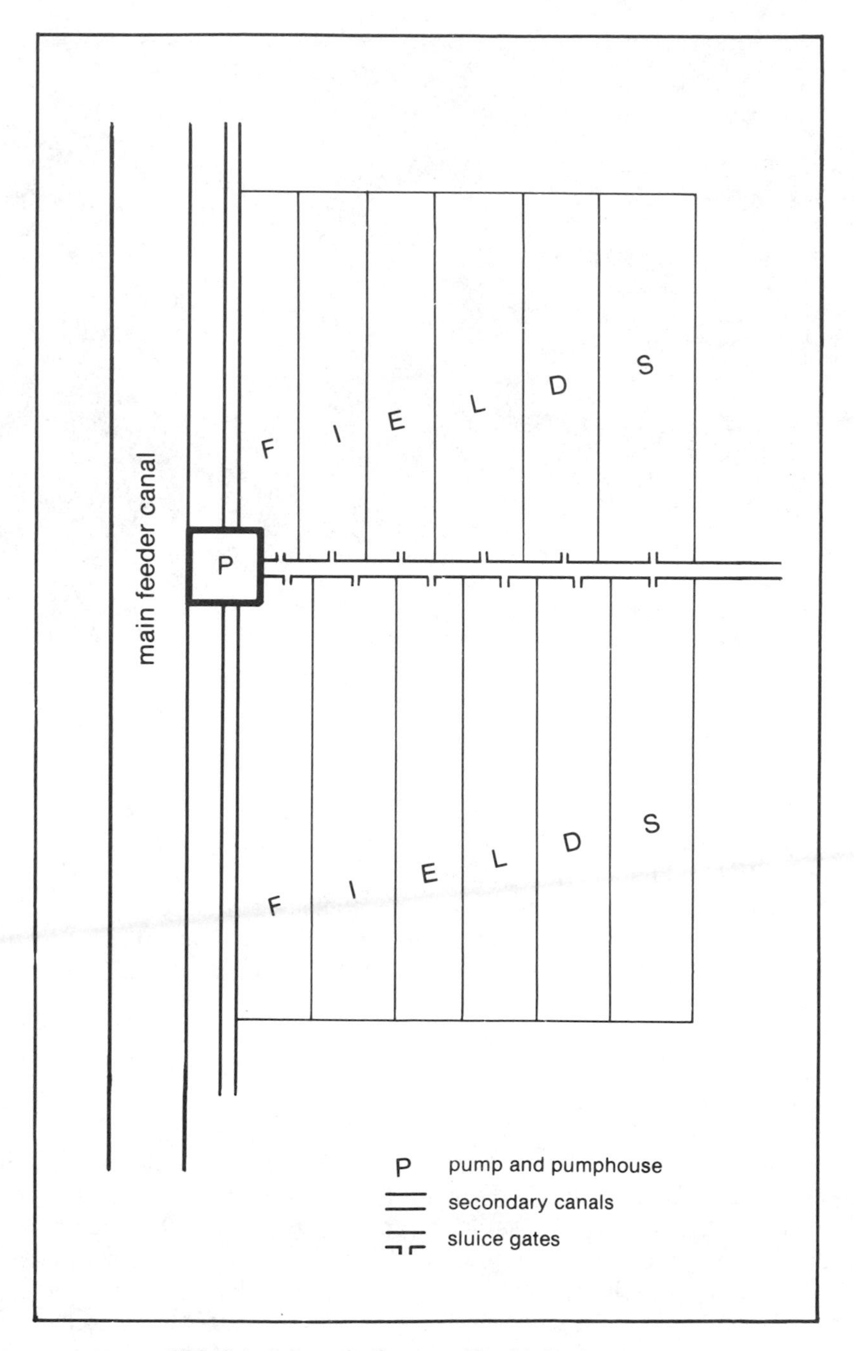

FIG. 8-4. Schematic diagram of local irrigation system.

FIG. 8-5. Pump guard blocking the flow of water into one canal and directing it into another (N. Hopkins).

Maintenance of the feeder canals is the responsibility of the government, and a period in January each year is set aside to clean them. The irrigation ditches are the responsibility of the owner of the pumps that supply them, who must see to their maintenance from time to time. The organization of irrigation water within the field is the responsibility of the farmer and, of course, varies from crop to crop.

The Household and the Labor Process

The household remains a central institution to the understanding of the labor process in the village. Household organization reflects the needs of subsistence cultivation, but the household's role has changed because of the penetration of rural areas by capitalism. The household is the principal organizer of labor, whether from among its own members, by hiring extra laborers as needed or by seeking off-farm or farm labor employment to supplement the direct income from agriculture, and so it serves as part of the process of control of labor.

Most households are nuclear (71%), and the average size is 7.5 persons. There is a tendency for the larger households to have access to more land and to raise more buffalo and cattle, or perhaps there is a tendency for poor households to be small. Households with at least 1.2 ha have an appreciably higher average household size, number of couples, number of male workers, number of cattle and buffalo, and total crops grown than do those holding less. In the 1981 survey, landless households had an average of 6.6 members, while households farming 2 ha or more had an average of 10.6 members. The number of buffalo and cattle per household was 0.83 for those farming less than 0.4 ha, and 1.82 for those farming between 1.2 and 2 ha.

Not all households are primarily involved in agriculture. In the 1981 sample, 29 (27%) derived their principal source of income from farming; 27 (25%) from day labor, not necessarily in agriculture; 24 (22%) from government jobs; 15 (14%) from crafts or trade; 4 (4%) from specialized agricultural jobs such as transport or winnowing. Three (3%) relied on migration income, and four (4%) relied on rent or pension. Some people combine agriculture with other jobs, so the phenomenon of the *worker-peasant* is beginning to appear. Many of the landless work outside agriculture, and a few merchants and government workers do very well.

The household embodies the sexual division of labor. Women are responsible for housekeeping and animal care. Caring for animals, milking them, and preparing cheese are also women's work. Women do not work in the fields in Musha, although in Lower Egypt they commonly do. Field work is strictly an activity for men and for children.

Children's jobs are clearly specified. They include cutting clover for the animals, cotton harvesting, and insect control. The latter is organized by the village cooperative. Cotton picking is organized by each farmer on a piecework basis. All except the smallest farmers, however, employ a labor contractor to recruit children for their picking gangs.

The head of the household plays a key managerial role. He must be sure that things are done on time—the inputs purchased, the labor hired, and the machinery scheduled. The household itself may supply labor for many of these steps, depending on size of the crop and household labor force. Since only about one-third of the households have more than two adult males (and they are not always available for agriculture), virtually every household hires labor once in a while, and some rely on it. In a sample of 58 farmers, 19 (33%) used mostly or only hired labor, while 29 (50%) used mostly or only family labor. Those that hire some labor constitute 67%. This figure should be taken as a minimum, since in most cases households do not count the labor hired to work with machines. It is noteworthy that when farmers give crop budgets they are likely to cost the labor automatically (even if they really use unpaid family labor).

Households decentralize the control of hired labor. Only the household head follows the crop throughout the cycle, even though others do the work. The small size and shifting composition of work groups enable the ability of the household head to make microadjustments and decisions. Thus the small farmer household is a relatively more efficient organizer of labor than the capitalist farmers would be if they used their control of machines to mobilize and control labor.

Exchanges between households are generally unequal. For example, labor needs are met much more by hired labor than by direct exchange. Those households that control the mechanized means of production and more land are in a favorable position. The pattern of labor transfer between households could be on an exchange or a reciprocal basis, but is most often between households that only hire in, and those who only hire themselves out. Thus the organization of work fosters a stratification system in the village, even though the household remains at the core of the organization.

It is hard to judge how profitable farming is. Most figures supplied by farmers are contradictory and unreliable, showing profit margins far too small to justify the attention given to farming. Overall, about 40% of the cost of growing a crop goes to labor and another 40% to seed, fertilizer, and other inputs, while 10% each goes to hiring of machinery and to rent and tax (the cost of land). Income depends on yields, which are highly variable, and on market price. Net income from a hectare is around U.S.$250 (1981). With workers earning around $1.75 for a 5-hour day, there may be some truth to the belief that they are better off than small farmers. Farmers generally consider cotton their least profitable crop because of the high labor costs involved; they prefer crops with high market value such as chickpeas. Yet so far they have been slow to move into vegetable gardening for the nearby city of Asyut.

The Labor Process for Wheat and Cotton

Two of the most common cultivars in Musha (and in Egypt as a whole) are wheat and cotton. In the 1981 sample, wheat was grown in the previous year by 84% of the farmers, and cotton by 79%. These two cultivars exceeded all others, with the possible exception of bersim. Wheat can be taken as representative of the winter crops, and to some degree of such summer cereals as maize and sorghum. Cotton is significant both economically and because the crop rotation system is built around it.

Wheat

Wheat is normally planted in November and harvested in May and June. It can follow cotton in the rotation but cannot precede it because of the relatively late harvest date. It is an important crop for the farmer for both grain and straw. The value of the straw is about 30–40% higher than the value of the wheat harvested from the same area, because of price controls on the wheat. Short-stemmed wheat of the Mexipak variety (HYV) is not popular because it produces less straw.

To grow wheat it is necessary to register with the government cooperative, and comply with the government crop rotation. The village cooperative records the amount of land the farmer holds, and the amount he will plant in wheat. The farmer must then get the village bank to authorize his loan – in the form of fertilizer and perhaps seed or insecticide – which he will later pick up from the bank's warehouse. The next step is to irrigate the land. Each time, the farmer has to deal with the pump mechanic, the pump guard (to determine when water will be available), the owner (to pay his fee), his neighbors (to ensure that there will be no damage to adjoining fields), and perhaps a worker.

Plowing the land involves hiring a tractor and a driver (Figure 8-3). The owner must be paid and the driver tipped. Someone from the family will generally supervise. Supplementary tasks, like harrowing and leveling, which are common in the delta, are rarely done in Musha. Fertilizer and seed must be hauled from the village bank to the home, and then from the home to the field. This is done by donkey, camel, or tractor, depending in part on the amount involved. Households lacking transportation must hire it. Wheat is generally sown by the broadcast method. The worker can be either a household member or hired. He may work in conjunction with a tractor so that the seed is covered. Fertilizer is broadcast by hand. Spreading manure (not used much for wheat) requires a great deal of shoveling, loading, and transporting before it is plowed under. Farmers do not use insecticide or weeding for wheat.

The wheat is harvested using a small sickle. The plant is cut at its base or simply pulled up by the roots. Most farmers hire labor who are paid piece rates according to the area harvested, although some farmers pay by the hour or the day. Family labor is also occasionally used, especially by small farmers. Machines have been introduced but are not widely used because they cannot cut low enough to preserve the whole plant. The reaped wheat must be bundled into sheaves. This job, slightly more specialized job than reaping itself, is done by older men. The worker is paid piece rates.

The sheaves are transported to the threshing ground at the perimeter of the village. The transport is done either by a camel or by a tractor and wagon. Costs are set by trip, regardless of the distance, and are paid in cash. Camels are easier to load and unload, and can go places tractors cannot, but are slower. After the loading, the field is opened to gleaners.

The grain is threshed using a tractor and a drum thresher. Threshing wheat is a lengthy and tedious job, and the usual gang has five members – one to feed the machine, one to shovel away the threshed crop that has passed through the machine, and two or three to hand sheaves to the feeder. The farmer must arrange for the tractor and for the hired labor. The threshed grain is winnowed and sifted. Winnowers are specialists, and because they are paid piece rates they require little supervision. The winnowed grain is measured and sacked by the winnower. At this critical juncture the owner will be present. The winnower is paid a share (typically 1/48) of the harvest for the combined operation. The final step in producing is to haul the grain and the straw from the threshing ground back to the storeroom in the house. Animal transport is likely to be used, either donkeys or camels. Farmers estimate the average yield for wheat at around 8 ardebs per feddan, or 2.86 t/ha.[4]

The grain is sold to merchants who funnel it into the city. In the immediate past, the government was also a major purchaser of grain. There are also dealers who buy straw to sell to other villages. In either case, the farmer must correctly time the sale to maximize profit. Some grain will also be used at home, milled into flour and baked into bread. Farmers may also use their own straw and bran (a by-product of milling) for animal fodder at home.

Cotton

This cultivar is normally planted in Musha in April or May and harvested in October. Approximately one-third of Musha's land is planted in cotton each summer. The cotton planted in Musha is not the famous Egyptian long-staple cotton, but a more ordinary variety used domestically.

The government issues seed and fertilizer for the cotton crop through the cooperative and the village bank. Plowing for cotton is always by tractor, which is also used to prepare small ridges that raise the cotton plants above the field ditches. Workers use hoes to prepare low bunds across the ridges, which help direct the flow of irrigation water. The labor here is likely to be hired unless the field is very small. Irrigating for cotton follows the same procedure as for wheat, save that it must be done more often. Cotton is planted by hand, usually by a crew of two to five workers. Working as a team, they move down one furrow after another. The worker chips a small hole with his hoe and drops several seeds into it. Weeding cotton is carried out by the same kind of work gang, and with the same short-handled hoe. While weeding, the workers repair ridges and bunds to prepare for the next irrigation.

Pest control is handled by the cooperative, which organizes gangs of children and young men to hand-pick the eggs of the cotton worm off the leaves. The government also organizes aerial spraying of most fields, those that are grouped together, and charges $66.25/ha.

Cotton is picked by gangs, mostly adolescents, who are paid per pound. They are frequently recruited by a "labor contractor" with whom the farmer makes an arrangement. The large farmers are likely to deal with a labor contractor from another village who brings in gangs of children from that village. Small farmers deal with a local labor contractor who recruits the children in his street. The labor contractor receives a share in the children's wages. When small farmers recruit their own labor, they pay the parents rather than the children. Farmers estimate their average cotton yield at about 7.5 qantars per feddan, or 2.8 t/ha.

The cotton is transported in plastic bags to the house of the farmer, either by camel or wagon. There it is unloaded and packed into large burlap sacks, usually by two men. The sacks are suspended from a strong beam, and one man uses his weight to pack the cotton down while the other man hands more cotton to him. Full sacks weigh around 180–200 kg.

The burlap sack is transported to the government grading and weighing station, again either by camel or by wagon. At the weighing station, government cotton graders grade the cotton, and the sacks are weighed. The workers at the station must move the sacks around for these tasks. They are paid both by the government and by the farmers.

This review of the labor process demonstrates the complexity of the agricultural cycle from the point of view of the individual farmer. The farmer is the only one who follows a single crop from beginning to end, and he must constantly negotiate with a wide variety of other people—government officials, machine owners, day laborers, labor contractors for child labor in cotton picking, neighbors, contractors for transport animals, and merchants who are eager to purchase his crop at a low price. The farmer needs to know enough of the traditional skills of farming to supervise the hand and machine work, but he is also dependent on the owners of the machines and on the government if he is to grow a successful crop. The principal role of the farm household is as manager of a wide range of external inputs rather than as the mobilizer and foreman of internal household labor. This conclusion has important implications for control and for the extraction of surplus.

Marketing

Musha farmers grow both for direct consumption and, increasingly, for sale. The cultivars used for each objective have been discussed. In addition, dairy products, poultry, and small amounts of vegetables may be sold if surpluses exist. As long as dairy products and poultry involve modest sums of money, they remain the province of women (Zimmermann, 1982). Prices are a never-ending topic of conversation. Both the government trade network and private merchants are important for marketing. The small market in Musha serves village consumers, and the periodic markets are concentrated on livestock trading.

There is considerable government intervention in marketing, which principally takes the form of "forced deliveries" for some of the basic field crops. In 1981 in Musha this included cotton, beans, and lentils, and wheat was on the list until 1979. Farmers were obliged to sell their entire cotton crop to the government. They were required to sell a set amount per feddan of other cultivars under cultivation, leaving some for home consumption or for sale at the higher market price.

Cotton illustrates government involvement in trade. At delivery the farmer is paid a base price. The village bank calculates the farmer's indebtedness to the bank for inputs and services, and deducts that from the remainder of the value of his cotton. The farmer may then receive a further sum. The cotton price is fixed each year by the government, relative to the world market price. The government sets a price intended to motivate the farmer to cultivate properly (i.e., just enough over the costs inputed to the farmer), and to generate the desired level of state income. Farmers are generally aware of the world market price and feel cheated by the notably lower price the government offers them. Harik (1984:70; cf. Korayem, 1982) figures that the government profit from cotton during the 1960s was always at least 30%, and that it reached 181% in 1966–1967. The low cotton price is passed on by the farmer in low wages to hired workers. The farmers thus squeeze the workers, and the state squeezes the farmers.

There are around 100 to 150 "centralizing merchants" in Musha who trade in grains. Merchants purchase grains from the individual farmer and sell them to urban-based merchants. To enter the market, they must have some capital, as well as a reputation for trustworthiness. They must also know how to judge the quality of grains and how to judge measures, and they must keep abreast of price fluctuations.

Arbitraging merchants trade straw, livestock, and vegetables between regions or villages, seeking profit due to local price variations or differences in availability. A network of local merchants in each village or region deal with each other. For instance, a straw dealer from Musha may buy up straw in Musha and transport it to another village. In that village, however, he must have a local partner who knows the people and their relative trustworthiness because delayed payment is common.

Livestock dealers usually meet in weekly markets scattered throughout the area. They range from small dealers working on a commission basis, who bring an animal belonging to a neighbor, to large dealers with large amounts of capital who may actually buy up animals in a village and sell them to butchers in the markets. Again, the reluctance to deal with strangers encourages this system of professional brokers built on personal networks.

Fruits (grapes and pomegranates) are sold in the field or in the village to merchants who come to seek them out. The farmers themselves never transport these fruits to market. Farmers who grow onions and other vegetables may sell either in bulk to a merchant or in small amounts to neighbors. In the latter case they rely on word of mouth to spread the information. Most of the small-scale peddlers who retail fruits and vegetables in Musha supply themselves from other villages, from Asyut, or less frequently from local farmers. Musha farmers are reluctant to grow perishable vegetables, apparently because of the uncertain marketing possibilities.

Role of the State

The state is a major partner in Egyptian agriculture. State policies set the framework for most if not all agricultural activity and structure the situation within which individual farmers must make choices. The state has made major investments in agriculture, such as those concerned with the irrigation system and other infrastructural projects. Generally these are carried out with a minimum of consultation with local farmers and village leaders, although there is some evidence that major landowners were influential in directing government attention in the past to particular projects of interest to them.

The state also provides agricultural credit for inputs, in effect financing the practices of agriculture and relieving farmers of the necessity to finance each year's crop from the preceding year's harvest. At present this job is done by the village bank branch of the Principal Bank for Development and Agricultural Credit. The village bank provides credit for the inputs to those farmers whose status and size of landholding has been guaranteed by the village cooperative. At the high point of the socialist period, in the early 1960s, state intervention extended to the creation of a state-controlled crop rotation system, intended to ensure that large blocks of land would be cultivated in the same crop (especially cotton) and that nationwide quotas of certain crops (cotton and food crops) would be met (El-Shagi, 1969; Ibrahim, 1982).

The state is also involved in marketing certain crops and in administering policies (such as the import of wheat on concessionary terms from the United States and elsewhere) that may have a depressing effect on the prices paid to domestic producers. The motivation for such state policies is generally to provide certain basic consumption goods to the urban population at a politically acceptable price.

The state has a role in setting land tenure policy and the rules governing tenancy. These rules determine access to this essential resource and also indirectly determine the quality and size of the labor force, by affecting the match between personnel and resources at the household level. A certain proportion of the costs of the reproduction of the labor force are pushed back onto the households. The presence of landless households, and the small scale of most near landless household enterprises, means that excess labor with poor bargaining power is available for the large farmers.

The state also directly affects the labor market by its control or lack of control of migration inside or outside the country. It allows the safety valve of moving from the village to the city or outside the country, and to some degree encourages migration because of the benefits the state hopes to draw from an urban labor force or from the remittances sent back in hard currency by those working abroad.

Conclusion

The farming system of the Upper Egyptian village of Musha responds to a certain number of imperatives. One of them is the landscape itself, especially the combination of rich soil, no rainfall, and a large river, creating what is in effect a very large oasis. A second one is the role of the state, which has remodeled the landscape and certain administrative regulations with a combination of economic and political goals in mind. A third is the division of local society into classes, and in particular the existence of a class of large and wealthy landowners who occasionally have taken the initiative in mechanizing irrigation and traction.

The essence of the social organization of agriculture is the formation of unequal relations between the state and the large farmer, between the large and small farmer, and between the farmer and the laborer. The farming system and its evolution depend on the ways in which surplus value is transferred among these various partners, tending to end up with the state or the wealthier farmers. The momentum that perpetuates the farming system stems from each party's attempts to improve its own position. The nature of these opposed interests is apparent in the disposition of surplus and the control of labor.

The Extraction of Surplus

The state has two methods for extracting surplus from the agricultural population. It taxes the land, albeit lightly, and controls the terms of trade for certain crops, presumably in its favor. It sets the "farmgate" price for cotton below the world market price, intending to reap a profit on the transaction. The relatively low fixed price for food crops helps subsidize the urban populations who buy them through consumer cooperatives. It is not clear whether the state comes out ahead in these transactions, not least because the state also assumes responsibility for providing, financing, and maintaining the infrastructure necessary for agriculture (irrigation works, subsidized inputs, roads, and the transportation system).

There is also private extraction of surplus, some of which reflects differential ability to profit from state policies. The relatively wealthy make use of their control of resources to generate income from others. These wealthy include the owners of machines, the owners of large amounts of land, the merchants, and the hirers of large amounts of labor. Those who control capital in the form of farm machinery rent it on a piecemeal basis to others in exchange for a share of the income from agriculture. The profitability of machinery is enhanced by state subsidies for fuel, oil, and the purchase price of the tractor. Rental income is generated particularly from free market rentals, usually for one agricultural season rather than a full year or longer. The share of the agricultural income that goes to the owner of the land rather than the farmer is much higher than in the case of legal rentals. Some surplus value is extracted through the wages paid labor, although Egyptian rural wages have risen markedly since around 1970. Wages reflect the relative bargaining strength of the workers and the employers, particularly the large employers who set the price. Since they set it fairly low, this also benefits the small farmers, who then pay the standard wage.

Control of Labor

Control of labor operates first through the household. The sexual division of labor is visible in the household, and the head of the household has some ability to recruit male kin for labor, using kinship as a crutch. Second, the household hires labor from day to day, so the household head appears as a manager. Since the rental of land and machinery involves renting capital in small pieces to the direct producer, it does not lead to any direct effort to control labor.

Situations involving the hierarchical control of work are few. They include those moments when large gangs of workers are hired for a single task, but the implications of hierarchy are mitigated because such gangs are often recruited from outside the village, are composed of teenagers, or are paid on a piecework basis, encouraging self-exploitation. The working of a marketplace in labor, land, machinery, even water, enables people to avoid human situations that would sharpen people's sensitivity to class distinctions and to stratification, perhaps because the units people deal in are small. Workers are separated from the productive process, control of which remains in the hands of the individual farmers. There is some accumulation of capital by the larger farmers on the one hand, and a proletarianization of the labor force on the other.

Directions of Change

Many different forces currently affect Egyptian agriculture. One of the most crucial is the monetization of agriculture. Farmers consciously orient their production to the market, and become sophisticated in dealing with it. Accompanying the spread of money is the growth of capital in private agriculture, first of all in the form of machinery, and the continuing importance of wage labor. As wages rise, the antagonism between the partners increases. But the rural world is not a closed box, and many villagers find alternatives outside agriculture.

The state is continuing its efforts to increase the level of mechanization in agriculture, and continues to regard the agricultural sector as its private farm. It does not take into account the social effects of increased mechanization, particularly on stratification. Current policies on mechanization favor the larger farmer, both through the restriction of credit to this class and because attention is focused on large-scale machinery. Intensification of agriculture is sometimes articulated as a goal by national spokesmen, who advocate the use of improved seed and chemical inputs, the improvement of the irrigation system, and the encouragement of mechanization. On the other hand, the state is retreating from its ambitions to control the details of the cropping pattern and the marketing of a large number of crops.

Individual farmers intensify by seeking more profitable crops, switching from open-field crops to vegetables and fruits, and trying to avoid onerous crops such as cotton. (In some extreme case they have sold topsoil for brick manufacturing.) They seek to mechanize, more to avoid labor than to increase production. The partnership between the state and the farmer is an uneasy one. The state seeks to maximize production, whereas the farmer seeks to maximize profit while minimizing such time-consuming processes as controlling labor. The outcome for Egyptian agriculture is the result of these two forces.

Innovation in Musha lay in the hands of the large, capitalist farmers, who were seeking additional machinery to reduce their dependence on wage labor. In some delta villages the same process is occurring (Stauth, 1983). In a village near Cairo, the large farmers could no longer afford labor costs, while the small ones who relied on family labor were doing well (Taylor, 1984). In another delta village there seemed little prospect of agricultural advance (Zimmermann, 1982). These cases may indicate two of the possible lines of development in Egyptian agriculture—capitalist in any case, but with accumulation by large mechanized farmers in some areas and by small market-oriented farmers in others.

Notes

1. Field research in Musha was carried out between September 1980 and June 1981 with the financial support of the American Research Center in Egypt and the Population Council through its Middle East Awards program. For local support I am grateful to the University of Asyut and to the village authorities. Egyptian pounds are calculated into dollars at the rate of 0.83 = $1.00. Professor Robert Hunt made numerous helpful comments on this paper.

2. A survey of the current state of agricultural mechanization in Egypt was carried out by the Social Research Center of the American University in Cairo in early 1982 on behalf of the Agricultural Mechanization Project of the Egyptian Ministry of Agriculture, with funding from the U.S. Agency for International Development (USAID). Dr. Peter Reiss was instrumental in arranging this. The survey covered ten villages in the governorates of Beheira, Gharbiyya, Qalyubiyya, and Minya. The complete results are available as a report (Hopkins, Mehanna, & Abdelmaksoud, 1982).

3. This survey was carried out in Musha in June–July 1979 on behalf of the Catholic Relief Services office in Cairo. My thanks to Mr. Andrew Koval for making this possible, and for providing me with my first contact with Musha. The results appeared in a report (Hopkins et al., 1980).

4. Yields for other field crops in t/ha, as estimated by farmers, are: maize and sorghum, 3.33; beans, 2.58; lentils, 1.71; and chickpeas, 1.43.

References

Abul-Ata, A. A. The conversion of basin irrigation to perennial systems in Egypt. In E. B. Worthington (Ed.), *Arid land irrigation in developing countries*. Oxford: Pergamon Press, 1977.

Baer, G. *Fellah and townsman in the Middle East*. London: Frank Cass, 1982.

Barois, J. *Irrigation in Egypt*. Washington, DC: U.S. Government Printing Office, 1889. The Miscellaneous Documents of the House of Representatives for the second session of the Fiftieth Congress, 1888–1889 (Vol. 9).

Bösl, K. Musha: Struktur und Entwicklung eines ägyptischen Dorfes. *Geographische Rundschau*, 1984, 36, 248–255.

Brunhes, J. *L'irrigation, ses conditions géographiques, ses modes et son organisation dans la péninsule ibérique et dans l'Afrique du Nord*. Paris: C. Naud, 1902.

Butzer, K. *Early hydraulic civilization in Egypt*. Chicago: University of Chicago Press, 1976.

Clot Bey, A. B. *Apercu général sur l'Egypte*. Paris: Fortin, Masson et Cie., 1840.

Crary, D. Irrigation and land use in Zeiniya Bahari, Upper Egypt. *Geographical Review*, 1949, 39, 568–583.

El-Shagi, E. *Neuordnung der Bodennutzung in Ägypten*. Munich: Weltforum Verlag, 1969. (IFO – Institut für Wirtschaftsforschung, München, Afrika-Studien . 36.)

El-Tobgy, H. A. *Contemporary Egyptian Agriculture* (2nd ed.). Cairo: Ford Foundation, 1976.

Hamdan, G. Evolution of irrigation agriculture in Egypt. In L. D. Stamp (Ed.), *A history of land use in arid regions*, pp. 119–142. Paris: UNESCO, 1961.

Harik, I. *Distribution of land, employment and income in rural Egypt*. Ithaca, NY: Cornell University, Rural Development Committee, 1979. Special Series on Landlessness and Near-Landlessness No. 5.

Harik, I. Continuity and change in local development policies in Egypt: From Nasser to Sadat. In L. Cantori & I. Harik (Eds.), *Local politics and development in the Middle East*, pp. 60–86. Boulder: Westview Press, 1984.

Hopkins, N. S., Aboulmagd, N., Attia, J., Guindi, M. A., Hediah, L., El-Khadem, Y., Mobarek, S., Mohieddin, M., Shama, A., & Wassef, M. *Animal husbandry and the household economy in two Egyptian villages*. Cairo: Catholic Relief Services, 1980.

Hopkins, N. S., Mehanna, S., & Abdelmaksoud, B. *The state of agricultural mechanization in Egypt: Results of a survey, 1982*. Cairo: Ministry of Agriculture, 1982.

Hurst, H. E. *The Nile*. London: Constable, 1952.

Ibrahim, A. H. Impact of agricultural policies on income distribution. In G. Abdel-Khalek & R. Tignor (Eds.), *The political economy of income distribution in Egypt*. New York: Holmes and Meier, 1982.

Korayem, K. The agricultural output pricing policy and the implicit taxation of agricultural income. In G. Abdel-Khalek & R. Tignor (Eds.), *The political economy of income distribution in Egypt*. New York: Holmes and Meier, 1982.

Lyons, H. G. *The cadastral survey of Egypt: 1892–1907*. Cairo: Survey Department, Ministry of Finance, 1908.

Richards, A. *Egypt's agricultural development, 1800–1980: Technical and social change*. Boulder: Westview Press, 1982.

Rivlin H. A. B. *The agricultural policy of Muhammed Ali in Egypt*. Cambridge, MA: Harvard University Press, 1961.

Ross, J. C. *Notes on the distribution and maintenance of works in the basin system of Upper Egypt*. Cairo: National Printing Office, 1892.

Stauth, G. *Die Fellachen im Nildelta: Zur Struktur des Konflikts zwischen Subsistenz – und Warenproduktion im ländlichen Ägypten*. Wiesbaden: Franz Steiner Verlag, 1983.

Taylor, E. Egyptian migration and peasant wives. *Merip Reports*, 1984, #124, 3–10.

Willcocks, W. *Egyptian irrigation* (1st ed.). London: Spon, 1889.

Winkler, H. A. *Ägyptische Volkskunde*. Stuttgart: Kohlhammer, 1936.

Zimmermann, S. D. *The women of Kafr el Bahr*. Leiden: Research Centre Women and Development, Institute for Social and Culture Studies, State University of Leiden, 1982.

9

Intensification in Peasant Farming: Punjab in the Green Revolution

MURRAY J. LEAF

Distinct religious communities have often been associated with distinct ecological regions and distinct systems of farming. The Sikh religion developed in the Punjab, the "land of the five rivers," at the head of the Indus drainage. In this century, political events have increasingly concentrated Sikhs in the southeastern third of their traditional homeland—in what is now the Indian state of Punjab. Since 1965 this subregion has been transformed by the advanced technology referred to as the *green revolution*. This has produced what a decade ago might have been regarded as a contradiction in terms: a system of modern intensive peasant agriculture, depending at once on some of the most ancient technologies and institutions and on some of the newest.

This description of the revolution and the system that it produced is based on two studies of a representative village, Shahidpur (Figure 9-1). The first extended from October 1964 through 1965. A follow-up study was conducted in the summer of 1978. The two studies bracket the period of greatest agricultural change.

On the basis of a combination of historical and statistical associations with adherence to Sikhism, the village of Shahidpur (a pseudonym) was initially selected because it had a large Sikh majority, it was relatively old and well established and tightly nucleated, its farmers were independent peasants, and it was not near major towns or other such distorting influences. It also had substantial irrigation from mixed sources, high cropping and cattle densities (compared to non-Sikh areas), and high geographical and occupational mobility. All these features have been relevant to the way the green revolution has reshaped village farming.

Between 1965 and 1978 farming in this area became more intensive by all the usual measures: area under irrigation, number of harvests per year, input and output per hectare or per person, capitalization per hectare and per person, population densities per hectare, energy consumption per hectare and per person, and total tonnage and calories produced per hectare and per person.

An emphasis on such measures has marked much of the work of conven-

Murray J. Leaf, School of Social Science, University of Texas at Dallas, Richardson, Texas.

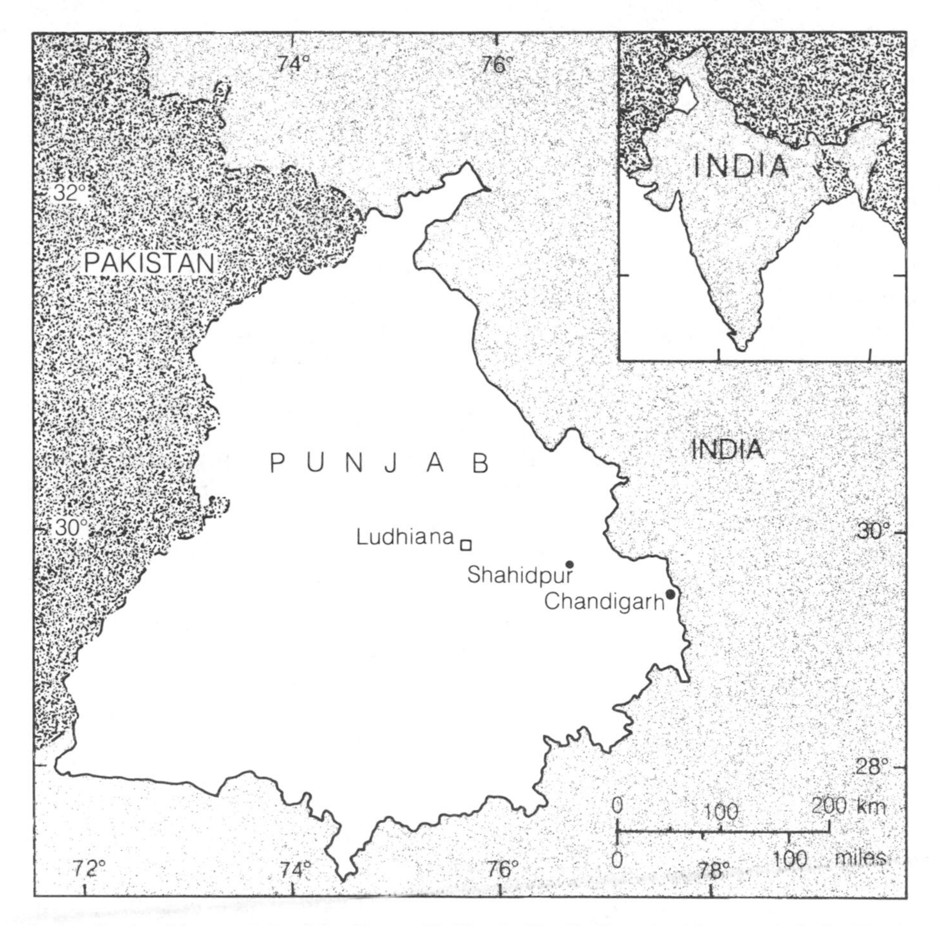

Source: Reprinted by permission from Eleanor M. Hough, *The Co-Operative Movement in India*, 5th ed., revised by K. Madhava Das. London: Oxford University Press, 1966.

FIG. 9-1. Shahidpur and Punjab.

tional farming systems analysis. A prominent version, termed farm systems research (FSR), has recently been codified for development studies (Haines, 1982; Shaner, Philipp, & Schmehl, 1981). FSR has two main aspects. The first is the dual nature of development projects both to design innovations and to provide an ongoing analysis of their acceptance or rejection. In this aspect, the method depends on the use of interdisciplinary teams of designer-analysts. The second aspect is the use of some type of input-output schema to depict the farming system by representing the activities of a single ideal or typical farm, often ignoring exogenous (nonfarm) variables. Neither of these methods is exactly appropriate here. This study describes inputs that were adapted in the village; it does not prescribe new inputs. It does not describe a typical farm because these farms are far too complex to be uniform, particularly with regard to the array of cultivars, animals, and combinations of farm and nonfarm activity. Finally, exogenous variables cannot be ignored, especially over time and in a situation of externally introduced change. The green revolution largely depended on villagers transferring reliance from local resources to resources produced regionally, usually by technologies that no village alone could develop or sustain. Also, a village farm does not typically control all the resources it needs to operate. Essential services and equipment are not found in the farm unit but at the village or regional level. For these reasons, the model of farming in Shahidpur must be a village model. The model should show which resources exist in the village and how each village farm family uses them to construct its respective farming system.

I have argued elsewhere that people organize their activities by means of cultural information systems (Leaf, 1972, 1984). A cultural information system is a consensually established network of ideas, related by learning and habit to a set of conceptualized objects. Six such systems are present in substantially universal consensus in this village and Punjab villages generally, although many others are shared by specialized groups such as the village masons or tax assessors. The six cultural information systems are (1) the village ecology, (2) the management system, (3) the economy, (4) the kinship system, (5) the system of political ideas, and (6) the system of religious ideas. Individual or family farming systems are built up when the decision makers draw on those parts of the general information systems that define their specific resources and aims. A family with cattle makes decisions using the various conceptions of cattle: as biological objects, as economic goods, as requiring a specific management routine, and perhaps as part of a patrimonial property. A family with land makes decisions using the various definitions of land, and so on. A decision maker's final action in any instance always depends on the interplay of all relevant frameworks. A landowner with crops will have to strike a balance between the agronomic practices that will produce the best plants, the allocation of economic resources that will yield the greatest profit, and the use of tools and helpers that can be most easily secured and controlled. Each framework provides a basis for optimization, but they may not be mutually compatible; the farmer has the final problem of constructing his farming system by balancing them as best he can.

The activities that may be grouped together in describing farming systems are defined mainly in the ecology, economy, and system of management. These are described here in some detail, with major emphasis on the ecology. The systems of ecology, economy, and management appear in use to be more concrete; their ideas are closely tied to the specific objects manipulated and considered in producing farm products. The networks of religious, kinship, and political ideas appear in use to be sets of conventional definitions imposed by social sanctions. They define modes of interpersonal relations, virtues and vices, social groups, and gods. Bayliss-Smith (1982:23–24) is right to insist that no farming system can be fully explained without taking them into account.

The Sikh religious organization has been a major forum for debate on social and agricultural policy (Leaf, 1976). Within this forum, the religious ideas of the importance of the individual, the idea of self-sacrifice (the subordination of individual and family interests to the survival of the community), and the rejection of caste in favor of respect for all kinds of work, framed proposals for policies to support the efforts of small private farmers such as were typical of the Sikh tracts. In 1967, when former Punjab State was divided into Punjab and Haryana, the new Punjab was centered on the Sikh area, and the policies that had first been articulated in the Sikh religious organization became the main program of its government and the basis of a new ethic of public service and governmental respect for the governed.

At the same time, the ideas of kinship, as sanctioned by courts and supported by the de facto absence of any effective state system of social security, are the basis of the continuing importance of the family as the most reliable basis of long-term personal security and therefore of socioeconomic decision making in the village, including agricultural decision making.

Finally, the political ideas that define parties, usually called "factions" in the Western literature, provided the only widely recognized framework for conceptualizing and massing support around conflicts of interest – defining the hidden power blocs underlying the overt debates phrased in religious and kinship terms. Such blocs were instrumental in obtaining the 1967 division of the state, and they continued to play a vital role in village affairs, as will be described, relating formal bodies like the *panchayat* and credit cooperative to village family interests and linking village interest groups to regional and even national political coalitions.

Without such frameworks – each in its own way linking individuals and guaranteeing that a person who makes a promise now can be held to it later – there would be no overall social organization, no allocation of tasks, and no way to debate such allocations. The work would never get done, technology would never be developed, and no one would act.

Agricultural intensification is often accompanied by structural change, change in the kinds and proportions of agricultural inputs and outputs. The relationship between such structural change and intensification is a major practical and theoretical problem, but it is usually approached only inferentially. A great advantage of looking at a single village over time is that the relationship can be seen directly, for example by viewing *intensification* as increases in the numbers of things that the ideas and terms of the information systems refer to and *structural change* as change in the content and organization of the systems themselves.

Ecology

The basic idea of an ecological system is expressed in many ways in the village. One of the simplest is that what comes from the soil must go back to the soil. In saying this, villagers understand three general points. The first is that living things must be kept alive; death is irreversible. Plants, animals, and family members must be sustained throughout the year, even though they may only be needed at some points in the year. The second is that if at any point in its growth cycle a crop undergoes stress that slows its growth, such as lack of water or nutrients, the effect cannot be erased by later supplying more of what was missing. The third point is that the needs of all plants and animals vary throughout their growth cycles.

Given these general understandings, the villagers' problem is to select the species that yield the maximum benefit and then to control the timing of their growth and their numbers in such a way as to obtain it. The main elements of the system are land and water, cultivars, cattle, the human population, and the various tools and shelters.

Land and Water

All available land in and around Shahidpur has been farmed since 1963. The total village land area in 1965 was 178.3 ha, of which 15.4 was nonagricultural; in 1978 it was 178.3, with 17.0 nonagricultural. Most of the nonagricultural land is concentrated in the nucleated housing area (*abadi*) in the center of the village lands. With the exception of four families who built houses in their fields between 1965 and 1978, farmers live in houses in the *abadi* and walk out to their fields (see Figure 9-2).

The village lies about 30 km southwest of the nearest point of the Himalaya Mountains, and about 3 km south of the Sutlej River, the southernmost of Punjab's traditional five rivers. The land surface is almost dead level. The soil is a fine sandy loam, with enough clay content to permit it to be baked into pots and bricks. Such soil can absorb large amounts of water, but water content markedly affects its physical properties.

Rainfall variation in Punjab is controlled mainly by distance from the Himalayas. Precipitation in the village area is about 83 cm a year, over half of which falls in July and August. This is about average for the state, and typifies the densely populated belt about 100 km wide that parallels the Himalaya Mountains at a distance of about 10 km throughout north-central India. Villages between this belt and the mountains receive more rain but are commonly subject to flooding or soil erosion. Villages beyond the belt, far from the mountains, have less rain, and their underground water is usually deep, hard to find, and of poor quality. Like most of the Indo-Gangetic Plain, the alluvium beneath Shahidpur is hundreds of meters deep. There is no layer of rock to collect water for wells. Because of this, available groundwater is limited to percolation from the surface supplies. Wells are sunk to whatever level allows them to collect by seepage a useful amount of water—traditionally, as much as bullocks could pump during a day, refilling overnight. The precise depth of the well varies with the amount of rainfall received.

Shahidpur has had some well-irrigated farmland for several centuries, but

FIG. 9-2. Farmer walking to newly transplanted rice field (July) (Photo by M. Leaf).

the water supply has changed substantially in the last 30 years. Before 1954, wells had to be dug to about 30 m. Subsequently a major canal system was built in the area, and in 1962 one of its minor distributories crossed the village lands. The channels are permeable and discharge water into the ground throughout their course. By 1965 the village water table had risen to 10 m, and by 1978 it was about 7 m. The costs of new wells decreased proportionally, and construction increased. There were 23 wells for agriculture in the village in 1955 and 24 in 1964; all were brick shaft wells fitted with Persian wheels and powered by oxen or camels. In 1965 two tube wells were sunk with mechanical pumps. By 1978 the number of tube wells had increased to 28, and all but about two of the older wells had also been fitted with pumpsets.

In 1965 the proportions of land types classified by water supplies was 11.8% canal, 4.4% well-canal, 61.8% well only, and 22% rainfed. By 1978 canal land had increased to about 14% well-canal to 9.7%, and well land to 75.3%. Rainfed land was negligible. Thus the village went from about 60% irrigated land in 1955 to 80% in 1965 to 100% in 1978. During the last period, total water production quadrupled.

The basic method of water distribution is unchanged, but the increased number of wells has made it much more efficient. There is no villagewide institution for water allocation. Wells are dispersed throughout the fields, and water is distributed from them, and from the canals, through small raised field channels. Farmers who share a well or a canal outlet take turns for periods of time that vary in proportion to their share rights, diverting the water to the fields where it is needed. Such sharing was usual for the older shaft wells; the new tube wells are almost always owned individually. Water is applied by flooding small rectangular pans (not rows) from these channels. The size of the pans is not fixed but depends on the rate of flow and the crop. In 1965 such pans were most often about 6 by 10 or 12 m, and were filled to the depth of a hand. Because of the greater rate of flow from the wells in 1978, they were commonly much larger: 5 to 8 by as much as 50 m. Depth was unchanged. Since crops needing the most water are planted nearest the water sources to minimize transport losses, a wider distribution of source points allows crops that require high water flows to be planted over more land without loss of efficiency. The larger flood pans also simplified plowing and reduced the needed channels and dikes.

Increased water supplies permitted increased use of chemical fertilizers. In 1965 fertilizer was used sparingly. It was relatively expensive, usually had to be bought with cash, and increased the risk of crop loss if the rains failed. Beginning in 1967, the state government developed a system of cooperatives to make fertilizer available in quantity on credit against the harvested crop. Local cooperative records show that in 1969–1970, total fertilizer loans were Rs. 14,000 (about U.S.$1,400 at the current exchange rate).[1] In 1974–1975, new loans were Rs. 50,701. In 1977–1978 they were Rs. 70,505. Allowing for price increases averaging about 150% from 1965 to 1978, the increase in volume must have been at least threefold. Fertilizer is always drilled into the furrows with the plow, either at the time of sowing or before.

To maintain good soil texture and drainage, fertilizer use is complemented with applications of compost. The total area set aside in the village for composting in 1978 was about four times that of 1965, thus corresponding closely to the increased fertilizer use.

Cultivars

Village farmers in 1965 grew 26-odd cultivars in two seasonal arrays. This diversity assisted the local diet, produced several environmental efficiencies (Beets, 1982:7–9), permitted more efficient use of local labor supplies, and raised total production. One array was sown from May to July and harvested in September–October. The second was sown in October–November and harvested in April and May.

Each array changed markedly between 1965 and 1978, mainly because of the introduction of new varieties that produce better with higher levels of inputs. The new crops were not limited to the "miracle" wheats and rices. In 1965 all cultivars were grown from seed or cuttings (cane) collected by the farmers from the best plants they had produced each season. In 1978 all but cane were grown from hybrid or improved seed obtained from commercial or state sources. Increases in yields varied greatly (Table 9-1).

The yields of wheat on average land increased most, almost doubling. The yield from sugarcane on the best (canal-irrigated) land, by contrast, did not increase at all. The changes in yields contributed to changes in areas of land planted to the various crops (Table 9-2).

Increased areas of maize and rice came primarily at the expense of cotton. The reason is directly related to the increased water supply. Two varieties of rice were grown, a tall *basmati* (*Oryza sativa* var. *indicum*, an improved indigenous type) and a shorter-statured *indicum* variety, created by genetic transfer, called *Jaya*. Both required more water than any other crop: 25 waterings a year (a watering is one filling of a flood pan). This was too much for the wells of 1965. The electric pumpsets introduced in 1970 greatly reduced the water cost because the government charged a low fixed monthly rate for the electricity. Rice gave a high financial return per hectare, required little weeding, was reliable in dry years, and was least vulnerable to lodging and disease in wet ones. Rice had not yet been integrated into the local diet, but its straw was used in fodder.

Cotton was the most important market crop in the state in 1965, and second only to cane locally. Villagers used it to make many of their garments out of homespun cloth, to make bedding, to make the webbing in cots, and to make all their finer string and twine. By 1978 almost all clothing was made by

TABLE 9-1. Yields (Q.) of major cultivars, 1965 and 1978.

		Land grade[a]		
Crop	Year	Best land	Average land	Poor land
Wheat	1965	42	26	13
	1978	47	41	24
Maize	1965	26	20	6
	1978	35	24	12
Desi cotton	1965	22	n/g	n/g
	1978	24	20	n/g
American cotton	1965	22	n/g	n/g
Sugarcane	1965	1,000	500	n/g
	1978	1,000	824	n/g
Rice (var. Jaya)	1978	74	47	n/g

Source: Farmers' estimates.

Note: Units are Quintals (Q.)/ha. 1 Q. = 100 kg.

[a] *Best* land is canal or well-canal land; *average* is ordinary well land; *poor* land is rainfed land, except for wheat and maize in 1978, where it indicates the poorest well-irrigated land.

TABLE 9-2. Summer crop pattern, 1965 and 1978: Percentage of total area under each cultivar.

	Year	
Crop	1965	1978
Maize	21	40
Mung and mash	2	1
Sugarcane	31	27
Chilis	1	1
Fodder	12	14
Desi cotton	8	1
American cotton	9	0
Rice	0	10
Other	14	6
Failed	2	0
Total:	100	100

Source: Farmers' estimates.

local tailors from commercial cloth. Raw cotton was needed locally mainly for bedding and cordage.

Cotton yields were highly variable according to weather conditions, but the crop was relatively drought-resistant. The *desi* (which means "local") variety was especially pest-resistant, needed only four waterings in the complete absence of rain, and matured earlier. The "American" type was less pest-resistant, but it ripened later and produced better if the rains were heavy or late. The two varieties were not expected to do well simultaneously.

Cotton produced no fodder and therefore had to be supported with additional land devoted to fodder crops. When water was scarce, there was no good alternative to accepting the risks of cotton as a market crop on some land; when the scarcity disappeared, rice was more reliable and useful. Planting of the American variety was especially reduced because its sowing and harvest times crowded those of the HYV (high-yielding variety) wheats. The *desi* cotton grown in 1978 was an improved variety, created from indigenous stock by state agricultural scientists.

The maize (*Zea mays* L. var. Ganga #5) grown in 1978 was a hybrid, the third in a series of replacements of the earlier local type. The 1965 maize needed no watering in years with good rainfall, and four if the rains absolutely failed, but a good crop only gave one ear on three plants. It was grown mainly for fodder, often on rainfed land. For this, the stalks and leaves were used. Grain was considered a windfall. In 1978 farmers planned on five waterings with ordinary rainfall. Grain production was more important, and grain yields were much more dependable.

Among the minor crops, mung (*Phaseolus mungo*) and mash (*Phaseolus radiatus*) are small peas, as important in the human diet as their dried stems and pods were for cattle fodder. Other crops included, in 1965, *arhar* (*Cajanus indicus*), sorghum millet (*Sorghum vulgare*), pearl millet (*Pennisetum typhoides*), hemp (*Cannabis sativa*), sesamum (*Sesamum indicum*) (for seeds and oil), some leguminous green fodders, and various vegetables grown in little garden plots near wells: onions, garlic, radishes, and herbs. *Arhar*, a drought-resistant legume used for food and fodder, was not mentioned in 1978.

Shifts in the areas under the winter crops were as dramatic as for the summer array (Table 9-3). Wheat dominated in both periods as the main dietary staple, a major market crop, and the principal base of fodder mixtures. Its increased cultivated area directly reflected the increased yields. Villagers experimented with new varieties in 1966–1967. By 1978 the two main varieties in use (*sonalika* and *kalyan sona*) were the last of a sequence of eight. The clear losers were the crops grown in pairs: wheat and chickpeas (*Cicer arietinum* L.), and barley (*Hordeum vulgare*) and chickpeas. They had been grown on rainfed land or marginal irrigated land, and were paired this way for several reasons. The legumes would provide some nitrogen to the grains, and legumes and grains grown together give better yields than either would alone on the same land.

TABLE 9-3. Winter crop pattern, 1965 and 1978: Percentage of total area under each cultivar.

	Year	
Crop	1965	1978
Wheat	21	69
Barley	1	1
Chickpeas	6	1
Wheat and chickpeas	24	0
Barley and chickpeas	1	0
Oil seeds	2	1
Fodder	10	1
Sugarcane	31	26
Lentils	1	1
Other	2	0
Failed	1	0
Total:	100	100

Source: Farmers' estimates.

Further, the crops were differentially drought-resistant (chickpeas more than wheat, barley more than chickpeas), so a complete loss of either pair was unlikely. By 1978 such insurance was unimportant, and mixed crops were more difficult to harvest.

Cane, though perennial, is only harvested for a few years. Raw sugar is essential in the diet, and sugar and cane are major market commodities. Its leaves are used for making coarse ropes, but it yields no fodder. Planted from cuttings in late spring, it grows through the summer and becomes dormant in winter, retaining its sugar content. It is then cut and taken to market or squeezed and boiled down, according to price and labor availability. In 1965 cane took more labor and water than any other crop. In 1975 its requirements were about average, but its price had not risen sufficiently to compensate for the improved returns on such alternative rotations as maize–wheat or rice–wheat. Therefore, plantings were reduced.

The winter fodder in 1965 was mainly *barsim* (also spelled *bersim*), a kind of clover (*Trifolium alexandirianum*), cut daily for green fodder as needed and mixed with dry matter. In 1978 it was largely supplanted by more dry maize and by millet grown between the seasonal arrays.

The main winter oil seed was mustard, whose leaves are picked daily as a vegetable before the mature seeds and pods are finally pressed for oil (and the residues then added to fodder). Mustard was often planted in the same fields as barsim. The legume aided the mustard, the latter growing above the former. After the mustard was pulled for pressing, the barsim beneath it would grow and be cut in turn.

Other cultivars included lentils and the garden crops: mainly peas, cabbage, onions, carrots, radishes, and cauliflower.

Increased water control permitted substantial plantings outside of the two traditional seasonal arrays. In 1965 fields were generally cleared by May, plowed, and leveled until sowing before the summer rains in late June. In 1978 many fields of irrigated millet were growing in June (for fodder) and some were planted in such new market crops as eggplant, melons, and okra grown for seed. In the fall, potatoes were a common catch-crop, along with more vegetables.

Cattle

For cattle too, intensification has been associated with structural change—change in both the proportions of types and the balance of uses to which they are put. In 1965 cattle were needed to meet two major needs: traction and milk production. There was not enough of either. More draft power was needed to increase production, and more milk was needed for the human diet, especially for children. Little if any was available to market. Yet both needs could not be met at once, because they required different kinds of animals. Power was provided by bullocks and camels, milk by buffalo and goats. Additional fodder could not be produced without more draft animals to operate wells (and perform other operations), but more draft animals could not be fed because the fodder was needed for milk animals. In 1978 the village produced much more fodder and needed much less draft power (Table 9-4).

By 1978 the bullock population had declined from 1965 levels by about 14%. Camels, whose major use had been on wells, were eliminated completely. The buffalo population, by contrast, increased by almost 50% and the female goat population almost doubled. There was abundant milk for the local diet, and milk marketed through a cooperative had become a major source of additional household income. Based on the village system of rating fodder consumption and work output by "plows" or "pairs," the total cattle population increased by about 24%. A *plow* is two bullocks; its fodder allowance when working is about 54 quintals of dry fodder equivalent a year; and its work output is rated at drawing a plow, operating a well, or pulling a fully loaded cart. A single camel is the equivalent of a pair, as is a single buffalo. A single cow, mule, or horse is rated at one-half of a plow, and a donkey or goat at one-quarter.

Rams are particularly significant because they are kept only for meat, a luxury in comparison to milk and power. In 1965 I was told that meat was eaten

TABLE 9-4. Cattle census, 1965 and 1978.

Type	1965 total	1978 total
Bullock	85	72
Ox cow	17	14
Ox calf (male)	13	9
Ox calf (female)	3	7
Guernsey hybrid calf	0	1
Camel	15	0
Buffalo cow	147	220
Buffalo calf male	6	14
Buffalo calf female	57	144
Ram	3	17
Goat (doe)	45	82
Kid	15	1
Mule	24	5
Donkey	1	13
Pig	13	16
Total count	444	615
Total adult pairs[a]	234.25	290.25

[a]A *pair* is a local unit for estimating work and fodder requirements, equivalent to two oxen.

only at weddings. Even then, the source was more likely to be an old doe than a ram. In 1978 meat was eaten five to seven times a month, and rams were slaughtered regularly to supply it. Obviously, the village ecological system had come to allow some latitude for comfort and taste in place of choices between hard necessities.

The Human Population

The human population figures as an ecological element both for the constraints its dietary and health needs impose on the other populations and for the energy pool it represents. Since the energy pool aspect is implicit in everything else, only the diet requires explicit description here.

In 1965 there were 812 resident and 59 nonresident villagers. These figures increased to 975 residents and 101 nonresidents in 1978. Residents are people who eat and sleep in the village; nonresidents are those reported as part of village families, but not at present residing or eating in the village regularly. Usually, they were away for work—in other villages or cities, and in distant countries including Bahrain and Germany. The reasons for their absence lie in the family management strategies described in the next section. Here, the important point is that the 20% increase in residents or the 23% increase in total villagers did not come close to the increase in overall agricultural productivity.

The diet is basically vegetarian, described locally in terms of "habitual" summer and winter meals. Each includes a bread made of the grain of the previous season, accompanied by a vegetable of the current season and tea made with milk and local sugar. In 1965 I estimated that the equivalent of about one-quarter of the village land was actually needed to provide it; in 1978 this portion would be about one-fifth. Yet the crops are so interdependent that the effect is pervasive.

Tools

The simplest way to describe the changes in village tools is by the main tasks they are used to perform: water lifting, soil manipulation, transportation, threshing, fodder preparation, and food preparation.

Water lifting, threshing, and fodder preparation, all quite separate tasks in 1965, were closely related in 1978. Water lifting, as noted, was done by Persian wheels. Threshing, for all grains but maize, was done in the field, usually with bullocks and a drag (although a village artisan of Mason caste had introduced a diesel thresher for wheat and owned a corn-cracking mill that the same engine was used to power). Fodder was chopped with hand-powered cutting wheels in houses or barns where the cattle were tethered. Since 1965 the motors that were introduced to drive the pumpsets have been applied to drive crushers for cane, centrifuges for sugar, and new chopper/threshers of several kinds. These are usually interconnected with systems of axles and belts around the well sites, and they save each farm family several man-months per year, greatly reducing the need for family members as a source of brute labor, but putting a new premium on mechanical knowledge and skill.

Soil manipulation includes plowing, leveling, and cultivation (Figure 9-3). In 1965 they were accomplished entirely by human and animal power. The tools, such as the bullock-drawn single-bladed steel plow, were generally improved commercial versions of traditional designs. In 1978, all the old tools were still in use and were much more numerous, but in addition there were three 50-hp tractors and one smaller one. Three were owned by families with holdings between 0.5 and 3.5 ha, who supplemented their farm income by hiring them out for custom work. The fourth was owned by a family who farmed just over 10 ha owned by a religious trust. They were generally equipped with a plow, a two- or four-wheeled cart for carrying bulky goods, and a set of vanes to be attached to the drive wheels for puddling rice paddies. Puddling is an entirely new operation involving flooding a field and agitating the soil to create a hardpan. Without it, fields will not retain the necessary

FIG. 9-3. Weeding maize (September) by landowner and laborer (Photo by M. Leaf).

water. After the rice is harvested, the hardpan must be broken up by deep plowing the dried field.

Tractors are officially considered to perform most agricultural operations at 5 to 7 times the speed of bullocks (Dhillon, 1978), although the owner of one of the 50-hp tractors in the village rated its speed for plowing at 12 to 14 times (cf. Leaf, 1984:130). There are many variables involved in such estimates, but even on the most conservative basis the four tractors represent an addition to the village land preparation capacity that could not have been obtained within the framework of the 1965 system. By 1978 the use of chemical herbicides had become a common replacement for hoeing and plow cultivation to control weeds.

Increased production requires increased transport. Apart from walking, the major means of transportation in 1965 were bicycles and oxcarts. There were about 120 cycles and 25 oxcarts. The latter were of the traditional type, locally made, with high wooden wheels. In 1978 there were about twice as many cycles, 11 of the traditional oxcarts, and 26 oxcarts of a new type built on an automobile or truck axle, with rubber-tired wheels. They could carry twice as much as the old type with the same bullock power and lasted twice as long. Villagers also owned four motorcycles, one automobile, and one 10-t truck, although the truck was not used locally. Considering all vehicles, including the four tractor carts, the increase in bulk cargo capacity was at least threefold. Again, this would not have been possible with the previous materials and energy sources.

In order to have more production in the village ecology, there had to be technological substitution: different cultigens, different inputs, and most important different energy sources. With them, the village ecology had, in a sense, opened up to include the state's electric power plants, its growing industrial system, and the plant-breeding laboratories of universities and research institutes. These structural changes within the ecology were closely related to changes in economic and social relations that make up its context. All were an inextricable part of the intensifying process.

The System of Management

The finely balanced relations among the populations in the village ecology are not the result of any kind of centralized action at the village level. Although it is convenient to speak of "the villagers" as doing this or that, the villagers as a whole are not a decision-making body and do not speak of themselves as one.

The Punjabi term for *management* is *parbandak*. Its root, *band*, literally means "bound" or "tied," and the whole word has the sense of "that which binds together." This conveys more of what is actually involved in de facto managerial control than does our idea of *ownership* alone, as a matter of legal rights. Legal rights are an element in control, but they are never the whole of it.

The core of the village system of management is consensus among villagers about the groups that have control of resources and about the patterns of cooperation among such groups. Any particular individual is accorded control over resources on the basis of his acknowledged place among these groups. In 1978 such groups were of three types: the *panchayat*, the cooperative committee, and households. Households were by far the most important. In 1965 the same three kinds of groups had been present in name, but the practical managerial importance of the cooperatives and *panchayat* were actually negligible.

Panchayat

The term *panchayat* means roughly a council of five, and *panchayats* have existed traditionally in many contexts: in villages, in castes, in religious communities, and in clans. These, however, generally appear to have been impermanent groups that gathered to express village consensus on some issue or problem and then dissolved. There is no indication in traditional accounts or current recollections that they had any clear sanctioning power or exclusive authority. Legally elected village *panchayats* became universal in Punjab only in 1952, the year Shahidpur's *panchayat* was established.

In 1965 the *panchayat* had an impressive array of legal powers. It could collect a household tax, and it was a court of first resort for minor civil and criminal matters with the power to levy fines and even impose jail terms. It could regulate local trade, and it maintained a village watchman. Yet in practice the powers were never exercised as the legislation envisioned. The reason was that members of the *panchayat*, among others, were official witnesses. This gave them the power to influence land transactions. The threat this posed ensured that the elections would be controlled by the village party alliances—groups of farm families who were engaged in active, often bitter, conflicts over land.

In 1964–1966, during the 17 months that I was involved in the village, *panchayat* members were never described as setting aside party ideas to act only in their legal capacity. They never represented the interests of village as a whole, only those of their respective parties. By 1978, however, there had been several occasions when the members of the *panchayat* did act collectively with the explicit purpose of representing the entire village community. They had raised money to pave a road through the village lands, erected well pumps at the points where the road entered the village, organized a substantial subscription to improve the local Sikh temple, and finally in 1976 undertaken a campaign to obtain government support and local money for a middle school in the village. The nearest such school was then about 3 km away. Teaching began in the village in 1978. For all but the road, the "taxes" the *panchayat* assessed were voluntary. Collections were not based on their legal powers but, rather, on a new application of the ancient idea of a group that worked informally by consensus. The effects were no less important. The road permitted travel all year, and by larger and heavier vehicles. The school provided access to the formal bodies of knowledge that were increasingly important in the village and that gave villagers access to new employment outside it. The *gurudwara* (Sikh temple) was an important symbol of the village's organizational strength and an important stage on which local and regional leaders met to inject village policy views into statewide debates.

Cooperative

The main change in the cooperative society has been suggested. It became the main conduit for the fertilizers on which the new crops depend. It also helped farmers obtain water-pumping equipment, other productive aids, and storage facilities. But this is not all.

The village cooperative is, legally, the local representative of two separate organizations. One, the Central Cooperative Bank, provided the short-term credit on a crop loan basis. The other, the Land Mortgage Bank, provided longer-term loans, such as for tractors or wells, which were secured by land. The increased importance of the first identity is indicated best by the increased volume of fertilizer loans, already noted. Mortgage loans rose from nothing in 1965 to Rs. 56,000 in 1974, the last year for which I have relevant records.

The changes in the cooperative came about as a direct consequence of government action and policy. From their beginning in the late 19th century through 1965, the declared purpose of cooperatives was to teach peasants sound economic practices by inducing them to put a small amount of their savings into a common pool and to lend to those in greatest need (Hough, 1966:55ff.). The money involved was insufficient to serve as serious agricultural venture capital. It was not available on a crop loan basis. The state government abandoned this conception when it restructured the cooperative system in 1967. Substantially increased funds were provided by the state and the Reserve Bank of India, and the expanded agricultural credit cooperatives were complemented by two new organizations, developed concurrently. The first was a system of marketing cooperatives at the state and national levels that undertook to buy the farmers' crops for central government stockpiles at stable and fair prices, announced before the planting season. Farmers were free to sell at higher prices if they could find them. The second was a dairy cooperative that provided loans to buy buffalo of certified quality and to improve dairy facilities. It also undertook to buy milk each day in the village, at prices farmers considered profitable. Since the milch cattle were fed mainly the residues of field crops, this produced a substantial added return to farming operations. These measures greatly reduced innovative risk, especially to small farmers, and thus accelerated the rate of agricultural expansion.

Households

None of the changes in the *panchayat* and the cooperative diminished the importance of households as the main locus of agricultural decision making.

The Punjabi term for *household* is *parivar*, defined in the village as a group of kin cooperating in a common property. The ideas of *kinship* and *property* have equal weight. Near kin who are not involved in the cooperation are not in the household; distant kin who do cooperate are in it. Nonkin who also may be in the house and cooperating, such as a full-time agricultural worker, are not in it.

Fundamentally, the basis of control of each household is its physical house. Houses are built solidly, with strong gates and tight internal storage rooms. Almost all have hand pumps for water, and most of the landowners' houses include substantial barns and work areas. They are the site of the support activities of all village occupations. They typically have more than one physical unit—sets of rooms, storage areas, and work areas. Those of farm families were from five to seven times as large as those of landless families, an average size under roof being about a hundred square meters. In 1965 four families (of Sweeper caste) had lived in mud-walled houses, but in 1978 these were gone and all structures were of baked brick.

The only major capital goods not secured in houses are buffalo of some poorer families, tractors (which are generally kept in sheds in the fields), well pumps (secured in brick buildings in the fields), and land itself. Legal rights in land are well defined, but land itself is of necessity unwalled and unfenced because field demarcations vary seasonally and farmers often have to reach their plots by crossing the land of others.

The main managerial trend between 1965 and 1978 in village houses was professionalization. As households acquired more capital, they tended to concentrate more intensely on a narrower range of occupational activities, evaluated in a wider regional framework. Two results of this were a decline in family size and more even family sizes across the different village occupational groups (Table 9-5). There were 118 families in 1965 and 152 in 1978; average family size declined by 6.5%. In 1965 the family sizes of Jats and of Harijans were near opposite extremes of the village range; in 1978 they both had moved closer to the average.

There are three main occupational groups. About half the villagers belong to landowning families, about one-third are in landless families relying primarily on agricultural labor, and the remainder are specialized artisans or professionals. In 1978 all landowners were of the Jat or Sadhu caste, whereas in 1965 five Harijans and one Sweeper had also been farming marginally. The total number of farms was about 70. The average size of a holding was then 1.97 ha, the largest 9.55 ha. In 1978 the number of farms declined to 64. The average size increased to 2.31 ha, and the largest was 9.70 ha. In both periods, families with larger holdings had greater agricultural income per capita (Leaf, 1984:225ff.).

Most landless laborers are of Harijan caste, although Sweeper, Potter, Barber, and Goldsmith families also do such work for the same pay. The professionals and specialized artisans include three Brahmin families (two doctors' families and a village shopkeeper); the Masons (who do masonry, carpen-

TABLE 9-5. Household population by jati, 1965 and 1978 (residents only).

	Population		Number of families		Average family size	
Caste	1965	1978	1965	1978	1965	1978
Brahmin	24	19	3	4	8.00	4.75
Jat	332	373	45	57	7.38	6.54
Mason	42	60	7	9	6.00	6.67
Goldsmith	16	13	4	4	4.00	3.25
Barber	2	8	1	1	2.00	8.00
Water-carrier	2	1	1	1	2.00	1.00
Potter	78	98	5	11	15.60	8.91
Cotton-ginner	10	12	1	3	10.00	4.00
Sadhu	10	16	1	1	10.00	16.00
Harijan	242	310	41	52	5.90	5.96
Sweeper	48	67	9	9	5.33	7.44
Total	806	977	118	152	6.83	6.43

Source: Personal village censuses, 1965 and 1978.

try, blacksmithy, and milling); two families of Potters (who make clay vessels); and several families of Weavers (included among Harijan caste).

Almost all of the village population increase occurred among laborers and artisans. The reasons for this, as well as for the shift in family sizes, lie in shifting managerial responses to the perceived relationships between education, farming, and other forms of employment in the region. These have led landowners to migrate out of the village at greater rates than others. Among Jats, the ratio of immigrants to emigrants was 1 : 1.87. That is, almost two people left for every one who came in (or returned from being out in 1965). Among Harijans, the ratio was 1 : 1.21, also a net outflow but much smaller than among Jats. By contrast, the ratio among Sweepers was 3 : 1, among Potters 1.8 : 1, and among Masons 1.67 : 1.

Landowners have been leaving this and similar villages to seek salaried work for generations in order to obtain cash to expand or improve their holdings (Kessinger, 1974:94–96). Concomitantly, they have always invested relatively heavily in education. As the green revolution developed, this old pattern, along with the greater per capita income on larger farms and increasing need for managerial sophistication in place of simple labor power, induced farmers with marginal holdings or less interest in agriculture to sell their lands to others. Those who stayed in farming concentrated more on supervising work teams, administering herbicides and insecticides, maintaining equipment, arranging sales, repaying loans, purchasing seed, and controlling water distribution. Such tasks, unlike direct labor, allow significant economies of scale.

Some families who sold their land left the village. Others modified their houses and commuted to salaried employment in nearby towns and villages as teachers, clerks, and the like. The labor capacity thus removed from the land-owning group was replaced by poorer people returning or immigrating from other areas to work as intensive field hands, usually in well-organized teams. A laborer's income in 1978 was about ten times what it had been in 1965.

Harijans are emigrating to take advantage of a legal reservation for them of 25% of all public jobs. This gives them a considerable advantage in realizing salaried returns from education compared to equally poor families of other groups.

The bases of managerial control did not change greatly between 1965 and 1978, although there was a movement away from strictly local security measures to greater reliance on regional police, as indicated by a willingness to leave tractors and well equipment in the fields. There were, however, definite changes in the relative importance of different resources, especially in the increased importance of formal knowledge compared to land, muscle power, and tools. Each family, of every economic class, had to balance the difficulty of financing formal education with the need to obtain local knowledge and short-term income. Once education was completed for one or more family members, there were constant problems of striking a balance between using external employment to build up the family's position in the village agricultural system, versus using it to find a way to move out of agricultural work entirely.

The Economy

The *economy* is the system of buying and selling, and of evaluating objects for buying and selling. Its core is the system of ideas that define money, the concept of ownership, and the conventions of market transactions. The major changes in this system between 1965 and 1978 were the introduction of new kinds of credit and the liberalized security requirements.

As money values are applied to objects in the context of market conventions in any area, a pattern of known and expected prices is established. This forms a dynamic secondary system of information of more immediate interest to decision makers than the general conventions themselves. Between 1965 and 1978 there was an overall price inflation in Punjab of about 400%.

A *market*, in ordinary usage (as opposed to formal economic usage), is a set of priced goods whose uses are mutually interrelated. In this sense, there were markets in the village for credit, crops, land, capital improvements to land, labor, housing, cattle, and transportation. A few examples will explain how they worked.

The Market in Credit. Credit in Punjab is not really an autonomous commodity available for general purchase by anyone able to pay the price. Rather, it is an aspect of purchase arrangements for some specific types of goods. Nevertheless, credit increased in importance throughout the period of the green revolution.

The increased availability and liberalization of security requirements for agricultural credit have already been indicated. The remaining point is that credit became absolutely cheaper after 1968 than it had been in 1965, and rates subsequently remained low or declined. The main source of credit in 1965 had been private commodity brokers who had offices in designated market towns. Unlike banks and the cooperative, they lent without material security, usually charging 12–18%. When the cooperatives were reorganized, they charged between 10.5 and 12.5%, subject to an absolute credit limit of Rs. 9,000 per farm for fertilizer and related inputs. This amount was adequate for the small farms in this area, and forced the brokers into other lines of work.

In addition to cooperatives, branches of some major private banks had been established widely in small towns and large villages throughout the state in response to government urging and were also offering credit for investments such as shops, small factories, tractors, and wells.

The Market in Crops. It is often suggested that only market crops, and not subsistence crops, are evaluated for their economic profitability. In reality, all crops are so evaluated. The culturally established algorithms used in making such evaluations are one of the most important keys to the pattern of farming behavior. They tell farmers what resources to bring together for each crop, and allow them to estimate their possible gains. Ultimately, calculations based on the algorithms structure competition between farmers and provide the logic that underlies both familial and village cropping strategies. Examples of the cropping algorithms for the major crops of 1978 on land of the "best" grade are given in Table 9-6.

Given this information, farmers evaluate each crop from two main perspectives. If there is a local need, they ask if the cost of production is significantly less than the price they would pay in the market. If it is, they will grow enough for the house as a minimum. Beyond this, if the cultivar can be placed in a rotation that will be as profitable as or more profitable than the rotations that have yielded the greatest profit recently, then more will be planted. Exactly how much depends mainly on ecological, rather than economic, factors: available labor, water, the need for by-products (such as fodder), and the exact parcels of land at the farmer's disposal. Cotton, for example, shows a very high difference between cost and return, but could not be planted in rotation with wheat and only grew well after fallow or legumes.

The Market in Land. The price of land serves as a benchmark for evaluating all other prices. In 1965 farmers quoted the price of the best grade at Rs. 17,650/ha (about U.S.$3,716 at the legal exchange rate). In 1978 the price was Rs. 76,000/ha (about $7,600). For average land the prices were Rs. 9,400 and Rs. 47,000 for the two respective years. In addition to outright purchase, land could be obtained by mortgage, rent, and holding in trust; in the order stated, these methods were progressively less expensive and less secure. In a *mortgage*, a creditor lends the landowner one-half the market value of the land and occupies the land, farming it, in place of collecting interest. Land farmed on a mortgage arrangement must be returned whenever the owner repays the money,

TABLE 9-6. Principal economic characteristics of major crops, 1978.

Crop item	Gain[a]		Cost[a]	Net[a]
Wheat				
Grain: 4 quintals (Q) @ Rs 130	520			
Fodder: 13 bundles @ Rs 2	26			
Seed: 6.3 kg @ Rs 2 each, once every three years.			4.00	
Fertilizer:				
Urea: 1 quintal			30.00	
or Can: 21 kg and superphosphate: 32 kg		or	(13.00 + 18.00)	
Waterings, 6.[b]				
@ Rs 2 each (electric)			12.00	
or @ Rs 6.75 each (diesel)		or	40.50	
Weeding, chemical			10.00	
Harvest, 1/25th of yield			21.84	
Transportation to Uncha Pind, Rs 1/Q			4.00	
	546	min.:	81.84	460.16
		max.:	103.34	438.66
Maize: Ganga #1				
Grain: 3 Q @ 130	390			
Fodder: 10 bundles @ 2.60 ea	26			
Seed: 1.7 kg			–	
Waterings: 5				
@ Rs 2 each (electric)			10.00	
or @ Rs 6.75 (diesel)		or	33.75	
Fertilizer				
Compost: 3–4 reris @ Rs 10 ea			30.00	
Can: 42 kg			28.00	
Superphosphate: 33 kg			18.50	
Hoeing: 2 @ Rs. 5 and food (Rs. 5)			20.00	
Transportation: Rs. 1/Q			3.00	
	416	min.:	109.50	306.5
		max.:	133.25	282.75
Rice: Jaya				
Grain: 6.25 Q @ Rs. 175	1,094			
Strat: 20 bundles @ Rs./2	40			
Puddling:				
with own oxen			0.00	
or with hired tractor		or	6.50	
Seed 2 Kg @Rs 2 each			4	
Transplanting			10	
Waterings, 25 times: @Rs 2 each (electric)			50	
or @Rs 6.75 ea (diesel)		or	168.75	
Fertilizer				
41 kg NH3			29	
28 kg superphosphate			12	
Transportation to Uncha Pind Rs. 1 per Q			6.30	
	1,133.75	min.:	117.80	1,024.70
		max.:	236.55	905.95

(continued)

TABLE 9-6. Continued.

Crop item	Gain[a]		Cost[a]	Net[a]
Sugarcane				
Cane 85/Q @ Rs. 13.50 each	1,147.50			
Fertilizer				
50 kg urea			80	
Waterings: 10:				
@ Rs. 2 (electric)			20	
or @ Rs. 6.75 (diesel)		or	67.50	
Hoeing: 2–3 times, Rs. 10 each			30.0	
Transportation to Uncha Pind			85	
	1,147.50	min.:	215	932
		max.:	262.50	884.50

[a]Per bigha (1 bigha = .084 ha)

[b]This is the rate to purchase water if the farmer has no well. With a well, the cost is Rs.37/bigha for diesel, and variable but less with electric motor.

immediately after the harvest of whatever crops are standing at the time. Annual rental rates for land were consistently about 3% of its market value, which, as farmers worked it out, was half the value of the net gain in crops it would produce (Leaf, 1984:125). Land held in trust was free, but had to be surrendered on demand.

In 1965 only 57% of the land was farmed by its owners; 31% was farmed on tenancy and 9% on mortgage. In 1978 over 88% was farmed by those who owned it, 5% on tenancy, and 3% on mortgage. There were two main reasons for the shift. Rainfed land had been rented often for fodder production to supplement irrigated production on well land. This strategy disappeared with rainfed land and persian wheels. In addition, new laws made it easier for a tenant to claim to be a hereditary tenant, whom the owner could not remove at will. This made owners less willing to rent.

The Market in Capital Improvements. Throughout the period of the green revolution, the capital investment that yielded the greatest return was well construction. The rising water level had reduced the cost of building a traditional shaft well to Rs. 2,000 in 1965. Such a well could upgrade 2.5–5.0 ha of rainfed land to well land, with a gain in market value of up to Rs. 12,000. Building a well thus yielded an instantaneous economic return up to six times its cost, whereas buying developed land did not.

A tube well with pump and motor was Rs. 7,000 in 1978. There was no more rainfed land to upgrade, but as an index to expected gains in productivity, the differential between canal land and well-canal was about Rs. 16,000/ha. One well could upgrade about 2.5 ha. The savings in draft animals and the ability to run additional machines with the pump motors were added gains.

The economics of tractor ownership after about 1970 were similar to those of well construction. It was widely held in official circles in 1978 that farmers with less than about 12 ha could not use a tractor economically. A tractor cost between Rs. 55,000 and Rs. 85,000. It could plow or cultivate 4 ha in 2 days at most and could be used for custom work the rest of the season. A reasonable gross return from this was Rs. 50,000 a year, equivalent to the gross income from about 3.5 ha of good land, although the tractor cost only about the equivalent of 1 ha. By 1980 six more tractors had been bought, and in 1986 there were no more bullocks in the village.

An important complement to crop production aids were new means of storage. In 1965 there were two main methods: sealed bins made of brick or adobe, and sacks on a pallet, preferably over a concrete floor in a tight room. These could reduce storage losses to 10–12% so long as the roof did not leak. Galvanized storage drums had been available in 1965 but were relatively expensive and small, so they were used only for such remunerative purposes as holding sugar for crystallization. By 1978, air-tight galvanized drums were available that could hold 1 mt of grain. The price was Rs. 325, but the government would issue a 50% subsidy. Cooperative buying had reduced annual grain price fluctuations to about 10%, but even so such a drum could pay for itself in 2 or 3 years. Most farmers had two.

The Market in Labor. There are three main points to make about the relationship between the market in labor in the village and the overall process of intensification. The first is that the basic structure of the market and its conventions did not change. The second is that wages did not change as much as many other prices. The third is that nevertheless laborers' incomes, both monetary and real, increased greatly.

The village followed a general practice in all types of hiring arrangements whereby a provision was made for maintenance plus a fee over and above it. In 1965 the artisans' wage rate was about Rs. 5/day plus food; in 1978 it was Rs. 25 plus food. Ordinary laborers received Rs. 2/day plus food in 1965, and Rs. 5 plus food in 1978. Food was the same for both types, in both years. When the 1978 rates were multiplied by the larger number of workdays/year, the gains were dramatic. Annual wages, including the value of food received for an ordinary laborer, would have been about Rs. 490 in 1965; in 1978 the lowest-paid agricultural worker would have earned about Rs. 1,600, a skilled worker about twice that (Leaf, 1984:132).

The Market in Housing. Between 1965 and 1978 housing costs increased more than most others, and this undoubtedly accounts for the relatively low rate of increase in housing area compared to the growth in population. Houses are so seldom bought and sold outright that there is no price for houses as such. The relevant price is for construction, the work of the masons. This is set on the basis of materials plus labor. Labor is paid at the rate just described, with the understanding that each mason gets the skilled rate and has two assistants paid at the lower rate. Materials are bricks, cement, iron bars for windows and reinforcement, and iron beams for ceilings. Between 1965 and 1978, the price of bricks increased tenfold, while the other components quadrupled.

The Market in Cattle. There were well-established markets for all cattle, and the way prices changed between 1965 and 1978 illustrates particularly well the relation between the ecology and other cultural systems in forming the overall farming system (Table 9-7).

Bullock prices increased fourfold and buffalo prices somewhat less. Camel prices declined. Buffalo are in demand because of improved credit for buying them, more discretionary income for buying milk locally, improved facilities for marketing beyond the village, and sufficient fodder. The prices have not gone up more, evidently, because the supply has increased. Buffalo must first produce calves in order to produce milk, and the government's artificial insemination program assures that the calves will themselves be good milk animals. In one year a buffalo can produce milk whose value exceeds its purchase price. Three-quarters of its fodder consists of crop by-products that have no incremental cost and very little net market value. This is the *economic* reason for the great increase of the buffalo population. The new camel price is presumably sustained by other villages in the area still converting to mechanical pumping.

The Market in Transportation. The increased flow of goods in and out of the village was supported economically by a relative decline in transport costs. The improved oxcarts of 1978 sold for between Rs. 2,000 and Rs. 2,500, as compared with Rs. 800 for the old type in 1965. Since the new carts can carry twice the old load and last twice as long, the prices represent a decline in the average cost per unit of capacity. Costs were further reduced by a state government policy of opening many more market centers. From 1965 to 1978, round-trip time to the nearest market was reduced from a full day to one-third of a day.

Transport costs in 1965 usually amounted to slightly more than 2% of the price of high-value commodities like wheat or brown sugar, but could be as much as 50% of the price of low-value goods like fodder or manure. By 1978 they had declined to less than 1% of the value of the major grains and less than 20% of the value of the low-value goods. This directly supported the opening up of the village ecology to regional resources. Transportation by tractor cart and truck was competitive in price with that by oxcart.

Between 1965 and 1978 there was a marvelous fit between what was profitable economically and what was productive ecologically. Yet prices did not cause ecological relations, or the reverse. Rather, people referred to both prices and ecological information to make their decisions, bringing the two factors into alignment. This fact was used by wise administrators to speed change, as in decisions to hold down the prices on well pumps and other key inputs, and to support the prices of commodities. By the same token, however, different pricing policies could have precisely the reverse effect. Since 1978, in fact, this has been the case.

TABLE 9-7. Economic characteristics of cattle, 1965 and 1978.

Type	Life span	1965 prime price	1978 prices	Annual fodder consumption	Yield
Buffalo	6–10 yr	Rs. 800	Rs. 2,000–3,000	54.0 quintals	Milk at 10 kg/day max., 1 kg/day min.
Bullock	10–12 yr	Rs. 500	Rs. 1,500–2,500	27.0 quintals	One pair can pull 20 quintals in a new-type oxcart.
Camel	12–15 yr	Rs. 1,000	Rs. 500	54.0 quintals	Can carry approx. 4 quintals in pack or replace two bullocks lifting water
Goat (doe)	N/R	Rs. 30–60	Rs. 200–350	13.5 quintals	Milk at 0.5–1.0 kg/day, then meat
Donkey	12 yr	Rs. 200	Unknown	13.5 quintals	Carries 0.75 quintals/pack
Mule	20 yr	Rs. 3,000	Unknown	27.0 quintals	Carries 1.5 quintals/pack
Pig	N/R	Rs. 2.25/kg.	Rs. 8–14/kg	N/R	Bristles, then meat
Ram	N/R	Rs. 2.75/kg	Rs. 12.00/kg	3.5 quintals	Carries 0.75 quintals/pack

Source: Farmers' estimates.

Shahidpur and Punjab

The changes in Shahidpur's agriculture were repeated throughout the central Punjab. Between 1965 and 1978, total food grain production increased from about 3,370,000 mt to 11,676,000 mt.[2] The greatest part of this came from increased productivity per hectare. For example, the yield for wheat over this period increased from 1.24 mt/ha to 2.73 mt. In rice the increase was from 1.00 mt to 2.55, making rice the second most important crop in the state in total tonnage after wheat. Subsequently, 1980 was a disastrous year because of widespread hail in the winter crop season, followed by government policies designed to force sales to government stores at prices far below normal market levels. Concurrently, oil price increases generated upward pressure on the prices farmers paid for inputs that the government was unable to resist, resulting overall in increasingly unfavorable terms of trade for agriculture through at least 1984. The political turmoil in the state over this same period, fueled in part by these price relationships, has had its economic consequences, but the basic pattern in place by 1978 has held. The revolution has moved on, not back.

Punjab's 1964 multicrop agriculture, with high levels of investment and irrigation from mixed sources, became dramatically more intensive, and this was supported by new institutional arrangements. Levels of investment, irrigation, and productivity per capita all increased. These changes were associated with changes in specific methods of farming. Yet growth in agriculture at the state level did not result in any simple way from a directly parallel expansion of the farming systems of the millions of households of the state—as a sack of balloons might expand because each individual balloon expanded.

Households changed in many directions. Some large farmers sold off land and sent family members out for salaried work as teachers, officials, and technicians. Others bought more land and equipment to go with it. Laborers who had done some farming moved back to more complete reliance on wages, but in a more specialized form, and farmers relied on them for more of their total farm labor requirements than previously. Village artisans invested in expensive equipment to repair the farmers' new tools and mill their grain, village cooperative secretaries and *panchayat* members expanded their field of action, doctors and other medical specialists spread through the villages, and numerous new town and regional specializations developed. Each supported the other; each contributed to the structural change that was both a concomitant and a basis of the increased productivity.

From the farmers' viewpoint, there was never one typical farming system or even a reasonably limited number of them. There were, rather, the general information systems and the uses that could be made of them. Household farming systems were like individual games, and the general information systems were the rules. In Punjab's green revolution, both changed, in an ongoing and ordered interaction.

Notes

1. The rupee was exchanged legally at about Rs. 4.75 to the U.S. dollar in 1965, and illegally at about Rs. 7.50 to the dollar. In 1978 the legal rate was about Rs. 8.00 to the dollar, and the illegal rate was about Rs. 12.00. None of these rates gives an accurate impression of the rupee's buying power and psychological importance. Generally, at both times it was more analogous to a dollar than to any part thereof. In 1965 a workshirt cost R. 8, a bicycle Rs. 30, a kilo of meat Rs. 2.75. In 1978 the prices for the same items were Rs. 20, Rs. 130, and Rs. 12. Accordingly, to convey most clearly the sense of the values involved, prices have been left in rupees.

2. The figures for 1965 are from *The Statistical Handbook, Punjab 1976–1977*. Those for 1978 are from Gill (n.d.).

References

Bayliss-Smith, T. P. *The ecology of agricultural systems*. London: Cambridge University Press, 1982.

Beets, W. C. *Multiple cropping and tropical farming systems*. Boulder: Westview Press, 1982.

Dhillon, A. S. *Package practices for the kharif crops of the Punjab state, 1978*. Ludhiana: Punjab Agricultural University, 1978.

Gill, M. S. *Punjab maintains lead: A review of agricultural production in Punjab*. Chandigarh: Agricultural Information Service, Department of Agriculture. (Undated, about 1982.)

Haines, M. *An introduction to farming systems*. New York: Longman, 1982.

Hough, E. M. *The co-operative movement in India* (5th ed.). Revised by K. Madhava Das. London: Oxford University Press, 1966.

Kessinger, T. G. *Vilayatpur: 1848–1968*. Berkeley: University of California Press, 1974.

Leaf, M. J. *Information and behavior in a Sikh village*. Berkeley: University of California Press, 1972.

Leaf, M. J. Economic implications of the language issue: A local view in Punjab. *Journal of Commonwealth and Comparative Politics*, 1976, 14, 197–203.

Leaf, M. J. *Song of hope: The green revolution in a Punjab village*. New Brunswick, NJ: Rutgers University Press, 1984.

Shaner, W. W., Philipp, P. F., & Schmehl, W. R. *Readings in farming systems research and development*. Boulder: Westview Press, 1981.

10

Intensive Paddy Agriculture in Shyampur, Bangladesh

ABU MUHAMMAD SHAJAAT ALI

The more the mango, the more the paddy, and the more the tamarind, the more the floods.
—Bengali proverb

A popular explanation of agricultural intensity and its change involves the relationships between forces of demand, especially as characterized by population pressure, and the input–output character of the farming system (Boserup, 1965; Brookfield, 1972; Turner, Hanham, & Portararo, 1977). Most case studies of this relationship have focused on "unsaturated" agroecosystems (Barlett, 1976; Brookfield, 1962; Lagemann, 1977; Norman, 1977) in which increased demand can be countered by the areal expansion of agriculture, intensification of production, or both without radical changes in the farming system. Studies of near-saturated agroecosystems, in which expansion and intensification of agriculture are highly constrained, have shown mixed results in regard to the demand theme (e.g., Metzner, 1982). Some indicate that continued population pressures lead to increased output per unit area, but at very high rates of diminishing returns to inputs. Others find that stasis in production has been reached (Parrack, 1969).

Farming systems throughout much of Bangladesh operate in near-saturated conditions (e.g., Chapman, 1984; Stoddart & Pethick, 1984). Given the current state of the farming systems, highly intensive, wet-rice agriculture seems to have reached its production capacity, and production is not keeping pace with demand resulting from increasing population pressures. This study examines the wet-rice farming system of the village of Dakshin Shyampur in Dhaka, in the central part of Bangladesh, as a case of a near-saturated agroecosystem. The central issue is the role of pressures of demand on agriculture.

Abu Muhammad Shajaat Ali. Department of Geography, Jahangirnagar University, Savar, Dhaka, Bangladesh.

The Village Setting

The village of Dakshin Shyampur is situated in the northern part of the Dhaleswari River floodplain in the central region of Bangladesh. Everywhere the land is flat and moderately drained. The floodplains of the Turag and Bansi rivers, tributaries of the Dhaleswari, are located to the east and west of the village, respectively (Figure 10-1). The only significant elevations on the floodplains are the low artificial ridges between the rivers, built primarily for settlement and associated nonagricultural uses.

Shyampur typifies the villages of the region. It is small in size, with 482 people on 22.5 ha, and overcrowded, 21.4 persons/ha (2,164 persons/km^2). Virtually every family is Islamic, illiterate (ca. 12% of population is formally educated), and engaged in some facet of agriculture. Intensive wet-rice cultivation is carried out on 13.2 ha of the village's land. But even with three harvests a year of various crops, insufficient food is produced to feed the population.

Three classes of farmers exist in the village. The *landless* (22 families) own only a homestead and/or less than 0.2 ha of land. *Smallholders* (20 families) own from 0.2 to 1.0 ha of land, including a homestead. *Largeholders* (17 families) own more than 1.0 ha but less than 3.3 ha, including a homestead. Because Dhaka City is only 8 km to the east, day-labor employment can be found. Also, a small number of villagers engage in business or are lower-level civil servants.

Shyampur was established in the late 1940s by 50 people (17 families) from neighboring villages. The settlers occupied the higher ground between the Turag and Bansi rivers, and began to raise this ground by 3–5 m above the rivers to ensure the safety of homes from floods. The population has grown steadily through 1980 (Table 10-1). In 1980 there were 59 families in the village, with an average size of 8.2 persons per family. There were 274 males and 208 females, and a working population of 114.

Physical Setting

The environment is wet, and the villagers have elevated the interfluvial ridges to create a flood-free living zone. These ridges and the minor relief of the floodplain create five levels or elevations relative to the rivers and permanent water bodies. Level 1, *chala* or *tek*, rises some 5 m above the rivers. It comprises some 6.9 ha or about 30.7% of the village land. It is free of flood and used mainly for homesteads, orchards, and gardens (Figures 10-2 and 10-3). Level II, *nama*, constitutes the upper portion of the floodplain, covering 1.8 ha or about 8.0% of the village land. It is infrequently flooded and is used for the cultivation of aus paddy (*Oriza sativa* L. var.), wheat (*Triticum sativam*), winter vegetables and pulses, and jute (*Corehorus olitorious*).[1] Levels III and IV, the middle and lower portions of the floodplain respectively, are called the *baid*. Baid lands

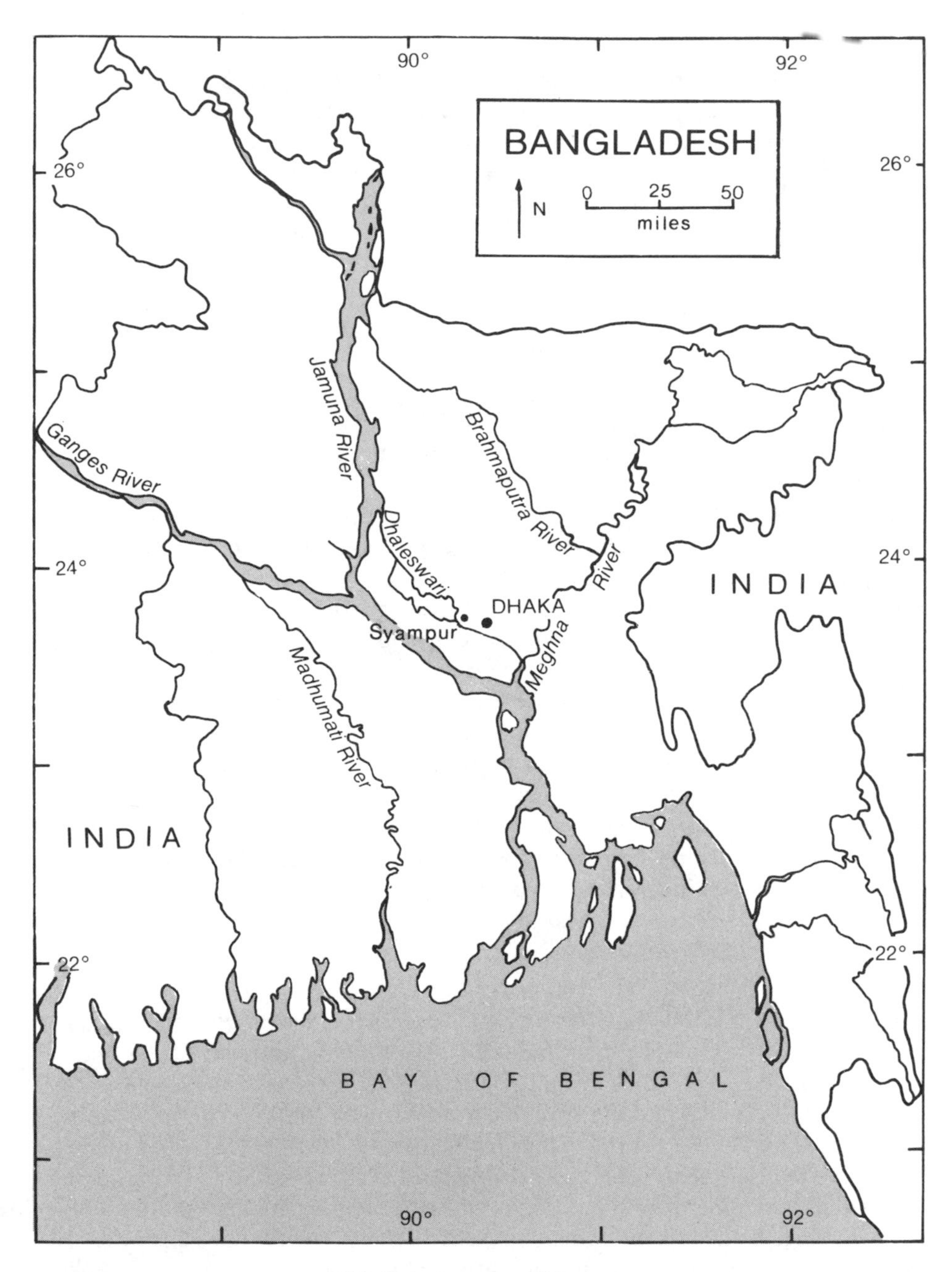

FIG. 10-1. Location of Shyampur.

TABLE 10-1. Population and land use, 1950–1980.

		Area (ha)					
Year	Population	Settlement	Cropland	Perennial trees	Idle land[a]	Water bodies	Roads, footpaths
1950	50	4.1	15.3	0.4	0.8	0.6	1.3
1960	175	4.8	15.0	0.5	0.5	0.4	1.3
1970	290	6.1	13.6	0.5	0.6	0.4	1.3
1980	482	6.7	12.8	0.8	0.5	0.3	1.4

Source: Ali (1982).

[a]By 1980, most idle land was owned by urban developers.

occupy 13.5 ha, nearly 60% of the village. They are regularly flooded during the monsoon. The upper and middle *baids* (Level III) are suitable for aman paddy(*O. Sativa* L. var.), winter vegetables, wheat, and jute, while the lower *baid* (Level IV) is suitable for boro paddy (*O. sativa* L. var.).[2] About 1.3% of the total area of the village is composed of permanent water (Level V), including small ponds and depressions, as well as drainage canals linked with the Turag River.

The soils in the village are of two distinct types: (1) red or mature soils, related to the *chala* lands; and (2) various alluvial soils associated with the *nama* and *baid* lands (Figure 10-4). The red soils are sandy loams of poor agricultural fertility. They are well drained and easy to till, but are poor in lime, phosphoric acid, potash, and organic matter. The alluvial soils are immature floodplain deposits ranging from sandy loams to clays in texture and from blackish grey to black in color (Government of Bangladesh, 1972). These soils are difficult to till but tend to be fertile for agriculture if properly drained.

Based on environmental characteristics, particularly soil, drainage, flooding, and suitability for the growth of particular cultivars, the land of the village can be categorized into three broad agricultural land types.

1. *Low agricultural constraints* (68% of the village, including all of land Levels II, III, and IV): This land is gradually sloping, infrequently to regularly flooded, and suitable for year-round cultivation. Annual inundation nourishes the soil, but high floods sometimes damage crops.
2. *Medium agricultural constraints* (30.7% of the village, including all of land Level I): This is the highest elevated land that is level and flood-free. It is suitable for garden cultivation if irrigated and fertilized. Productivity is lower than that of land Levels II, III, and IV.

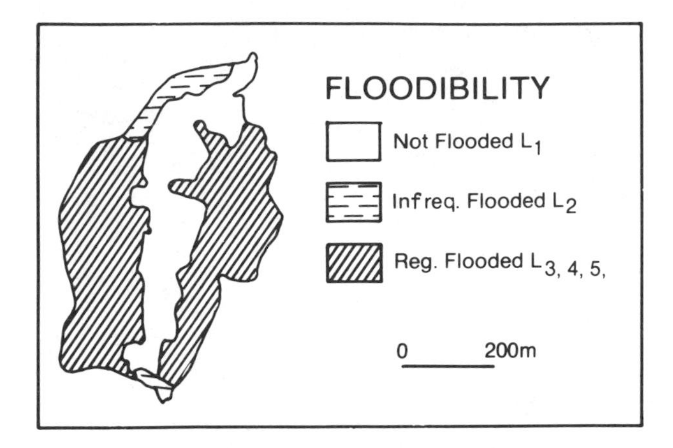

FIG. 10-2. Flood characteristics, Shyampur.

FIG. 10-3. Diagram of land levels and uses.

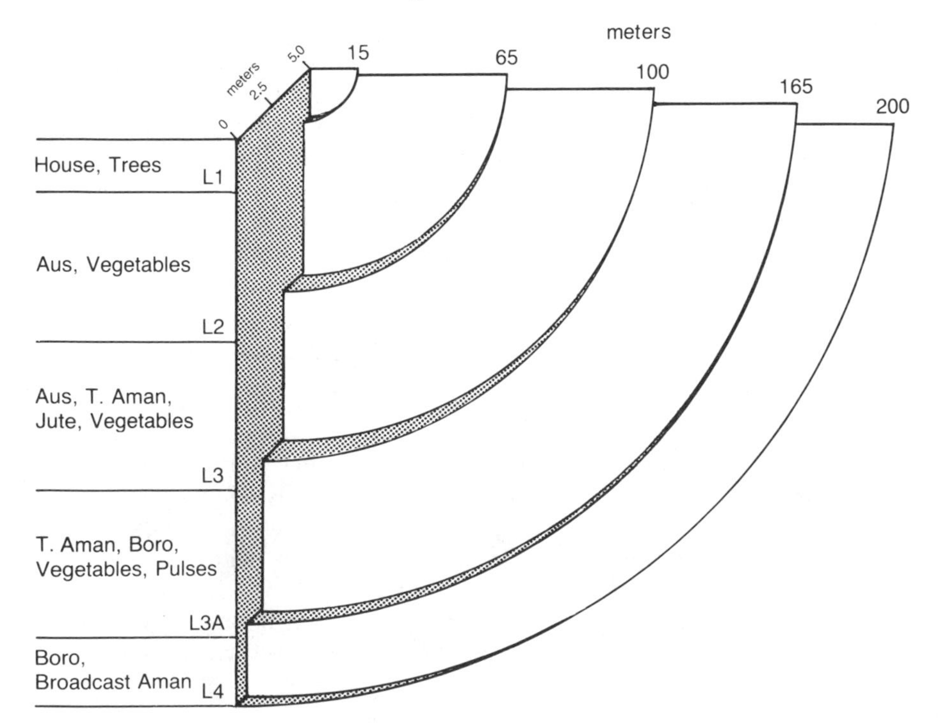

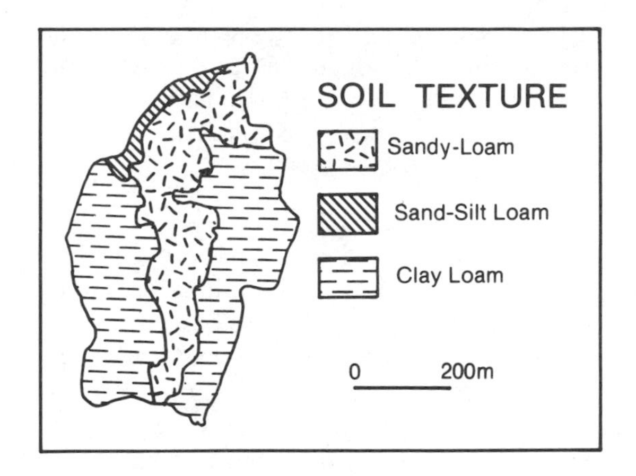

FIG. 10-4. Soil textures, Shyampur.

3. *High agricultural constraints* (1.3% of the village, including all of land Level V): These lands mainly comprise water bodies and severely flooded depressions; the latter inhibit cultivation.

The village enjoys a tropical monsoon climate. Three distinct seasons exist in the year: the nor'wester (March–May) characterized by *kalbaisakhi*, or hot weather, violent thunderstorms and cyclones, and low rainfall (an average of 64.07 cm); the monsoon (June–October) characterized by hot weather (+35°C) with heavy rainfall (averaging 146.39 cm); and the winter season (November–February) characterized by cold (18°C average minimum temperature in January), dry weather (average rainfall of 8.33 cm). At no time does temperature impede cropping.

The monsoon rains cause floods in the area. Floodwaters deposit large amounts of silt and blue-green algae to the *baid* lands, providing excellent nutrients to the soil when the water recedes. During the winter and part of the nor'wester season, the soils of Levels II–IV dry out, and irrigation is necessary for cropping.

Agriculture

Though small in area, Shyampur is endowed with relatively good quality agricultural land. The climate and soils favor the cultivation of double and triple cropping, with the exception of some low-lying fields that are flooded for long periods. Average cropping intensity in recent years ranges from 200 to 277%, or a frequency of cultivation of 2.0–2.7 harvests/plot/year. The principal type of cultivation is rainfed paddy (wet-rice) agriculture, supplemented by irrigated paddy in the dry season and by the cultivation of vegetables, jute, wheat, and legumes. The system can best be characterized as a modified traditional one, in which production has been mainly for subsistence (consumption) but is increasingly changing to commodity or market production. Paddy is the principal food crop, jute and vegetables the principal cash crops. Farms are very small in size and are arranged in fragmented plots.

Land Use

In 1980 Shyampur had 12.8 ha of land under field crops. There were also 0.8 ha of economic trees and about 0.5 ha of land in idling (see Table 10-1). Economic trees include fruit and timber species, as well as bamboo. A drainage canal and two ponds account for 0.3 ha. The dirt roads and paths leading to the local market, neighboring villages, and fields (*ail* or footpaths) cover about 1.4 ha. Houses and associated structure take up 6.7 ha of land.

During the past 30 years, the village has experienced rapid population growth and significant changes in its land-use patterns. Most notably, land under homesteads has increased by 63% in order to accommodate the addition of 42 families or 432 people. Consequently, arable land has declined by 16.3%, reflecting a conversion of 2.5 ha of arable land into settlement and nonagricultural uses. Among other land uses, the area under perennial trees has increased, while the area of idle land and water bodies has decreased (see Table 10-1).

Cropping Schedule

The monsoon rainfall and associated floods play a major role in various aspects of cultivation, particularly in scheduling. Approximately 30% of the annual precipitation falls during the nor'wester months (March–May), facilitating rainfed aus paddy, jute, and other crops known as *bhadoi* or *kharif* crops.[3] From June through October, 67% of the annual rainfall occurs. All cultivars grown during this time are called cold-weather or *haimantic* crops.[4] The winter (November–February) is dry, with only 3% of the annual rainfall. Cultivation during this time involves some irrigation. All cultivars grown at this time are called *rabi* crops.[5]

The agricultural year begins in late February (*falgoon*) with the preparation of plots in the Level II (*nama*) and Level III (*baid*) lands, which are usually not flooded. Fields are plowed three to four times. Plowing is generally done from 6:00 A.M. to 2:00 P.M. and 4:00 P.M. to 6:00 P.M. It takes about one working day (8–10 hours) for one male with a wooden plow and a pair of oxen to plow 0.2 ha (1.5 *pakhi* or 0.5 acres). Therefore, plowing takes an average of 15–20 worker-days/ha. While waiting for the rains, the farmer breaks large earth clods with a wooden hammer (*mugur*), iron spikes, and wooden beams. Immediately after the rains begin, the field is plowed again, weeded, and leveled. An overlapping, oval-shaped plow is the basic instrument used (Ahmad & Khan, 1963:22). Plowing continues until mid-April, by which time the field has been plowed three or four times.

Field preparations also include the construction and maintenance of ridge-like plot boundaries (*ails*) and bunds (*bandhs*), both of which retain water in the plots. *Ails* are about 0.5 m in height and are made of earth and weeds collected from the field. About 12.35 worker-days are required to construct all *ails* around a 1-ha plot of land. Ails also serve as footpaths used to walk into the field without disturbing the crops. Large *ails* or bunds (*bandhs*) enclose a number of large and smaller plots. Bunds are about 1 m in height (sometimes 1.5 m), are made of earth and weeds, and are rather permanent features. Bunds are normally made in order to protect fields from floodwaters and to impound water on the higher level of land. Since agriculture depends on either the availability of water or the protection of land from floodwater, farmers take care to ensure that all the *ails* and *bandhs* are well maintained.

Land preparations for the rainy season and cold weather harvests are done from late February through March; land preparation for the dry, early-summer harvest is done in November and early December.

Rice seeds are cleaned and kept in a closed mud pitcher full of water to facilitate germination before sowing. There are differences in terms of seed, fertilizer, and labor requirements among local and high-yielding varieties; broadcast and transplanted paddy; aus, aman, and boro paddy; and jute. Difference also exists in scheduling (Table 10-2; Figure 10-5).

In the early part of the agricultural year (March), the fields of land Levels II and III a and b (*nama*, and upper and middle *baid*) are cultivated to *bhadoi* crops, while the lower *baid* fields are sown to *haimantic* crops such as broadcast aman paddy. Aus paddy is sown in late March and continues until mid-April. At the same time, jute is sown on some Level III (middle *baid*) plots, and broadcast aman is sown on Level IV (lower *baid*) fields. Transplantation of aus paddy takes place in late April through late May, depending on the full outbreak of the nor'wester rains, which deposit 5–8 cm of water on the fields. The high-yielding variety (HYV) of aus paddy is a dwarf plant that is normally sown on irrigated land and can be transplanted earlier, in late March to mid-April. Both aus paddy and jute are harvested more or less at the same time, in July–August.

Immediately after the harvest of aus paddy and jute, the fields of Level II are cultivated with vegetables, while both the upper and middle *baid* fields (Level III) go under transplanted aman paddy. Aman paddy covers nearly all plots in Level III from July to December, and on Level IV from March to December. The full onset of monsoon rain in June facilitates this cultivation. Various vegetables, such as spinach (puisak, *Spinacia oleracea*); balsam apple (Karala, *Momordica* spp.); rigged gourd (*Lagenaria* spp.); taro (*Colocasia* escolenta); and okra (dhyaras, *Hibiscus esculentus*) are grown on Level II fields during the rest of the year.

Broadcast aman paddy is generally long-stemmed, survives in deep water, and can withstand heavy floods. The plant is grown in the rising flood waters and survives as "floating" paddy. Yields are low, but farmers cultivate it on the

TABLE 10-2. Current scheduling, input, and output for leading cultivars in Shyampur.

Crop	Time of sowing–transplanting	Time of harvest	Labor requirements (work days/ha)[a]	Seed requirements (kg/ha)	Fertilizer requirements (kg/ha)			Average yield (kg/ha)[b]
					Nitrogen	Phosphorus	Potassium	
Broadcast aus (local)	March	July–August	150	80	35	35	30	825
Transplanted aus (HYV)	March–May	July–August	250	22	180	110	110	1,700
Broadcast aman (local)	March	November–December	150	80	25	25	16	1,008
Transplanted aman (local)	June–July	November–December	190	27	35	35	30	1,850
Transplanted aman (HYV)	June–July	November–December	250	22	180	110	110	2,500
Boro (local)	November–December	April–May	190	27	35	35	30	1,900
Boro (HYV)	December–February	April–June	250	22	180	110	110	2,945
Jute	April	June–July	300	9	35	30	35	8 bales (each bale = 180 kg)
Vegetables[c]			40–300					3,394 kg–4,500 kg

[a]Work day = 8 hours by an adult male. The labor requirement for all crops are shown up to the point when the product is ready for marketing.

[b]All measures are for the cultivation of one crop per plot per year. Yield of paddy indicates the average yield of unhusked paddy. Yield decreases if the plot is cropped with paddy subsequently.

[c]Vegetables can be grown almost year round, including labor-intensive ones. Each cultivar has different seed, fertilizer, and labor requirements as well as different rates of production.

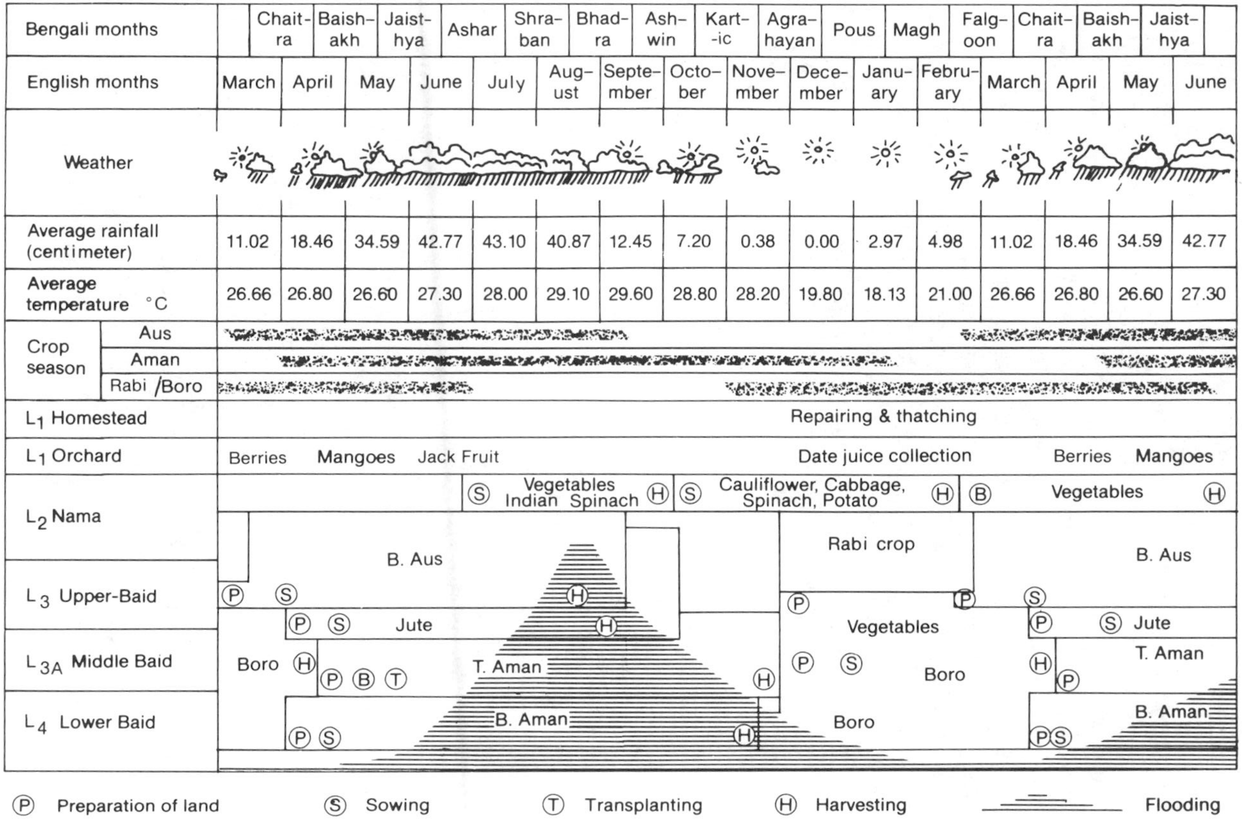

Bengali months: Chaitra, Baishakh, Jaistha, Ashar, Shraban, Bhadra, Ashwin, Kartic, Agrahayan, Pous, Magh, Falgoon, Chaitra, Baishakh, Jaistha

English months	March	April	May	June	July	August	September	October	November	December	January	February	March	April	May	June
Average rainfall (centimeter)	11.02	18.46	34.59	42.77	43.10	40.87	12.45	7.20	0.38	0.00	2.97	4.98	11.02	18.46	34.59	42.77
Average temperature °C	26.66	26.80	26.60	27.30	28.00	29.10	29.60	28.80	28.20	19.80	18.13	21.00	26.66	26.80	26.60	27.30

Weather

Crop season: Aus; Aman; Rabi /Boro

L_1 Homestead: Repairing & thatching

L_1 Orchard: Berries, Mangoes, Jack Fruit, Date juice collection, Berries, Mangoes

L_2 Nama: (S) Vegetables Indian Spinach (H); (S) Cauliflower, Cabbage, Spinach, Potato (H); (B) Vegetables (H)

L_3 Upper-Baid: (P) (S) B. Aus (H); Rabi crop; (P) (S) B. Aus

(P) (S) Jute (H); (P) Vegetables (P); (P) (S) Jute

L_{3A} Middle Baid: Boro (H) (P) (B) (T) T. Aman (H); (P) (S) Boro (H); (P) T. Aman

L_4 Lower Baid: (P) (S) B. Aman (H); Boro; (P)(S) B. Aman

(P) Preparation of land (S) Sowing (T) Transplanting (H) Harvesting Flooding

FIG. 10-5. Cropping schedules and related information.

lowest-level fields, which otherwise would be uncultivated except for the dry season.

Transplanted aman paddy is grown on Level III fields, where the degree of inundation favors this variety. Monsoon flooding (both rainfall and river) supplies the water and nutrient supply in the fields. Damage to transplanted aman paddy can result from excessive rainfall at the time of germination, uprooting the young plants. Low monsoon rainfall (drought) can result in insufficient water as the plant goes to ear, creating an empty ear.

As the monsoon flood waters recede, in late November to mid-December, Level II and III fields dry. The water leaves large amounts of silt and decomposed blue-green algae, which add nutrients to the fields. Shyampur farmers use the dry fields to grow various individual *rabi* (dry, early summer harvest) crops which include boro paddy, pulses, vegetables, and wheat. Recently, boro paddy and vegetables have become the dominant cultivars.

Boro paddy grows well in the low-lying *baid* (Level III and IV) fields, which comprises about 60% of the total village area. The requirement for good harvests of boro paddy is adequate irrigation facilities, because the fields dry out during the winter months (November–March). Boro fields need about 6–8 cm of water for 2–3 months. The major task for the farmers is to provide water to the dry fields. This is done by small-scale irrigation, in which water is lifted by traditional methods or by low-lift electrical or diesel pumps. If the water supply to these fields is ensured, a high paddy yield is also ensured. Boro paddy is normally sown in mid-November to mid-December, transplanted in late December and January, and harvested from mid-April to mid-June, depending on the time of sowing and on the variety of seeds.

The farming system of Shyampur is an intensive one in which cultivation is virtually continuous on many fields. The specifics of the field system are related to land level, floodability, rainfall, and water availability (irrigation facilities). The critical time of the agricultural year is March–April (the early months of the Bengali year), when the majority of the arable land goes under preparation, sowing, weeding, and so on. At this time, farmers must decide which fields to cultivate. Paddy is the consumption crop, but it involves considerable labor to produce and is vulnerable to loss by flood. Vegetables are somewhat less labor-intensive to produce than is paddy and are not so dependent on nor'wester rains. Moreover, vegetables have a good market value in the nearby cities. Paddy is the riskier venture ecologically because of flood hazards. In contrast, jute's market value has been low or too fluctuating in the past, creating production risks. The decision as to which of the two cultivars to grow hinges on perceptions of the price of jute or on the needs for paddy.

Field Characteristics

Fields in Shyampur are small in size and rectangular in shape, with minor exception. Family size and inheritance laws are the major causes of the fragmentation and small size of plots. Over 86% of the plots in the village are smaller than 0.2 ha; the largest plot is 0.6 ha. Sometimes larger plots are broken up into smaller ones by the use of temporary *ails*, which make it easier to retain irrigated water to cultivate different crops. On average, there are three plots of land per household in Shyampur.

The rectangular shape functions in relation to the types of cultural implements used and crops grown (Ahmad, 1961:24). Most of the cultivars (for example, transplanted rice, sugarcane, cauliflower, potatoes, and bananas) are sown in either rows or ridges. Irregular shapes result from a crossing footpath or a meandering channel.

Agricultural Implements

Traditional implements are dominant, including the plow, hoe, sickle, harrow, leveler, matchet (rake), wooden hammer, and wooden hoe. People and oxen are the principal sources of energy. In recent years a few largeholders and richer farmers have used small power tillers and spray machines, and low-lift irrigation pumps. Smallholders and poorer farmers cannot afford these tools.

Irrigation

Cultivation during the dry winter and early summer (November–March) months is dependent on irrigation facilities. Major crops grown at this time are high-yielding boro paddy and various kinds of vegetables, both of which require irrigation for adequate yields.

Both traditional and modern irrigation methods are practiced in Shyampur. Three common traditional methods are the pitcher (*kalash*), the swing basket, and the *don* (a pivoting log hollowed out in the shape of a dugout canoe and fitted with a counterbalance). Modern methods include the low-lift pump. In smaller plots of vegetable gardens, the pitcher is used to lift water from the wells and ponds. For larger plots, a swing basket and *don* are used to lift water from canal and marshes. Low-lift pumps are used on HYV boro in large plots because this rice requires more water than the local variety and because the technique becomes cost-effective on large plots.

Low-lift pump irrigation is generally organized by a group of farmers interested in cultivation of HYV boro paddy. Interested participants discuss the fields to be cultivated, probable cost, and management devices, and distribute the responsibility of installation and management of the pump. Sometimes government credit is available to farmers interested in purchasing a low-lift pump, but in most cases HYV cultivators bear the cost themselves.

Tenancy

Virtually every occupant of Shyampur is a farmer of some type. Five percent rent land; 12% are landowners and rentors; about 76% are landowners who neither rent nor lease; nearly 2% are landowners who both rent and lease; and 5% own and rent land.

Farm Production and Farm Income

Only 15% of the farmers in the village produce enough food to meet their annual requirements. Fifty-five percent harvest enough consumption crops to last from 4 to 9 months; and 30% produce an amount that lasts for less than 3 months. Largeholders (>1.0 ha) with small families may grow a jute cash crop. All farmers produce vegetables for market sale, and some farmers must also sell their consumption crop in order to meet immediate family demand for nonfood commodities; to repay debts; to purchase fertilizer, irrigation water, and cattle; or to meet the expense of marriages.

Despite the apparent depressed conditions, farm income in Shyampur is comparatively higher than in other "typical" villages in Bangladesh because of the proximity of the Dhaka City market and its high demand for vegetables. In 1980, 29% of the farm families reported that their annual farm income in addition to subsistence was below U.S.$93 or *taka* 2,500 (in 1984, U.S.$1.00 U.S. = *taka* 27.00). Another 40% had an annual farm income ranging between U.S.$93 and $278 (*taka* 2,500–7,500). The rest, 31%, had an income exceeding U.S. $278 (*taka* 7,500).

Nonfarm Income

Approximately 61% of the Shyampur farmers considered farming as their major occupation, although they may engage in off-farm work. The frequency and duration of off-farm work depends on household need, family size, time required for farming, and availability of off-farm work. Twenty percent work as wage laborers but also engage in small-scale farming. Businessmen with small farms constitute 7% of the male work force, while 12% are salary earners who have limited involvement in farming.

The majority of farmers (84%) are engaged in off-farm work more than 100 days a year, and only 16% worked fewer than 90 days a year off the holding. Off-farm employment, including small business, trade, industrial, urban, and agricultural work, is profitable. About 7% of farmers reported off-farm income below U.S.$75 in 1980. Seventy percent earned between $75 and $295 per year, and only 23% earned more than U.S.$295 per year.

Female employment outside the household and in agriculture is insignificant because a surplus of male laborers exists and because tradition encourages the female members of the family not to seek out-of-home employment. Females do raise poultry, work kitchen gardens, husk paddy, collect fuel, and perform household work for neighboring families.

Household Decision Making

In general, the decision-making process among farm households in Shyampur is similar. Male heads of households discuss with other male adult members, and with other laborers, issues involving the conditions of plots, the level of expected consumption and market demand, the cost of production, and the availability of resources. "Heads" then make decisions concerning agriculture and distribute responsibilities to the various household members. Social decisions are also made by the head but in consultation with his parents, wife, and eldest son.

Variables that strongly influence cropping intensity are the household's consumption needs and its monetary needs for clothing, education, ceremonies, and so on. If consumption needs are not met because of crop failure, decisions are invariably made toward the intensification of market crops.

Largeholders tend to keep informal accounts of their general income and expenditures, including costs of production and sales of market output. The household head and his eldest or another educated son keep these accounts. At the end of the cropping season, the cost and net benefit are calculated, either for each crop and or for all crops grown during the season. Farmers do not normally keep track of the family labor inputs, but they carefully record the hired labor inputs and costs. They also keep accounts of the cost of fertilizer, seeds, pesticides, irrigation, plowing, and harvesting. The total sale of products and net profit are calculated (Table 10-3). Farmers do not calculate marginal returns for labor and other investments separately. But after 2 or 3 years of cultivating the same cultivar and of observing the response of each to specific inputs and the market price fluctuation and net returns, they know which cultivar is more profitable and what its response is to labor and other inputs. Farmers also often discuss input–output relationships and market prices with one another.

In summary, agriculture in Shyampur is dominated by paddy cultivation supplemented by vegetables and other cash crops. Because of high population pressure and high market demand for cash crops, the cropping intensity in the area is very high. The frequency of cropping is influenced by the floodability of the land and availability of irrigation facilities. Boro is the leading paddy crop. The majority of the farmers are landless and/or smallholders, and most of the fields are less than 0.2 ha in size. Traditional farm implements are dominant, although diffusion of technological innovations and neotechnical inputs are evident. Sixty-nine percent of farm families earn less than U.S.$278 annually from the farm, and 77% earn less than $295 from off-farm work.

Population Growth and Agrarian Change

From 1950 to 1980 the population of Shyampur increased almost ten times. This high rate of increase can be attributed to both in-migration and natural growth. During the period 1950–1960, population growth was mainly due to in-migration of 23 families from neighboring villages. This migration related to the growth of Dhaka City, which created land pressures and also offered job opportunities. Since 1960, population increases in Shyampur have been primarily related to natural growth.

Population growth had placed pressures on agriculture, resulting in an

TABLE 10-3. Cost–benefit data of rice, jute, and vegetables in Shyampur, 1979–1980.

Crops	Cost per hectare (U.S.$)[a]					Total cost/ha (U.S.$)	Total production/ha (t)	Farmers' price/t (U.S.$)	Total price of products	Net return/ ha ($/crop)	Farmers' preference level
	Labor[b]	Seed	Irrigation	Fertilizer	Pesticide						
Boro (HYV)	82.0	5.6	74.0	18.5	2.8	183.0	2.3	166.7	383.3	200.3	High–very high
Jute	78.0	1.1	–	2.6	.6	82.3	1.26	101.9	128.0	45.7	Low–medium
Vegetables[c]	11.4	.4	2.8	3.7	–	18.3	9.2	24.9	229.0	210.7	Very high

[a]One U.S. dollar is equivalent to 27.0 taka in 1984.

[b]Excluding family labor. Wages of agricultural labor were approximately 15.0 taka in 1980.

[c]Calculated for spinach and amaranth.

intensification of cultivation. Fallowing has virtually disappeared (Figure 10-6). Annual cropping (one harvest/year/plot) has been restricted, and double and triple cropping have increased (Table 10-4, Figure 10-7). This change has been associated with increased labor inputs and with the adoption of new technologies such as HYV seeds, chemical fertilizer, pump irrigation, and crop diversification.

In the past, annual cultivation was devoted mainly to boro paddy, whereas double cropping involved jute, broadcast aus or aman, and vegetables. Triple cropping included local aus paddy or jute, or summer vegetables, and broadcast or transplanted local aman paddy, winter vegetables, or boro paddy. These patterns have changed. Now there is no annual cultivation; all fields are used for double and triple cropping. Double cropping on the low-lying fields includes broadcast aman and high-yielding boro paddy. Triple cropping involves HYV aus paddy, jute, or summer vegetables, followed by transplanted aman paddy, followed by various kinds of winter vegetables or HYV boro paddy. Normally, two paddy crops and one vegetable crop are grown on high-quality land.

The level of cropping frequency reflects the need for extremely high levels of output induced by pressure (or demand) for production in the village. Those fields that are used for two to three harvests per year are good-quality land but are constrained by the regular monsoon flood, such that cultivation during the monsoon months may be risky. Farmers have had to adjust to this constraint.

The growth of population, in terms of both number of families and family size, has also been associated with reduced per capita or per family land holding. In 1950 the average per capita and per family land holdings were 0.45 and 1.32 ha, respectively, including land used for homesteads and roads. In 1980 these figures had been reduced to 0.05 and 0.38 ha, respectively (Table 10-5).

The impact of population pressures on landholdings can be viewed in at least three ways. First, population growth led to the expansion of land for housing, reducing the amount of land for agriculture. Between 1950 and 1980, about 2.5 ha of land was diverted from agricultural to settlement use, a 16.3% loss of arable land. Second, population growth has resulted in the fragmentation of agricultural plots through changes in ownership by sale and hereditary transfer (see Table 10-5). Third, plot sizes have been reduced, affecting production and output intensification. The rate of adoption of neotechnic inputs is positively related to the increase in size of holding in Shyampur. For example, the proportion of a cultivated area fertilized and the amount of fertilizer used per unit of land are positively related to an increase in farm size. All farmers with large holdings (≥ 1.0 ha) have been using chemical fertilizers for at least 8 years (in 1980). Only 14% of smallholders have used it for 3 years, although they are all aware of higher yields from fertilizer use (Ali, 1982:82–84). Most of the smallholders in the village complain that they cannot use fertilizer because of insufficient land.

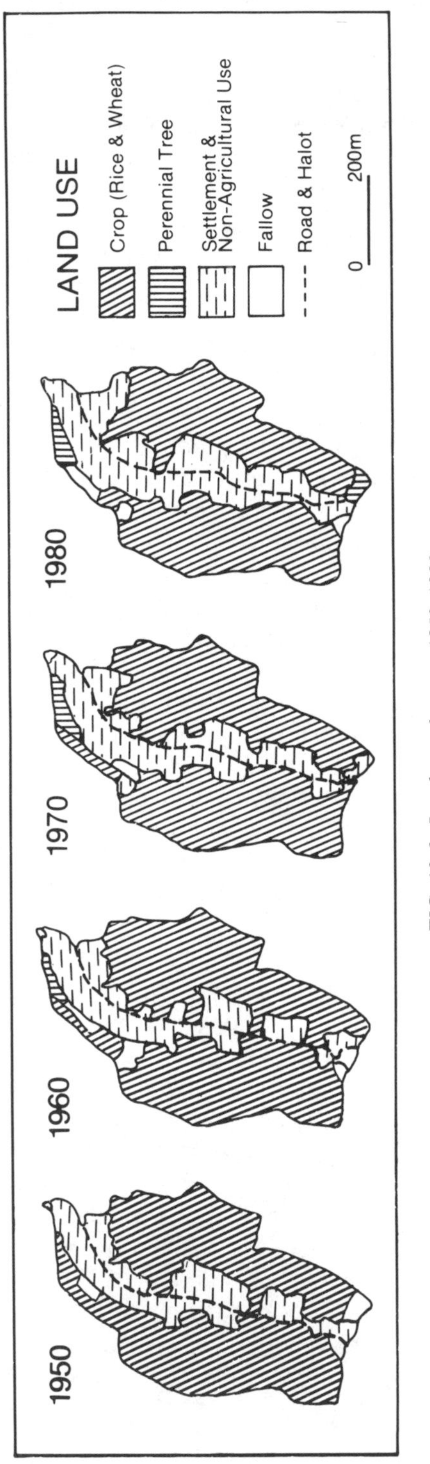

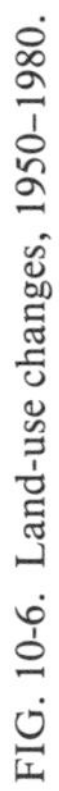
FIG. 10-6. Land-use changes, 1950–1980.

TABLE 10-4. Relationships between population growth and cropping intensity in Shyampur, 1950–1980.

Year	Total population	Average family size	Number of families	Population density/km^2[a]	Cropping intensity (%)[b]	Area (ha) under: Annual cropping	Double cropping	Triple cropping
1950	50	3	17	224	188	3.0	11.1	1.2
1960	175	4.4	40	786	206	2.5	9.1	3.4
1970	290	5.8	50	1,302	207	6.3	–	7.3
1980	482	8.2	59	2,164	177	–	2.9	9.86

Source: Ali (1982).

[a]The density of population represents the ratio of the total population of the village to the total area of the village and does not include lands owned by the residents elsewhere outside the village boundary.

[b]Intensity of cropping here represents the ratio of the total area of crops (per year) to the net sown area. It is calculated as (Total cropped area/Net cropped area) × 100. This ratio is somewhat similar to a frequency-of-cultivation measure of intensity. See Government of Pakistan, *Pakistan Census of Agriculture*, Vol. 1: *East Pakistan* (Bangladesh), 1960, p. xxx.

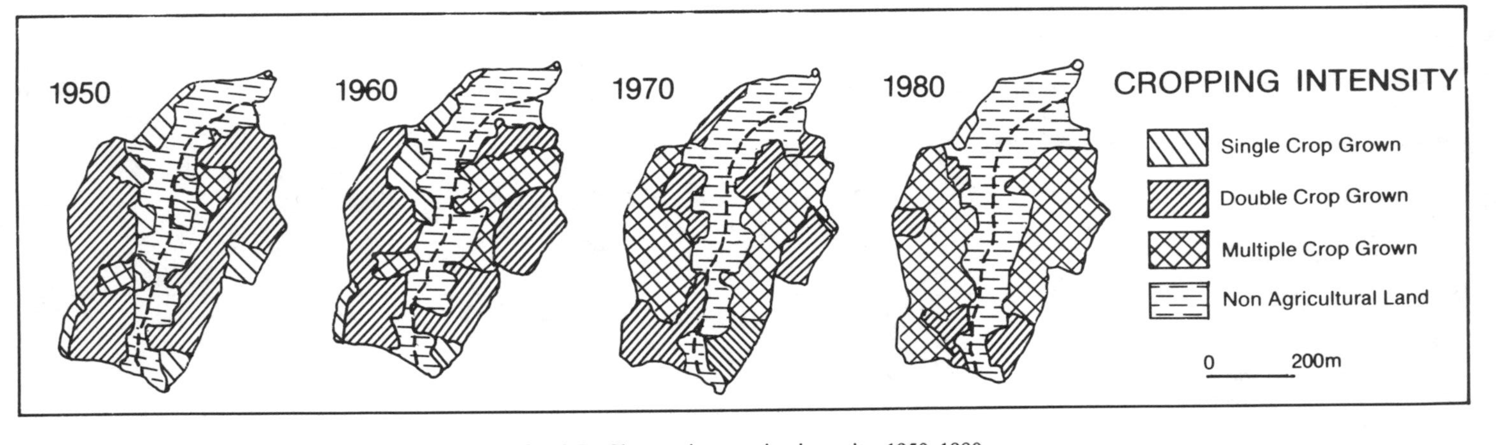

FIG. 10-7. Changes in cropping intensity, 1950–1980.

TABLE 10-5. Changes in per capita landholding and plot size in Shyampur, 1950–1980.

Year	Per family average holding size (ha)[a]	Per capita holding size (including homestead) (ha)	Per capita holding size (excluding homestead) (ha)	Number of plots by size (ha)				
				<0.2	0.21–0.4	0.41–0.6	0.61–0.8	≥ 0.81
1950	1.32	0.45	0.37					
1960	0.56	0.13	0.1	60	30	6	3	1
1970	0.45	0.08	0.06					
1980	0.38	0.05	0.03	129	15	9	1	0

Source: Ali (1982).

[a]Land owned within the village boundary are included here. If land owned elsewhere is included, the figure may increase somewhat.

As population has increased and agriculture has intensified, changes have taken place in family food requirements and production. The demands for cereal foods have risen so greatly that increases in agricultural production have not been sufficient to keep pace (Table 10-6; Figure 10-8). In 1980, for example, double HYV paddy cropping resulted in a slight increase in average production of rice/ha over that obtained previously. Despite steady increase in annual rice production over a 30-year period, food shortages are increasing.

Per hectare production of rice has increased since 1950 because of more frequent cultivation and technological change. From 1970 to 1980, however, total production of husked rice declined 0.9 t, even though HYV paddy was double-cropped and average per hectare production increased slightly, 394 kg/ha (Table 10-7). This decline in total rice production in the village is the result of the loss of 0.8 ha of cropland to homesteads, of the allocation of more lands to vegetable cultivation, and of the damage to crops by monsoon floods. The decline in total production has not been offset by the increase in production per hectare.

Also of interest is the apparent decline in the average productivity (per harvest per hectare) in both local and HYV rice (see Table 10-7). In 1970, 440 person-days/ha produced two harvests (one local and one HYV rice) per plot, yielding 4,419 kg/ha. In contrast, by 1980, it took 500 person-days/ha to produce two HYV harvests per plot, resulting in 4,813 kg/ha. These data (Table 10-7) indicate diminishing returns to inputs, including neotechnic ones. Total production has increased, but at the cost of lower productivity (Figure 10-9).

In 1950 the estimated total requirement of husked rice for the village was 10 t, and the village produced 16.7 t of husked rice in only one harvest. By 1960 the total demand for rice had increased to 35.2 t, while the estimated total production was only 29.8 t, a deficit of 5.4 t. This deficit was even higher if the costs of production, seed, and storage losses are considered. To meet this deficit, farmers grew jute and vegetables as cash crops, the return from which was used to purchase basic household needs and to meet higher production costs.

In 1970 total demand for rice was 58.3 t, and total estimated production was only 36.6 t, reflecting a total increase of only 6.8 t of husked rice over 1960 despite intensification of production. The estimated total requirement of rice rose to 97 t in 1980, but estimated total production was only 35.7 t. Both average and marginal productivity per unit of labor input per hectare have declined during the last 10 years, although total production per hectare increased slightly (Table 10-7; Figure 10-10).

It is impossible to triple-crop paddy because of the overlapping of cropping seasons. Moreover, some plots are flooded up to 3–4 months. Continuous cropping of land (no fallow), in the view of Shyampur farmers, results in land

TABLE 10-6. Food requirement, production, and technological change in Shyampur, 1950–1980.

Year	Population	Estimated requirement of clean rice (t)[a]	Estimated total production of clean rice (t)[b]	Difference between production and requirement (t)	Average total production of rice (kg/ha)	Rice varieties and frequencies of cultivation
1950	50	10.0	16.7	+6.7	1,817	1 paddy crop in the same plot/year
1960	175	35.2	29.8	−5.4	3,542	2 local paddy crops in same plot/year
1970	290	58.3	36.6	−21.7	4,419	2 paddy crops including 1 local and 1 HYV
1980	482	97.0	35.7	−61.3	4,813	2 HYV paddy crops[c]

[a]This estimate is based on per capita daily cereal requirement of 0.56 kg (equivalent to 1,850 calories).

[b]This estimate does not include cost of production, or seed and storage loss. Production rates and approximate total production estimated (for the cropland included within the village boundary) in the field survey were checked with government census records. Total areas under different varieties of paddy were estimated during the field survey and double checked with Settlement/Revenue Department of the government of Bangladesh. This estimate included land owned by the farmers within this village boundary and excluded those lands owned and or rented by the family elsewhere outside this boundary. Also, the production estimate included only farmstead rice production and not other cultivars. One kilogram of unhusked rice is equivalent to 0.6–0.75 kg of husked and clean rice, depending on the varieties of paddy.

[c]The cultivation of 3 HYV or local paddy crops on the same plot of land is not common because of overlapping cropping seasons and flooding.

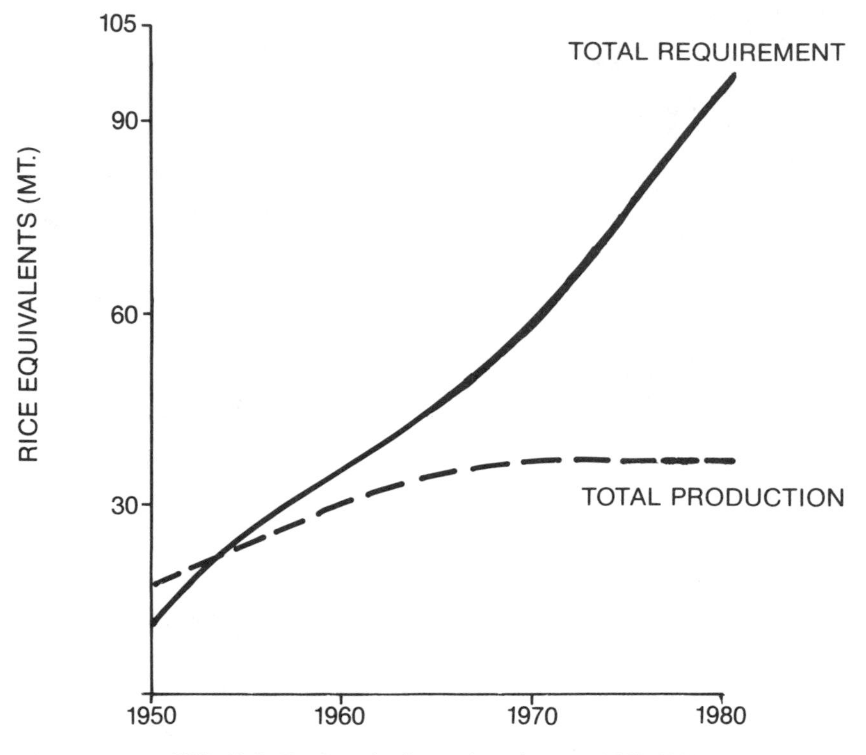

FIG. 10-8. Food production and requirement, 1950–1980.

exhaustion. The use of more fertilizer to offset this process can result in the formation of a chemical layer on the topsoil.

The increase in deficit production and access to the market and urban employment have led to a decrease in consumption production. The majority (70%) of the villagers in 1950 produced their annual subsistence needs (Table 10-8). Only 24% produced about half their needs, and 6% produced less than half. In contrast, only about 10% of the village cultivated their annual subsistence requirements in 1980, and 66% produced less than 3 months' annual subsistence.

Increased pressures for living space and demand for agricultural production over the last 30 years in Shyampur have resulted in increased input and output intensification of agriculture, at the cost of diminishing returns and lower total food production. The study of Shyampur did not directly address the question of the substitution of commodity production for consumption production, and data to ascertain the changes in standards of consumption are

TABLE 10-7. Returns to labor input in rice cultivation in Shyampur, 1950–1980.

Year	Rice varieties and number of harvests ( )	Total worker-days/ha/year used for rice production	Production of rice/ha/year (kg)	Average production rice/worker-day/ha/year (kg)	Marginal production of rice/worker-day/ha/year (kg)
1950	Local boro (1)	190	1,817	9.56	9.56
1960	Local boro (1) and transplanted local aman (1)	380	3,542	9.32	9.08
1970	HYV boro (1) and transplanted local aman (1)	440	4,419	10.04	14.62
1980	HYV boro (1) and HYV aman (1)	500	4,813	9.63	6.57

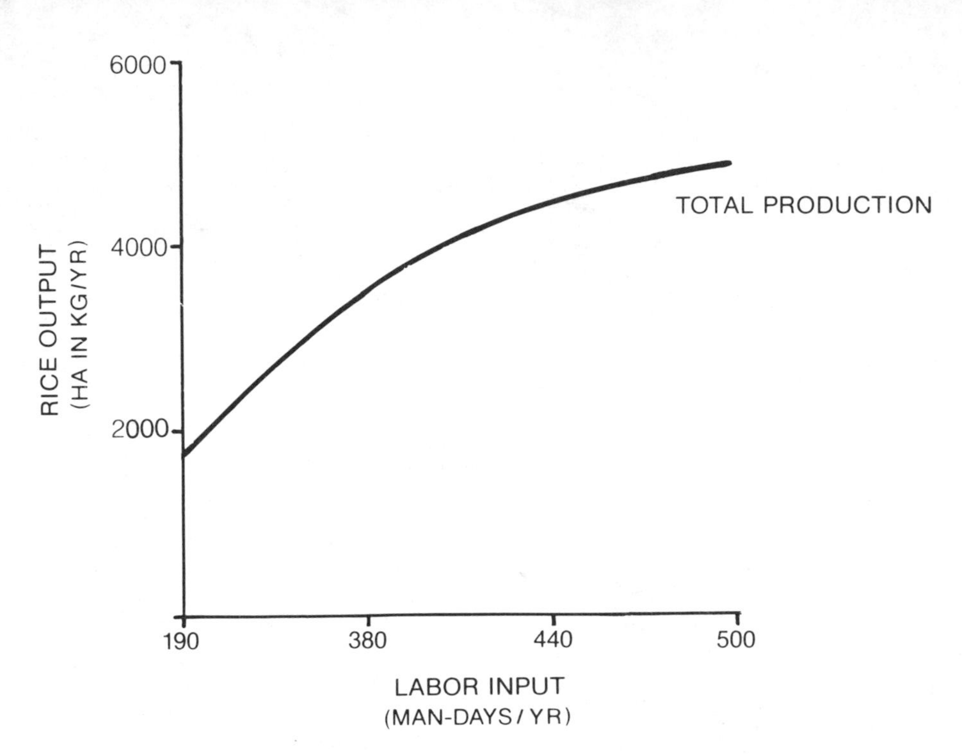

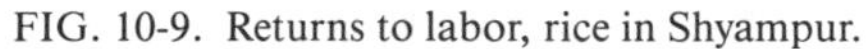
FIG. 10-9. Returns to labor, rice in Shyampur.

FIG. 10-10. Average and marginal production, rice in Shyampur (see Tables 10-6, 10-7).

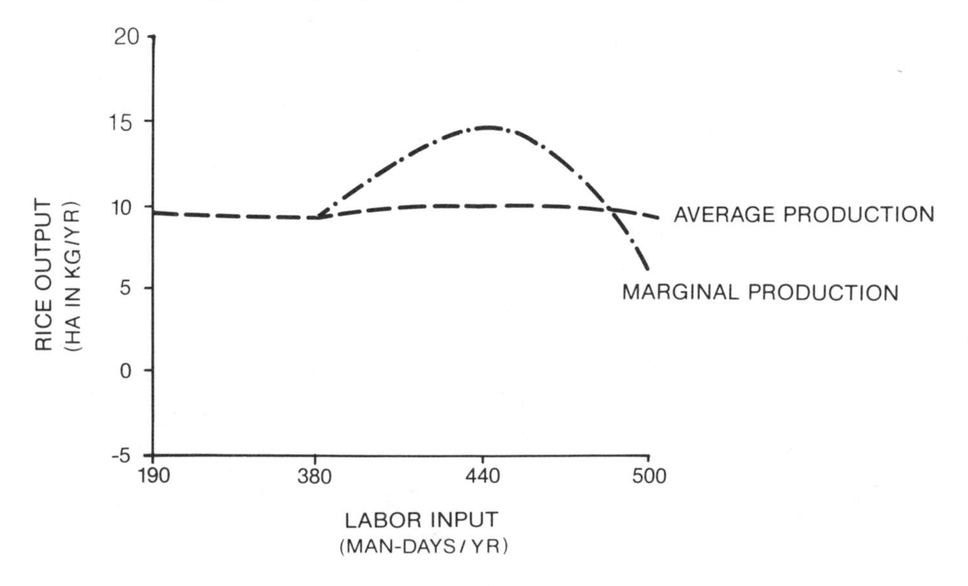

TABLE 10-8. Changing household food production in Shyampur.

		Percentage of farm families producing food for:			
Year	Population	Less than 3 months	4–6 months	7–9 months	10–12 months
1950	50	6	12	12	70
1960	175	15	20	30	35
1970	290	34	30	18	18
1980	482	66	15	9	10

not sufficient to make strong claims. I believe, however, that the income derived from increased commodity production has not yet offset the loss of foods resulting from the decline in consumption production. Constant intensification may have reached a threshold at Shyampur—given land, environmental, and technological constraints—that will seriously impede increases in both consumption and total crop production. In this sense, Shyampur appears to be a case of a near-saturated or involuted agroecosystem (e.g., Geertz, 1963).

Influence of Environment, Market, and Other Variables

The key environmental constraint on agriculture in Shyampur is flooding. Village farmers recognize this but have virtually no means to mitigate it. If one crop is damaged by floods, farmers attempt to compensate for the loss by producing another cash crop, for example vegetables, that yields a high market return (also see Paul, 1984).

The floods of 1970 were unusual, covering the southeastern and southwestern parts of the village for about 4 months (April–mid-August) and resulting in severe damage to paddy production. About 6.3 ha of aman paddy that had been sown was damaged and had to be resown. One harvest of HYV boro paddy was obtained (in December–April). To compensate for the overall losses, farmers used two paddy and one vegetable harvest on the remaining 7.3 ha of land that were not affected by the floods.

Severe floods reoccurred in 1980, when the southern part of the village remained waterlogged for about 3 months (mid-April–July). Moreover, heavy monsoon rains in the month of August led the farmers to anticipate severe damage to aman paddy. To compensate, immediately after the aman harvest, farmers of the village planted vegetables rather than HYV boro paddy. The reasons for this action were: (1) that vegetables have a short growing season (1–1.5 months) and two crops can be harvested in 3–4 months; (2) that the economic rent earned by vegetables was higher than that of boro rice; and (3) that boro rice requires high costs for irrigation and fertilizer, which the farmers could not afford. The result was that about 10.1 ha of arable land went under vegetable cultivation from September 1980 to March 1981. Under normal circumstances about 6.0 ha of arable land are used for vegetable cultivation during this time of the year (Ali, 1981).

This adjustment may indicate the future for Shyampur agriculture. The farmers are aware of the pressures on them, the intricacies of crop-environment relationships, and the changing nature of their agriculture. The village has emerged as a major source of vegetable production because of location and environment. Village agriculture may become a truck garden for Dhaka City. In terms of cost–benefit analysis of the three competitive crops—rice, jute, and vegetables—the last yields the highest net return, involves shorter growing periods, requires minimal irrigation and other inputs, and produces well at any time of year (Ali, 1981). Moreover, Dhaka City will be a constant source of demand for vegetables and will ensure relatively high market prices throughout the year.

The population of Shyampur is composed of traditional agriculturalists. Smallholder and landless farmers constitute 71% of the total farm families. Largeholder farmers are small in number but own about 67% of the village's total land resources. This unequal concentration of land resources affects agricultural production. Largeholder farmers are in many cases renters, whereas smallholder and landless farmers are rentees and sharecroppers. Sharecroppers are responsible for the inputs for cultivation, although they often have insufficient capital for them. Thus the proper use of fertilizer and irrigation cannot be maintained, and higher yields cannot be expected. Smallholder and landless farmers face great difficulties in case of crop failure or bad harvest. Their only alternative is to seek loans or land mortgages from the government or from rich farmers. In either case the loan repayment is impossible because of high interest rates and other terms of the loan. The result is reduction of smallholders and an increase in the landless.

Shyampur farmers respond to international fluctuations in market prices. They grew more jute in the 1960s, when the world market price was high, than they do now. At that time Bangladesh produced about 50% of the world's jute. When the market for Bangladesh jute collapsed in the 1970s, Shyampur farmers responded by severely decreasing land in jute. From 1975 to 1976 only 1.2 ha of land were in jute production. The international market for Bangladesh jute was restored in 1977, and the farmers promptly responded by increasing jute land to 3.5 ha by 1980.

The farmers of Shyampur respond to technological innovations in agriculture as well as they can. The first major technological innovation to diffuse to the village was chemical fertilizer in the early 1960s, followed by HYV paddy in the late 1960s. In the early to mid-1970s the irrigation pump, spray machines, and power tillers were introduced to the village.

In 1980 approximately 85% of the farmers had adopted some of these technological innovations, and there is reason to believe that more would do so if constraints were not so great. Field size and access to capital are the major barriers. Use of high technology is more common among the largeholders, who have sufficient capital and large enough fields to use mechanized technology. However, all farmers holding above 0.2 ha of arable land have adopted some neotechnic innovations, particularly chemical fertilizer (Table 10-9).

Conclusion

This study has examined a case of change in a wet-rice (paddy) farming system in Shyampur, Bangladesh. This system is located in a wet, floodplain environment in which monsoon flooding is a significant constraint to production. Despite these constraints, agriculture has intensified through the years in direct response to population growth and increasing demand on fixed land. Indeed, population growth has resulted in the loss of farmland to homesteads and other nonfarm uses; per capita land holdings and farm size have been reduced, and the frequency of cultivation has been raised to double and triple croppings. This frequency has led to increases in total production, at the cost of diminishing returns. Moreover, increases in production have not kept up with demand, and the gap between household food requirements and farm production is increasing, hinting that technological changes produce a stair-step function in input–output relationships (Boserup, 1965; Robinson & Schutjer, 1984). Apparently, however, declining farm size and environmental constraints have offset productivity increases related to new technologies. This situation seems to be a force in the movement of the farming system from consumption to market production. It remains to be seen whether this change will stabilize or increase the standard of food consumption (and the material wealth) of Shyampur farmers.

TABLE 10-9. Extent of adoption and period of experience with fertilizer use in Shyampur.

Landholding size	Percentage of cultivators using fertilizer	Percentage of adoptors using fertilizer for: Up to 4 years	4–8 years	8+ years	Share of total fertilizers use (%)
<0.2 ha	60	14.2	42.0	42.9	5.9
0.2–1 ha	100	11.7	47.1	41.2	26.9
≥1 ha	100	4.8	31.6	63.6	67.2

Source: Ali (1982).

Note: This use was associated with the cultivation of HYV paddy and vegetables, in both rainfed and irrigated lands, in order to increase cropping intensity and productivity.

Acknowledgments

This research was supported by the University Grants Commission, Bangladesh. The author acknowledges the guidance received from B. L. Turner II during the preparation of this chapter. He is also thankful for the detailed comments of Dr. Stephen E. Brush.

Notes

1. *Aus paddy* is a coarse, short-stemmed variety of rice (*Oriza sativa* L.), normally sown on high land during the summer months (March–May) and harvested in July–August. It is one of the *bhadoi* crops.

2. *Aman paddy* is a long-stemmed variety of rice that is sown during the rainy season (June–August) and harvested in November–December. It is one of the cold-weather or *haimantic* crops. *Boro paddy* is a short-stemmed variety of rice that is sown on low-lying areas during the winter months (December–February) and harvested in the summer (April–June). It is one of the *rabi* crops.

3. All cultivars harvested during the rainy season (July–August) are known as the *bhadoi* or *kharif* crops. The most important *bhadoi* crops are aus paddy and jute. The *bhadoi* crops are usually sown in early summer (March–May), and are dependent on nor'wester rainfall.

4. All cultivars harvested during the cold weather (November–December) are known as *haimantic* crops. The most important *haimantic* cultivar is aman paddy, which is sown and transplanted during June–August. A good harvest is dependent on the monsoon and late monsoon rainfall.

5. All cultivars sown and transplanted in the winter (November–February) and harvested in the early summer (March–May) are known as the *rabi* crops. These include boro paddy, wheat, pulses, oil seeds, vegetables, and tobacco.

References

Ahmad, N. *Land use in Rampal Union: A horticultural area*. Dacca: The East Pakistan Geographical Society, 1961.

Ahmad, N., & Khan, F. K. *Land use in Fayadabad area*. Dacca: The East Pakistan Geographical Society, 1963.

Ali, A. M. S. Farmers' perception toward economic benefits and changing cropping pattern in Savar area. *Environment and Region*, 1981, 1, 32–48.

Ali, A. M. S. *Dynamics of agricultural land use and socioeconomic infrastructure of Savar Thana: A case study*. Dhaka: Bangladesh University Grants Commission, 1982.

Barlett, P. Labour efficiency and the mechanism of agricultural evolution. *Journal of Anthropological Research*, 1976, 32, 124–140.

Boserup, E. *The conditions of agricultural growth: The economics of agrarian change under population pressure*. London: Allen and Unwin, 1965.

Brookfield, H. C. Local study and comparative method: An example from Central New Guinea. *Annals of the Association of American Geographers*, 1962, 52, 242–252.

Brookfield, H. C. Intensification and disintensification in Pacific agriculture: A theoretical approach. *Pacific Viewpoint*, 1972, 13, 30–48.

Chapman, G. P. A structural analysis of two farms in Bangladesh. In T. Bayliss-Smith & S. Wanmali (Eds.), *Understanding Green Revolutions*. Cambridge: Cambridge University Press, 1984.

Geertz, C. *Agricultural involution: The process of ecological change in Indonesia*. Berkeley: University of California Press, 1963.

Government of Bangladesh. *Soil survey report no. 1, Dacca District*. Department of Agriculture, Bulletin No. 1, 1972.

Lagemann, J. *Traditional African farming systems in Eastern Nigeria*. Munchen: Weltforum Verlag, 1977.

Metzner, J. K. *Agriculture and population pressure in Sikka, Isles of Flores: A contribution to the study of the stability of agricultural systems in the wet tropics*. Canberra: The Australian National University, 1982.

Norman, M. J. T. *Annual cropping systems in the tropics*. Gainesville: University of Florida Press, 1977.

Parrack, D. W. An approach to the bioenergetics of rural West Bengal. In A. P. Vayda (Ed.), *Environment and cultural behavior*. Austin: University of Texas Press, 1969.

Paul, B. K. Perception of and agricultural adjustment to floods in Jamuna floodplain, Bangladesh. *Human Ecology*, 1984, 12, 3–19.

Robinson, W., & Schutjer, W. Agricultural development and demographic change: A generalization of the Boserup model. *Economic Development and Cultural Change*, 1984, 32, 355–366.

Stoddart, D. R., & Pethick, J. S. Environmental hazard and coastal reclamation: Problems and prospects in Bangladesh. In T. Bayliss-Smith & S. Wanmali (Eds.), *Understanding green revolutions*. Cambridge: Cambridge University Press, 1984.

Turner, B. L., Hanham, R. Q., & Portararo, A. V. Population pressure and agricultural intensity. *Annals of the Association of American Geographers*, 1977, 67, 384–396.

PART IV

NEOTECHNIC AND COMMODITY-ORIENTED SYSTEMS

Introduction to Part IV

The four farming systems in this chapter are characterized by their large size and intensive mechanization, production for regional and international markets, and variable output intensities. Crop yields range from below 3,000 kg/ha/yr to over 12,000 kg/ha/yr (Figure IV-1). The former level is no greater than yields obtained by some of the paleotechnic systems examined in Part II. The higher yields are exaggerated as a result of the field weight of sugar beets (*Beta vulgaris*), a root crop. Omitting it, the higher yields for the neotechnic systems are similar to those exhibited by the labor-intensive mixed systems discussed in Part III. The range in yields is also a result of the differences in rainfed production from the semiarid environs of the Great Plains to the more mesic environs of Europe.

If these systems do not represent major advances in land productivity over those in Part III, they are related to major increases in labor productivity. In each system, labor is deemphasized, and high levels of neotechnic subsidies are employed. The types and levels of many subsidies are determined in conjunction with specialists hired directly by the farm or provided by some arm of the government. Some have argued that the level of subsidies for these systems is so high that the overall energetic efficiency is low (Pimentel & Pimentel, 1979), and therefore the long-term economic variability of neotechnic (mechanized) farming is in question. Others disagree, as noted in the case study of Great Plains wheat farming (Chapter 9). Although the role of mechanization as a subsidy is debatable at the energetics level, economists forecast economic inefficiencies if some types of North American systems do not mechanize more than they have (Martin, 1983). Such a shift, however, has significant labor and social consequences (e.g., Price, 1983).

Production in these systems is geared almost exclusively to the market, but as modified by sociopolitical controls (e.g., farm price support, production quotas, import restrictions). The farm has direct and indirect linkages to these controls, with the precise means related to the political-economic context in which the farm exists. The profitability of at least three of the systems seems to be critically linked to some form of governmental or multinational support.

The first case study in this section is that of rainfed, mechanized wheat farms in Sedgwick County, Colorado, part of the North American Great Plains "breadbasket." Agriculture here is not much older than 100 years and was developed largely *in situ* by farmers with little experience with grassland environments having major rainfall cycles. Farmers in Sedgwick County have always cultivated cereals, primarily wheat (*Triticum aestivum* L.), for a national

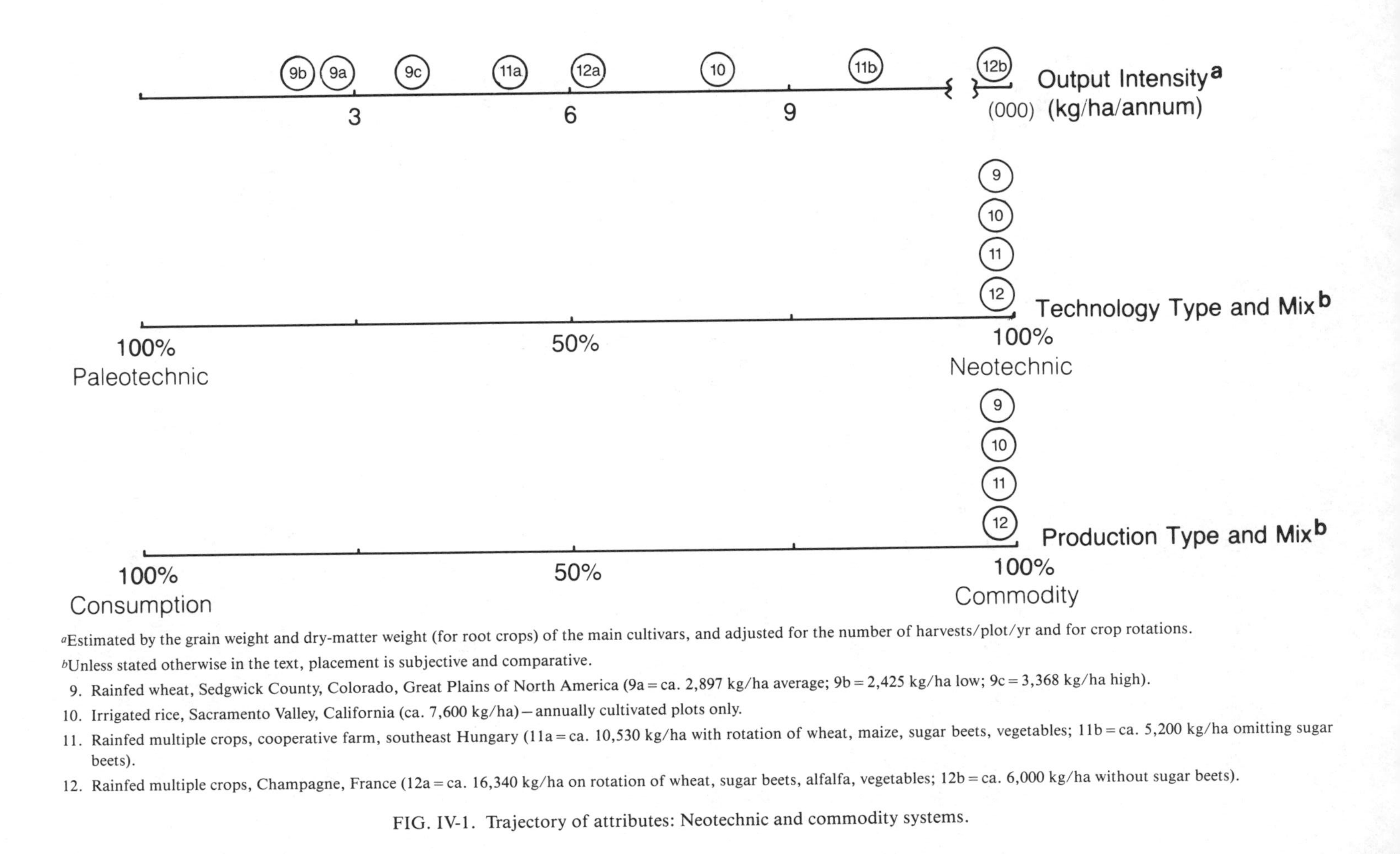

[a]Estimated by the grain weight and dry-matter weight (for root crops) of the main cultivars, and adjusted for the number of harvests/plot/yr and for crop rotations.

[b]Unless stated otherwise in the text, placement is subjective and comparative.

9. Rainfed wheat, Sedgwick County, Colorado, Great Plains of North America (9a = ca. 2,897 kg/ha average; 9b = 2,425 kg/ha low; 9c = 3,368 kg/ha high).
10. Irrigated rice, Sacramento Valley, California (ca. 7,600 kg/ha) – annually cultivated plots only.
11. Rainfed multiple crops, cooperative farm, southeast Hungary (11a = ca. 10,530 kg/ha with rotation of wheat, maize, sugar beets, vegetables; 11b = ca. 5,200 kg/ha omitting sugar beets).
12. Rainfed multiple crops, Champagne, France (12a = ca. 16,340 kg/ha on rotation of wheat, sugar beets, alfalfa, vegetables; 12b = ca. 6,000 kg/ha without sugar beets).

FIG. IV-1. Trajectory of attributes: Neotechnic and commodity systems.

and international market. Largeholder family farms have increased in size through time to their current average of 525 ha. These farmers have been subsidized by the national government both directly and indirectly (e.g. through sponsored research, particularly at state agricultural universities). Much of this research has involved proper tillage techniques and procedures—the focus of this chapter—to sustain yields. Currently, rainfed yields in Sedgwick County range from about 2,800 to 3,300 kg/ha of wheat on an annual cropping cycle, depending on tillage practices and rainfall conditions. Based on the method of estimation, the energetic efficiency of this system is higher than that which would result from the work of Pimentel and Pimentel (1979).

The second farming system is irrigated rice production in the Sacramento Valley of California. As in the previous study, the extensive use of agriculture in the valley is a relatively recent phenomenon, which, from its inception in the latter stages of the 19th century, has involved large-scale operations for the national and international market. The development of irrigated rice (*Oryza sativa* L.) has been primarily in the context of the international market and large family farms, many of them corporations involved with sharecropping. The system is almost totally mechanized from sowing to harvest, averaging a mere 4.4 person-hours/ha of labor. In this regard, the system has benefited from extensive publicly sponsored research and experimentation. Annual cultivation of rice alone yields up to 9,000 kg/ha, but the valley average is about 7,600 kg/ha. Some farms also cultivate a winter legume. Despite this productivity and the historical political influence and inventiveness of the valley rice farmers, this system is in some jeopardy because of the current low levels of demand for rice on the international market, the high costs of production, and changing government subsidy policies.

Rainfed production of multiple crops on a cooperative farm in southeastern Hungary constitutes the third case study. This farm, situated in the village of Kondoros, Békés County, has over 2,000 members working on 8,700 ha. The cooperative acts as a vertical production unit and as a social unit with responsibilities that extend to the community at large. Complex crop rotations, primarily involving cereals but also including sugar beets and vegetables, are produced in a highly mechanized way, complete with some activities determined by data analysis on the cooperative's computer. On an annual rotation of wheat, sugar beets, alfalfa, and vegetables, the farm averages some 10,500 kg/ha. This figure is reduced to 5,200 kg/ha if sugar beets are omitted. Cropping is integrated with livestock production, including dairy and beef cattle, pigs, and chickens. Coexisting with this system is private plot production focusing on labor-intensive cultivars and livestock. Management of the cooperative is organized in a complex fashion, particularly in comparison to the family farms characterized in the first two systems in this section. The trend has been for decision making to become increasingly the role of the cooperative as it responds to the market.

The final case study involves rainfed production of multiple crops in the Champagne region of France. As in the case of Sedgwick County and the Sacramento Valley, the development of extensive large-scale farming in the area is a relatively recent transformation, in this case beginning after World War II. About one-half of the "chalky plains" are consumed by farms of 100 ha or more, which rotate wheat, sugar beets, and alfalfa in an intensive system of cultivation, much like those employed on the Hungarian cooperative farm. Yields are high, some 16,000 kg/ha on an annual rotation with sugar beets, and 6,000 kg/ha without that cultivar. The high yields may be attributable in part to the mollisols that dominate the area and the fact that these soils had been rested until the last 35 years. High levels of mechanization and subsidies raise questions about the economic viability of production with increased energy costs. The Champagne system is a product of the European Common Market, with its guaranteed high prices for crops. Champagne farmers have developed political linkages to lobby for controls that benefit their operations (see Hill, 1984).

References

Hill, B. E. *The common agricultural policy: Past, present and future*. London: Methuen, 1984.

Martin, P. L. Labor-intensive agriculture. *Scientific American*, 1983, 249, 54–59.

Pimentel, D., & Pimentel, M. *Food, energy and society*. London: Arnold, 1979.

Price, B. L. *The political economy of mechanization in U.S. agriculture*. Boulder: Westview Press, 1983.

11

Dryland Wheat Farming on the Central Great Plains: Sedgwick County, Northeast Colorado

HANS-JOACHIM W. SPÄTH

Dry farming has evolved as the dominant agricultural system of the Great Plains of North America. Although this system has predecessors in both the Old and the New Worlds, these antecedents provided little guidance for the unique evolution of dry farming in the Great Plains. Beginning as an ill-adapted introduction of humid-zone techniques of soil and water management, it ultimately adjusted to environmental constraints through locally developed land resource management principles. The system also has adjusted to cope with dominant external economic and political forces. In less than 100 years, it has expanded over an entire continental landscape and has emerged as one of the major sources of global food supply.

The objectives of this study are:

1. To assess the agroecological setting, constraints, and potential of the central Great Plains as the backdrop to both settlement process and regional agroeconomic development
2. To explain the evolution of the physical and technical properties of the regional farming system from a semi-industrial to a high-technology, energy-intensive one within this ecological frame
3. To evaluate the performance of the present dryland wheat system in economic, energetic, and ecological terms, taking into consideration overall production conditions and development trends.

This analysis focuses on a set of farm operations in Sedgwick County in the northeastern corner of Colorado (Figures 11-1, 11-2). From 1979 to 1984, these farms participated in an agroenergetic regional analysis.

Physical–Geographical Setting

The constraints on and potential of agricultural production are related to the interactions among topography, soil resources, climate, soil moisture, and evapotranspiration. The central Great Plains are a tableland that is slightly tilted

Hans-Joachim Späth. Department of Geography and Institute for Dryland Development, University of Oklahoma, Norman, Oklahoma.

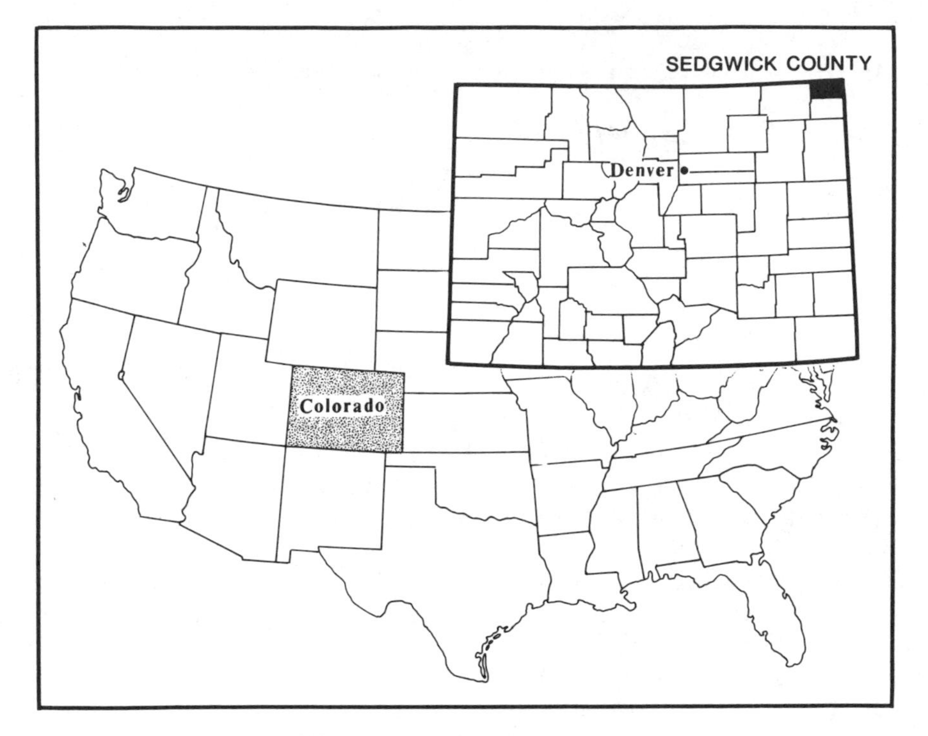

FIG. 11-1. Sedgwick County, Colorado.

(2 m/km) from an elevation of 1,830 m at the base of the Rocky Mountains toward the east. This generally level tableland is drained by major east-flowing rivers. Smaller tributaries create islandlike plateaus between major rivers. Dry farming is conducted on these lands.

The South Platte River passes through the northern part of Sedgwick County in a northeasterly direction, dissecting the tableland. The two parts of the tableland slope from about 1,250 m in the southwestern part of the county to about 1,050 m in the northeast, a distance of 93 km. Across these lands the South Platte is joined by numerous tributaries.

The cultivated area of the table land is covered predominantly by entisols and by light brown mollisols. Sandy soils are residual in origin; the loam, silt loam, and clay loam are loess deposits (Elder, 1969; Hoover, 1957). Both the sandy and medium-textured soils are deep and do not contain inhibiting rock layers. Nearly all soils are calcareous, with a B horizon lime layer varying in texture and depth with effective precipitation. The Soil Survey for Sedgwick County (USDA, 1969) identifies the soils in the upland wheat zone as Rago–Richfield–Kuma associations, which range from deep, nearly level to gently sloping, loamy soils.

Cultivated sandy soils average less than 1.0% organic matter; they are low in total N (0.05%) and available phosphorous (Greb, Whitney, & Tucker, 1953; Greb, Whitney, & Tucker, 1954; Olsen, Cole, Watanabe, & Dean, 1954). The mollisols usually average 1.4–1.8% organic matter, about 0.08% total N, and 17 kg/ha to 34 kg/ha available P in the surface 15 cm of soil. Considering the shallow A horizon of the brown and light-brown mollisols, any removal of topsoil by erosion seriously reduces nitrogen and phosphorous supplies.

The medium-textured soils possess relatively good water intake rates: 25 mm/hr for the first hour and 5–10 mm/hr subsequently. These soils have a water-holding capacity of 22–25% in the A and B horizons, and 18–22% in the silty parent material of the C horizon. Typical medium-textured soils hold 30–33 cm of plant-available moisture in a 180-cm profile (Greb, Smika, & Black, 1970; Smika, Black, & Greb, 1969; Späth, 1980a).

FIG. 11-2. Soil associations in Sedgwick County, Colorado.

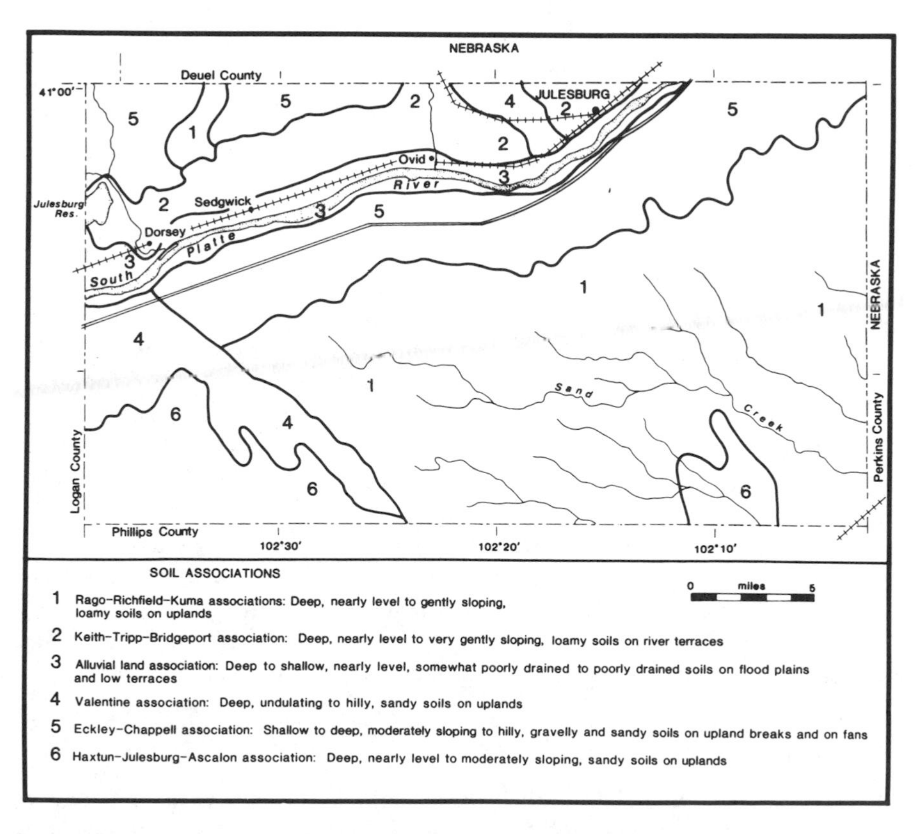

Source: U.S. Department of Agriculture, Soil Conservation Service, Denver, Colorado, personal communications, *Soil Survey for Sedgwick County, Colorado*, 1969.

A semiarid, continental climate dominates the county and region. Wide variations in temperature, precipitation, and humidity are characteristic. The annual range in temperature is from −35°C to 45°C. One year in 10 has either less than 28 cm or more than 50 cm annual precipitation; the average for 1941–1970 was about 43–46 cm. Approximately 75–80% of the precipitation occurs during the growing season from April to October. Average annual snowfall reaches 51 cm. If properly managed, snowfall can be an effective source of soil moisture, and there is evidence that wheat yields are highly correlated with snow accumulation (Black & Siddoway, 1971; Greb & Black, 1971; Späth, 1976, 1980a).

Evapotranspiration demand (precipitation minus potential evapotranspiration) is at a 300-mm level. Moisture deficit (potential evapotranspiration minus actual evapotranspiration) reaches 254 mm, assuming 35% overwinter moisture gain, a root zone of 180 cm depth, and a total storage capacity of 300 mm on silt loam (Späth, 1980a:22). Moisture deficit and drought are normal in Sedgwick County. It is a prime objective of agriculture to mitigate natural moisture deficits and drought by coverting as much precipitation into soil moisture as possible, by eliminating excessive weed growth, by conserving an optimal amount of soil moisture for crop growth, and by replacing continuous cropping by wheat summer fallow rotation. There are several soil and water management options in context with summer fallowing, each of which requires specific production inputs and results in specific ranges of food production potential.

Development of the Agricultural Landscape

Settlers initially halted their move into the semiarid West between 95° and 98° west longitude, where soil management techniques of the humid zone provided an economically acceptable balance between production inputs and food output. Periodic mitigation of climatic constraints, expansion of foreign markets, improvements in soil management techniques, and the slow westward expansion of population promoted the spread of farmers further west into the Great Plains.

Dry farming requires a match of technological ingenuity with ecological awareness in order to mitigate climatological and economic uncertainty. The initial farmer had to cope with these demands and begin the development of dry farming technology for the region. Moreover, the farmer had to deal with political and economic structures that affected this development.

The Boom-and-Bust History

Before the 1860s cattle barons had possession of the tableland of Sedgwick County (Figure 11-1). The first European agricultural settlers arrived as early as 1867, after the Union Pacific Railroad reached Julesburg, Colorado. The passage of the Homestead Act and the Morrill Act in 1862, and of the Hatch Act in 1887, permitted the development of the regional infrastructure, economy, and agrotechnology. In the early 1880s the railroad brought entire church congregations from Europe, especially from the Scandinavian countries, to settle on railroad land, to raise cattle, and to farm on a small-scale subsistence level. Severe shortfalls in annual rainfall in the 1890s, strong persistent winds, and dust storms put an end to the initial settlement drive in other parts of the Great Plains. Sedgwick County, however, being mainly rangeland, was little affected by these phenomena.

The introduction of heavy equipment and powerful steam tractors, as well as the rapid expansion of foreign markets in post–World War I Europe, brought about a significant settlement boom and a change in land use. The grassland of the tableland was plowed and turned into wheat fields. The moldboard plow, a humid-zone tool for soil, moisture, and residue management, turned over the topsoil, left field surfaces bare and unprotected, stimulated wind erosion, and minimized soil moisture harvest. Private experimentation and tests on research stations between 1910 and 1930 led to the development of shallow operating plows and harrows. These implements, however, allowed for a fallow efficiency (the percentage of fallow-season rainfall that can be stored as plant-available moisture in the root zone) of no more than 20% and produced an average wheat yield of about 1,075 kg/ha. Yet Sedgwick County's economy flourished on the basis of expanding wheat acreages in the 1920s.

This boom coincided with relatively high moisture supplies and contributed directly to the development of the Dust Bowl syndrome of the 1930s. High production was achieved at the cost of indebtedness and increasing exploitation of marginal and vulnerable cropland, all implemented in an attempt to avoid economic bankruptcy. Diminishing rainfall and wheat prices in the 1930s initiated a cycle of further expansion of acreage and production (which were supposed to provide the farmers with means for meeting their overextended financial obligations). In order to remain competitive, individual farmers had to acquire new machinery and further expand their holdings to make the investment in machinery profitable. Cash flow and risk increased. Crop prices and rainfall determined the solvency of individual operators. When domestic and foreign demands decreased in the 1930s, the new farm technology created enormous wheat surpluses and price depression. In 1931, for example, wheat prices were at $0.30/100 kg, which was only 18% of the 1919 level. When prices fell, individual farmers, paradoxically, expanded their acreage and increased their production. The effect, of course, was to increase the overall surplus. By this time tenancy rates and indebtedness had become so high that farmers were obliged to raise cash crops to meet their mortgage and rental payments.

The New Deal exacerbated this situation. The per capita agricultural loans and expenditures of the New Deal agricultural agencies were geared to absolute and percentage changes in farm income from 1929 to 1932, not to genuine economic need. The government indirectly encouraged the selling of land by small farmers, the eviction of small tenant farmers, and the introduction of more powerful tractors and equipment to the post-1934 Great Plains (Bowden, 1977).

The Homestead Act of 1862 was an early contributor to this pattern. Units of only about 65 ha (160 acres) were available to new settlers, regardless of intended land use or of the land's suitability for a specific use. Subsequent land policy did not provide for land-use regulations, which might also have served as conservation devices. Western railroads, new settlers, land speculators, local businessmen, state and local governments—all had different and often conflicting interests in both timing and volume of the general regional economic development.

In times of economic distress, the first to fail were settlers who were forced off their land. Land changed hands over and over contributing to the general growth of individual holdings far beyond the 65-ha unit originally intended by 19th-century land policy. Passage of the Federal Crop Insurance Act of 1938, which established the Federal Crop Insurance Corporation to write all-risk crop insurance on low yields (Ottoson, Anderson, Birch, & Henderson, 1966), seems to have come too late to halt the trend of farm aggregation. In the 1930s the average farm size in the central Great Plains was about 120 ha (300 acres). The production cost–return balance collapsed along with the ecological production potential during the erosion process of the 1930s, when early soil management practices left field surfaces bare and exposed topsoils were subject to severe wind erosion.

Another cycle of prosperous years began in the 1940s with high rainfall supplies and increasing demand for wheat from famine-stricken Europe. Small one-way disk plows were developed by the mid-1940s, and fallow efficiency levels improved to 20–24% with average yields increasing slightly to 1,142 kg/ha. Land-use restrictions were abandoned, and farmers' willingness to protect the soil diminished. For example, sandy soils, suitable only for grazing, were brought back into crop production. The introduction of the rod weeder in the late 1950s and of stubble mulching in the 1960s improved fallow efficiency levels to about 27%; average yields increased to about 1,747 kg/ha. By the time European agriculture was rebuilt in the 1950s, the Great Plains were producing an enormous surplus of grain. Diminishing rainfall and declining wheat demand again brought an end to economic prosperity; excessive soil erosion once again destabilized the regional agroecological balance. Population density again decreased, farm sizes increased, soil-conserving crop rotations and mixed farming systems were replaced by wheat monoculture, and specialization continued (Table 11-1). Farm employment declined.

Modern stubble mulching used large sweeps and rod weeders. In the 1970s fallow efficiency levels rose to about 33% and yields to 2,218 kg/ha. When practiced in combination with fall weed control, stubble mulching improved fallow efficiency to about 35% and yields to about 2,419 kg/ha. Minimum

TABLE 11-1. Population, number of farms, average farm size, and hired labor in Sedgwick County, Colorado, 1910–1982.

Year	Total population	Number of farms	Average farm size (ha)	Hired labor
1910	3,061	448	144	–
1920	4,207	487	195	399
1925	–	632	172	–
1930	5,580	560	222	–
1940	5,294	505	232	213
1945	–	466	266	107
1950	5,095	474	268	145
1954	–	417	285	161
1959	–	376	335	298
1960	4,242	–	–	–
1964	–	356	386	–
1966	3,600	–	–	–
1969	–	327	428	571
1970	3,405	–	–	–
1974	3,300[a]	302	458	536
1978	3,300[a]	316	434	346
1980	3,266	–	–	–
1982	3,219	253	520	390

[a]Population estimates and projections, U.S. Department of Commerce, Bureau of Census.

mechanical tillage with herbicide applications, as sometimes practiced since the late 1970s and early 1980s, can increase these performance levels to 40% fallow efficiency and yield levels of 2,688 to 3,360 kg/ha. (Greb, 1979). In spite of modern agrotechnology and training, soil erosion continued to threaten croplands throughout the Central Great Plains. Soil Conservation Service data show excessive damage to cropland in the late 1970s and again in the early 1980s throughout eastern Colorado (Table 11-2).

Increased production again has failed to mitigate economic hardship because of market conditions. Since the 1970s production costs have not been offset by wheat prices. Government programs of the 1980s (like Payment In Kind) have not helped either. Small family farms have continued to decline, as has the population (see Table 11-1). The trend is toward corporate farms.

The average farm size in the 1960s was about 340 ha (850 acres). The average hectarage of 45 farms surveyed in 1980, randomly sampled throughout Sedgwick County, was about 525 ha (1,290 acres). These operators felt that if dryland wheat production is to provide the only farm income, then 600–800 ha are needed for crop production. Considering technological advancements (e.g., high-yielding crop varieties, chemical fertilizers), this farm size compares well with the 1,040 ha that Webb (1931) identified as the Great Plains dry farming equivalent of a 65 ha farm in the humid east of the United States. A standard deviation of 280 ha for the sampled farms indicates that a large number of

TABLE 11-2. Wind erosion conditions for Colorado: Final report for wind erosion season (November–May).

Year	Land damaged (ha)				Crops or cover destroyed on land not damaged (ha)				Land not damaged due to emergency tillage (ha)			
	Cropland	Rangeland	Other land	Total	Cropland	Rangeland	Other land	Total	Cropland	Rangeland	Other land	Total
May 1969	10,988	279	4	11,271	56,356	0	0	56,356	29,525	0	0	29,525
May 1970	23,490	1,361	10	24,861	50,099	2,438	259	52,796	95,175	0	0	95,175
May 1971	20,756	405	16	21,177	111,902	41	20	111,962	51,767	0	0	51,767
May 1972	29,525	1,418	32	30,974	20,011	267	32	20,311	68,396	0	0	68,396
May 1973	21,425	5,358	486	27,269	9,315	790	16	10,121	25,394	0	0	25,394
May 1974	37,013	6,156	3,208	46,377	17,873	1,798	81	19,752	20,234	0	152	20,386
May 1975	261,737	321,671	6,806	590,215	388,189	139,636	7,857	535,681	201,737	0	81	201,818
May 1976	257,264	35,174	7,160	299,599	168,938	11,672	0	180,610	166,487	0	0	166,487
May 1977	695,102	438,291	8,821	1,142,213	168,245	151,648	118	320,011	319,788	688	0	320,396
May 1978	168,083	13,610	3,493	185,186	76,184	1,004	101	77,289	50,791	0	0	50,791
May 1979	45,988	3,038	1,004	50,030	10,004	0	130	10,134	23,867	0	0	23,867
May 1980	13,094	1,823	723	15,640	1,964	0	138	2,102	1,871	0	0	1,871
May 1981	925,830	6,926	932	933,688	19,213	1,377	32	20,623	35,843	0	0	35,843
May 1982	161,356	17,452	4,001	182,809	46,336	6,140	656	53,132	27,714	0	0	27,714
May 1983	147,187	2,373	1,458	151,018	31,003	284	16	31,303	14,345	0	0	114,345

Source: USDA-SCS, Denver, 1984.

substantially smaller farms have to compensate insufficient dry farming income by raising irrigated corn or livestock, or by off-farm wages.

In sum, the Dust Bowls of the 1930s and 1950s and the soil losses in the 1970s and 1980s were a result of complex technoecological and socioeconomic interactions that included unpredictable precipitation, highly variable crop prices, insufficient experience with new machinery, and ill-conceived land policies.

Land Tenure and Risk Management

A central issue is the relation of land tenure arrangements to the choice of tillage practices and to the cyclic recurrence of alternating economic and ecological success and failure. Absentee owners are commonly blamed for the conditions that led to the periodic environmental and economic disasters in the Plains. The evidence does not support this claim (Evans, 1938; Fischer, 1978; Hewes, 1973, 1977; Lee, 1983). Absentee ownership in the region increased between 1920 and 1933 and again between 1944 and 1950. Interestingly, the 1920s and 1940s were periods of rising wheat demand and expanding hectarage.

Early absentee owners came from the wheat country further to the east. They bought land along the western frontier of wheat farming in eastern Colorado or consolidated landholdings behind that frontier. For a few weeks each year, these owners harvested one crop and planted the next. They brought their equipment from their home farm in the east, spread their investment over a larger area, and distributed their risk more widely than local farmers. This practice increased their general stock of farming experience and economic well-being relative to local operators, enabling them to keep up with the latest improvements in farming equipment and to survive dry years. This allowed them to buy out hard-hit residents at low prices.

Absentee owners managed their straw residues and weeds less aggressively—and therefore controlled wind erosion more effectively—than did many local farmers who practiced clean or semi-clean tillage. They were the early adopters of conservative minimum tillage and chemical fallowing, practices requiring less work than traditional mechanical tillage systems.

A recent survey in Sedgwick County provided no evidence that small, family-owned and operated farms protect the environment more effectively than do corporate farms or tenants. In fact, varying tillage methods by field are used on a single farm, regardless of farm size or ownership.

As the total number of farms has declined, so has the percentage of absentee owners, from about 30% in the 1930s to 10% in the 1970s and less than 10% in the 1980s. These numbers vary considerably in the short run and are related to the absentees' commitment to wheat. These situations can be caused by the following: short-term high moisture supplies (which can attract investors to expand wheat land into sandy areas for a few years); number and size of foreclosed farms; the attractiveness of federal programs like the Conservation Reserve (which leads to the deactivation of physically present farms); and the availability of irrigation water and the potential for livestock production (which allows for potential diversification and independence from wheat monoculture).

Interaction between highly variable crop prices and precipitation has determined the scope of environmental and economic risk in Great Plains farming. Unsuitable land policies, lack of effective regulation, and land speculation contribute to this risk, even though these and other hazards both human and natural, are of secondary importance. Lockeretz (1981:19) concludes that the historic Dust Bowls were caused not by external economic, technological, and political factors but, rather, by "our attitudes and values." The Great Plains Committee and the Soil Conservation Service of the U.S. Department of Agriculture (USDA), both established by the federal government in the 1930s to develop and implement practical solutions to the Dust Bowl syndrome, as well as the congressional Great Plains Conservation Program of 1956, obviously have failed to have a significant impact on "our attitudes and values." Conflicting economic aggressiveness and environmental-ecological conservatory needs will persist and create future boom-and-bust cycles.

Economic incentives, provided by world grain demand or government programs, and relatively high supplies of precipitation over a number of years can initiate production intensification beyond that which is in balance with the long-term carrying capacity. This would lead to another bust phase and perhaps another Dust Bowl. On the other hand, modern tillage techniques and alternative crop rotations that can mitigate the impacts of environmental and economic changes may be widely adopted.

The Dryland Wheat-Farming System

Although a variety of technologies and procedures are practiced on the wheat farms of Sedgwick County, a typical system can be described. Winter wheat (*Triticum aestivum* ÄL.Ü) is the prime cultivar. It is planted early in September, provided enough moisture is available in the upper soil layer. Shoedrills and sometimes diskdrills (ranging in width from 6 m to 12 m) are used. After a growing period of about 300 days, the wheat is harvested by migrating commercial combine crews in early July. Only about one-quarter of the surveyed farms operate their own combines (with 4- to 7-m-wide headers). The majority of the farms either have on-farm storage facilities or own storage space in a cooperative grain elevator. Others deliver their harvest directly to central grain elevators.

Weed growth begins immediately after harvest. Postharvest weed control, therefore, often includes one 10-cm-deep sweep operation between July 25 and August 5 to destroy fall weeds in new wheat stubble and one 10-cm-deep cross-sweep operation between September 1 and September 10 to kill volunteer wheat if needed. These operations save up to 75 mm of available soil moisture, but they also reduce the rate of stubble on the field surface, which leads to increased soil moisture losses during the subsequent summer season. Spring tillage begins between April 20 and May 10 with a 10-cm-deep one-way disk or a sweep operation. Fallowed fields are cultivated in May–August with sweeps and rodweeders; seedbeds are prepared between August 20 and September 1, applying the rodweeder with tong attachments crosswise 75 mm deep. Stubble fields are worked four to eight times during the 14 months of the fallow season (Tables 11-3, 11-4). Barley (*Hordeum vulgare L.*), spring wheat (*Triticum aestivum L.*), and millet (*Panicum miliaceum*) are planted in February, March, or April; sorghum (*Sorghum bicolor L.*) in May, June, or as late as July. These crops are harvested in June or July and between September and December, respectively.

All field operations can be performed by one person using 150-to 250-hp tractors. Labor/ha/operation varies with the type and depth of tillage operation (Tables 11-3, 11-4). Harvesting requires at least two persons, one to drive the combine and the other the cart or truck that hauls the grain to storage. Yields range from 2,016 to 3,024 kg/ha, depending on soil and rainfall conditions.

The average cost per field operation was $1.42/ha in 1979–1980 and $1.62 in 1980–1981 (Table 11-5). In terms of tillage, the average production cost/return ratio is 1/7.5 at yield levels between 2,352 and 2,688 kg/ha and wheat price levels of $2.65–$3.03/kg ($3.50 to 4.00 per bushel). On an energy unit basis, the wheat farming system harvests 15–30 times the energy invested in the food production process.

Specific Strategies

This general pattern varies throughout Sedgwick County. The keys to this variation are how and to what extent soil moisture budgets can be optimized in terms of controlling weed growth, water intake, evaporation and runoff losses, water-use efficiency, and soil erosion.

The manufacturing of farm implements and all cultivation operations depend on fossil fuel inputs. Consequently, the performance evaluation of the systems includes an analysis of the energy input structure in order to determine the degree of dependency on fossil fuel input, and an analysis of the energy output/input ratios, which help to identify options for restructuring of cropping patterns, if needed.

Soil and Water Management

Soil management activities serve one goal: to minimize environmental constraints and to optimize crop production. Summer fallowing represents the single most important practice in wheat production under semiarid conditions. It was not universally adopted in Sedgwick County until after the 1930s Dust Bowl,

TABLE 11-3. Field-energy budget for fertilized wheat without herbicide/pesticide application during the crop year 1979–1980.

Field Number = 1.6 | Hectares = 32 | Year = 1979 | Land Use = SFAL | Legal Des. = 9N-45-10-NE 1/4 | Soil Des. = RCB-L

Tillage operation	Manufacturer + model	Width (m)	Tractor	Depth (mm)	Ha/hr	Hectares done	Load	Operation energy (kcal/1,000 ha)		Date
One-way	SC – null	10	VR700	127.0	6.00	32.00	2	Fuel energy =	8,406,409.2	5/30/79
								Labor energy =	12,555.0	
								Embodied energy =	6,623.3	
								Transportation energy =	384,203.9	
One-way	SC – null	10	VR700	76.0	7.00	32.00	2	Fuel energy =	7,005,341.0	6/05/79
								Labor energy =	10,462.5	
								Embodied energy =	6,623.3	
								Transportation energy =	384,203.9	
Rodweed	ML – null	15	VR700	51.0	12.00	32.00	3	Fuel energy =	3,399,084.7	6/30/79
								Labor energy =	6,277.5	
								Embodied energy =	8,647.1	
								Transportation energy =	384,203.9	
Plant	IH – null	7	MM G1000	51.00	4.00	32.00	4	Fuel energy =	5,886,551.9	9/03/79
								Labor energy =	18,832.5	
								Embodied energy =	176,659.9	

								Transportation energy =	301,874.5	
								Seed energy =	21,262,500.1	
	Crop variety = Baca			Seeding rate (kg/ha) = 39.00						
Rodweed + NH3 ATT	ML – null	15	VR700	51.0	12.00	32.00	3	Fuel energy =	3,399,084.2	8/01/79
								Labor energy =	6,075.0	
								Embodied energy =	17,597.5	
								Transportation energy =	384,203.9	
								Fertilizer energy =	89,317,728.3	
	kg/ha of N, P, K, S, Z, and manure, respectively			38.00						

Yield (kg/ha) = 2890			Hectares harvested = 32.00		7/08/80
Total fuel energy =	28,096,471.5 (19.96)	Remaining residue = 31.59%	Yield energy =	1,567,350,004.6	
Total labor energy =	5,440.5 (0.04)		Total input energy =	140,785,945.9	
Total embodied energy =	216,151.1 (0.15)		Total output energy =	1,567,350,004.6	
Total seed energy =	21,262,500.1 (15.10)				
Total fertilizer energy =	89,317,728.3 (63.44)		Output/input ratio =	11.13	
Total herbicide/pesticide energy =	0.0 (0.00)				
Total transportation energy =	1,727,689.9 (1.31)				

Abbreviations: SFAL = Summer fallow; Legal Des. = Legal description; Soil Des. = Soil description; RCB-L = Richfield Loam; SC = Schafer; null = no model number available; VR = Versatile; EN. = Energy; ML = Miller; IH = International Harvester; MM = Minneapolis Moline; NH3 ATT = NH_3–attachment (fertilizer spreader); N = nitrogen; P = phosphorus; K = potassium; S = sulphur; Z = zinc; Herb./pest. = Herbicide and/or pesticide.

TABLE 11-4. Field-energy budget for unfertilized wheat without herbicide/pesticide application during the crop year 1979–1980.

Field Number = 18.10 | Hectares = 28 | Year = 1979 | Land Use = SFAL | Legal Des. = 10N-45-5-NE-1/4 | Soil Des. = RAC-SIL

Tillage operation	Ma. + Model	Width (m)	Tractor	Depth (mm)	Ha/hr	Hectares done	Load	Operation energy (kcal/1,000 ha)		Date
Oneway	SC – null	10	MF 1150	76.0	4	28	2	Fuel energy =	10,708,454.7	4/01/79
								Labor energy =	18,832.5	
								Embodied energy =	4,528.4	
								Transportation energy =	533,263.4	
Oneway	SC – null	10	MF 1150	76.0	4	28	2	Fuel energy =	10,708,454.7	5/01/79
								Labor energy =	18,832.5	
								Embodied energy =	4,528.4	
								Transportation energy =	533,263.4	
Oneway	SC – null	10	MF 1150	76.0	4	28	2	Fuel energy =	10,708,454.7	6/01/79
								Labor energy =	18,832.5	
								Embodied energy =	4,528.4	
								Transportation energy =	533,263.4	
Rodweed	ML – null	10	MF 1150	76.0	4	28	3	Fuel energy =	8,866,442.8	7/01/79
								Labor energy =	18,832.5	
								Embodied energy =	4,521.1	
								Transportation energy =	533,263.4	

Harrow	IH – null	10	MF 1150	76.0	4	28	3	Fuel energy =	8,866,442.8	8/15/79
								Labor energy =	18,832.5	
								Embodied energy =	9,960.4	
								Transportation energy =	533,263.4	
Plant	IH 150	13	MF 2775	76.0	7	28	4	Fuel energy =	5,240,509.9	9/15/79
								Labor energy =	11,770.3	
								Embodied energy =	33,273.8	
								Transportation energy =	651,289.5	
								Seed energy =	19,440,000.0	

Crop variety = Scout 66 Seeding rate (kg/ha) = 36.0

Yield (kg/ha) = 2688 Hectares harvested = 21 7/14/80

Total fuel energy =	55,098,759.5 (70.62)	Remaining residue = 8.47	Yield energy = 1,458,000,004.3
Total labor energy =	105,932.8 (0.14)		Total input energy = 78,023,639.4
Total embodied energy =	61,340.7 (0.08)		Total output energy = 1,458,000,004.3
Total seed energy =	19,440,000.0 (24.92)		
Total fertilizer energy =	0.0 (0.00)		Output/input ratio = 18.69
Total herbicide/pesticide energy =	0.0 (0.00)		
Total transportation energy =	3,317,606.5 (4.25)		

Abbreviations: RAC-SIL = Rago and Kuma silt loam; MF = Massey Ferguson; ML = Miller; IH = International Harvester.

TABLE 11-5. Fiscal budgets for winter wheat, comparing conventional tillage versus ecotillage, for 1979–1980 and 1980–1981, on the basis of local prices per hectare in northeastern Colorado.

Inputs, conventional tillage				Inputs, ecotillage		
		1979–1980	1980–1981		1979–1980	1980–1981
July	Sweep	$1.42	$1.62	Herbicide	$5.07	$ 7.10
September	Sweep	1.42	1.62	No tillage	–	–
May	Sweep	1.42	1.62	No tillage	–	–
July	Rod	1.42	1.62	Sweep	1.42	1.62
August	Chisel	1.42	1.62	Sweep	1.42	1.62
				Rod (in Sept.)	0.70[a]	0.81[a]
Total Cost:		$7.10	$8.10		$8.61	$11.15

Conventional tillage	Ecotillage
1979–1980 output/input ratio 7.00 based on $2.65/100 kg ($3.50/bu) and 2,352 kg/ha	*1979–1980 output/input ratio 7.00* (based on $2.65/100 kg ($3.50/bu) and an additional 504 kg/ha over conventional tillage due to increased soil moisture harvest)
1979–1980 output/input ratio 8.00 (based on $2.65/100 kg ($3.50/bu) and 2,688 kg/ha	*1979–1980 output/input ratio 7.80* (based on $2.65/100 kg ($3.50/bu) and an additional 504 kg/ha over 2,688 kg/ha of conventional tillage)
1980–1981 output/input ratio 7.00 (based on $3.03/100 kg ($4.00/bu) and 2,352 kg/ha	*1980–1981 output/input ratio 6.20* (based on $3.03/100 kg ($4.00/bu) and an additional 504 kg/ha over 2,352 kg/ha of conventional tillage)
1980–1981 output/input ratio 8.00 (based on $3.03/100 kg ($4.00/bu) and 2,688 kg/ha	*1980–1981 output/input ratio 6.90* (based on $3.03/100 kg ($4.00/bu) and an additional 504 kg/ha over 2,688 kg/ha of conventional tillage)

Source: Anderson (1981).

Note: Planting costs at $2.00–$3.20/ha are not included.

[a]One operation figured at half of custom cost due to an average 2.5 operations saved with chemical fallow.

although sufficient scientific information about the practice was available as early as the turn of the century (see Widtsoe, 1911). Summer fallowing in Sedgwick County implies extending the season between crops to about 14 months (from about July 14 to September 9 the following year), in order to accumulate sufficient soil moisture to reduce the risk of crop failure during the next crop cycle.

The objectives of summer fallowing in the winter wheat–fallow rotation are: (1) to reduce weed growth from harvest to planting to an absolute minimum; (2) to keep wheat stubble upright as long as possible in order to increase snow accumulation and reduce wind velocities; (3) to preserve a straw mulch of 1,120 ka/ha (1,000 lb/acre) to 1,680 kg/ha (1,500 lb/acre) until planting time to reduce wind impact, evaporation, and runoff; and (4) to maintain a rough field surface with large clods during fallow and after planting. The results are (1) improved moisture penetration, (2) reduced evaporation and transpiration losses and maximized soil moisture storage, (3) improved plant nutrient availability (nitrogen), (4) improved seed germination, (5) minimized erosion hazards, and (6) minimized energy and financial inputs.

Tillage techniques have changed constantly as a result of improved understanding and equipment. As the capacity to retain straw mulches and control weeds throughout the fallow season has improved, fallow efficiency rates have gradually improved and have led to significant yield increases. Acceptance of new techniques was initially slow, however, hindered by financial constraints, lack of information, and imported soil management traditions.

The tillage systems in use today in the county are maximum tillage (or black fallowing or clean tillage), conventional tillage (or bare fallow or semiclean tillage), stubble mulching (or conservation tillage), minimum (mechanical) tillage, and ecotillage (or chemical tillage). No-till fallow has been introduced on experimental plots on a few farms. Each system has distinct mechanical operations and implements, residue rates remaining on the field surface at the end of the fallow season, fallow efficiency and crop yield levels, and energy output/input ratios (Table 11-6).

If used, deep tillage is the first operation on wheat stubble. It breaks up clay horizons or tillage pans created by successive shallow tillage, buries straw and weeds, and produces large soil clods necessary to conserve snow melt. It can also serve as an emergency tillage for wind erosion control. Shallow tillage requires less time, fuel, and labor inputs, and preserves plant residue on the field surface. Seedbeds are prepared by shallow-tillage implements.

Clean tillage is used on roughly 5% of the land today. Ordinarily it starts with postharvest or early fall plowing, leaving almost no stubble on the field surface. Low-speed, short-throw moldboards ensure that the stubble will not be turned over entirely, and that subsequent sweep or harrow operations will restore 25–35% of the preplowing field surface stubble rate. Frequent and deep disking, mostly after significant rains during spring and summer, and harrowing for seedbed preparation renders the field surface completely bare, smooth,

TABLE 11-6. Tillage systems and related agroecological parameters for northeast Colorado.

Tillage system	Activities	Number of mechanical operations	Remaining residue (%)	Fallow efficiency (%)	Yield (kg/ha)	Energy output/input ratio
Clean tillage	Moldboard plow; one-way and offset or double-offset disk (deep) 2–3 times; harrow	5–8	0 (0)[a]	20–24	1,680–2,352	5–15
Semiclean tillage	Disk plow once; chisel plow; rodweeder	4–6	8–24 (5–10)	24–27	2,016–2,688	10–15
Stubble mulching	Large V-blades or sweeps; rodweeder	4–5	40–55 (15–30)	30–33	2,352–2,688	15–20
Minimum-tillage	Large V-blades or sweeps; rodweeder	2–4	50–60 (25–35)	30–40	2,688–3,360	20–30
Eco-tillage	Herbicides in fall; large V-blades; rodweeder	1–3	60–80 (35–50)	33–45	3,024–3,696	20–35

Source: Späth (1984).

Note: Average data for unfertilized and fertilized wheat combined and generalizes for all soil types, farm sizes and wheat varieties.

[a]Chemical decomposition and weathering accounted for.

and pulverized. Weeds are destroyed by five to eight operations, but so are straw mulches and erosion-resistant soil clods. Fallow efficiency levels reach 22–25%, yields reach 1,680 to 2,352 kg/ha.

Semiclean tillage uses a disk plow either in the fall or subsequent spring and reduces residue rates to 10–20% of postharvest straw rates. Remaining residue rates depend mainly on disk diameter, working depth, and traveling speed. Subsequent chiseling (with sweeps and rodweeding) maintains a low rate of straw mulch, and provides some erosion control and better fallow efficiency than clean tillage. But both fallow efficiency (24–27%) and yield levels (2,016–2,688 kg/ha) compare badly with the performance of the more conservative stubble mulching and minimum tillage (see Figure 11-3).

Stubble mulching with four to five operations is practiced on about 50% of the wheatland. Large V-blades or sweeps undercut stubble and weeds in late spring. Percentages of residues and fallow efficiency improve significantly. With increasing mulch rates, moisture intake improves and evaporation losses and erosion rates decrease. Upright stubble is more effective than leaning stubble; leaning stubble is more effective than flat stubble. Fallow efficiency levels from 32 to 36% provide for yield levels between 2,352 and 2,688 kg/ha. If fall weeds can be controlled, an additional 1.27–3.81 cm of soil moisture can be conserved, and yields can be increased by 134–470 kg/ha, depending on local moisture/crop yield functions.

Sedgwick County's farmers are quick to adopt new technology such as minimum tillage—a variation of stubble mulching, using the same implements but performing only two to four mechanical operations. It is practiced on about 5% of the wheatland. The presence of fall weeds and the frequency of summer rainfall, however, determine the feasibility of this management option. There may be trade-offs between increased transpiration losses and increased soil moisture storage. If implemented successfully, this system achieves a fallow efficiency of 30–40% and yields range from 2,688 to 3,360 kg/ha.

Ecotillage replaces one to three mechanical operations by contact and preemergence weed control herbicides. The objective is to kill all weeds, leaving stubble undisturbed and reducing mechanical operations to one to three per season. Plant and soil residual effects vary greatly from field to field. If implemented successfully, fallow efficiency ranges from 35 to 45%. In contrast to minimum tillage without herbicide application, ecotillage can conserve between 2.54 and 5.08 cm more available moisture and about 22–34 kg/ha more available nitrogen per fallow season. Fuel savings are enormous; yields can be increased by 200–1,000 kg/ha (Smika & Wicks, 1968; Späth, 1984), and erosion impact can be minimized, since harvested straw rates are subject only to weathering impact and chemical decomposition. On those fields that have been treated with herbicides for a number of years, broadleaf weeds decline and annual grassy weeds increase.

The overall agroecological advantages of ecotillage are significant. Present trends toward higher fuel and equipment costs; improved herbicide adaptation

FIG. 11-3. Dryland wheat stubble-mulch after disking, Sedgwick County (Späth).

to specific weed, soil, and weather characteristics; shorter straw wheat varieties; higher wheat yield potentials; and increased concern for overall energy conservation and efficiency improvement will help ecotillage to become a widely established alternative in the future.

All these tillage systems are designed to reduce water consumption by weeds. Average annual water requirements of broadleaf weeds of roughly 500 kg of water per kg of dry-matter production match the water requirement given for winter wheat. Two and one-half cm of water can grow 515 kg of wheat per hectare. Assuming a rather conservative 1 : 1 straw/grain ratio growth input near the heading stage, about 268 kg/ha of grain are lost for each 2.5 cm of water that is used by weeds. These data from regional experiment stations correspond well with an average soil moisture/grain yield function (equation 11-1)

$$Y = -2{,}000 + 10X \tag{11-1}$$

(where Y = wheat yield in kg/ha, X = available moisture in millimeters; $r = 0.83$), which describes performance levels on the farms under investigation in terms of their water-use efficiency, averaged for all wheat varieties and soils in the range of 1,000–2,700 kg/ha of wheat per hectare.

Uncontrolled weed growth also consumes soil nutrients, otherwise available to crops. Postharvest weed growth from mid-July to late September uses about 34 kg/ha available nitrogen and up to 7.5 cm of moisture. If fall weed control is successful, wheat grain yields can be increased by 330–950 kg/ha and straw yields by 560–1,100 kg/ha. However, a disk operation is commonly used in fall, which destroys a substantial portion of the surface cover, thus sacrificing the potential advantages of fall weed control. In saturated wheat markets, flexible crop rotations of fall and spring grains might provide a means not only to improve economic performance, but also to cope with weeds that establish themselves under monoculture conditions.

Conservative wheat cropping also reduces natural nutrient supplies. Fertilizer applications in 1 out of 4–8 years restore desirable fertility levels. Applications of 34–50 kg/ha of N can increase wheat yields from 5 to 15% over unfertilized wheat under favorable rainfall conditions on sandy soils, where natural supplies of total N are initially low, and on heavier soils under less than favorable rainfall conditions (Späth, 1984). Yields decrease when N applications reach 67 kg/ha. Phosphorus is the only other nutrient to which wheat shows a consistent response. This response is limited to medium-textured or eroded soils with less than 28 kg/ha of available P in the main root zone.

Contour terraces are installed mechanically on some of the wheatland in order to control runoff and increase water infiltration rates. In general, however, sloping terrain is used as rangeland only, and water erosion is a localized minor problem.

Wind erosion is a significant problem on sandier soils and on intensively tilled field surfaces. It occurs primarily after wheat has been planted, when ground cover is at a minimum. A second maximum occurs during spring and early summer of the fallow season, when initial spring tillage reduces straw rates substantially and when wind speeds are at a maximum. Long-term erosion causes a change in soil texture (mainly a reduction of those medium-diameter soil particles that have the capability of forming erosion-resistant soil clods), and a significant loss in organic matter, phosphorus, and potash. The result is a reduction of the potential to store plant-available moisture. The increase in local soil aridity and reduction of soil fertility are related to the process of desertification.

Field observations show that optimal fertility is the easiest and most effective profitable way of increasing wheat water-use efficiency. Nutrient-deficient wheat uses water at approximately the same rate as nutritionally balanced wheat, but with a much lower yield. Eroded fields, therefore, require fertilizer supplements not only to offset the nutrient losses caused by erosion, but also to optimize reduced moisture supplies. Individual operators can control erosion more readily than they can resort to these delicate management options after the damage has occurred.

Wind erosion is controlled on fallowed land by creating large clods on the soil surface, which roughen the field surface, reduce wind speed and trap drifting soil, by reducing field length along the main wind direction by strip cropping, by establishing vegetative barriers to reduce wind speed and soil avalanching, and by maintaining a vegetative cover to protect the soil. Stubble mulching incorporates most of these conservation measures.

Performance Evaluation

The overall performance level of the dry farming systems in Sedgwick County can be expressed in terms of water-use efficiency, "soil moisture–soil erosion–crop yield" systems control efficiency, cost–return budgets, and in terms of cultural energy input–food energy output relationships.

WATER-USE EFFICIENCY

Production risks have been greatly reduced because of improved tillage technology. Based on data from North Platte, Nebraska, for 1940–1970, the average regional yield probability at the 1,344-, 2,016-, and 2,688-kg/ha level for wheat–fallow was about 95%, 70%, and 40% (Figure 11-4). Between 1980 and 1983, average fallowed soil contained between 2.5 and 33 cm of available moisture in a 1.5-m deep soil profile at planting time. These moisture levels are sufficient to produce a good stand in fall and to prevent winter-kill damage (Johnson, 1964). Equation (11-1) points out that at least 20–25 cm of total available moisture (stored available soil moisture at planting time plus crop season precipitation) are consumed before any grain is produced. Such yield functions, however, vary greatly, mainly with soil texture, wheat variety, and local precipitation/temperature relationship, but also with technoprocedural strategies such as planting time, planting equipment, seeding rate, and planting depth.

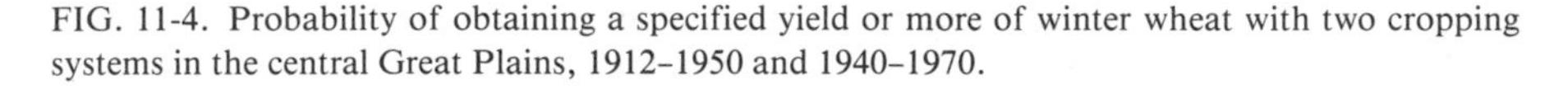

FIG. 11-4. Probability of obtaining a specified yield or more of winter wheat with two cropping systems in the central Great Plains, 1912–1950 and 1940–1970.

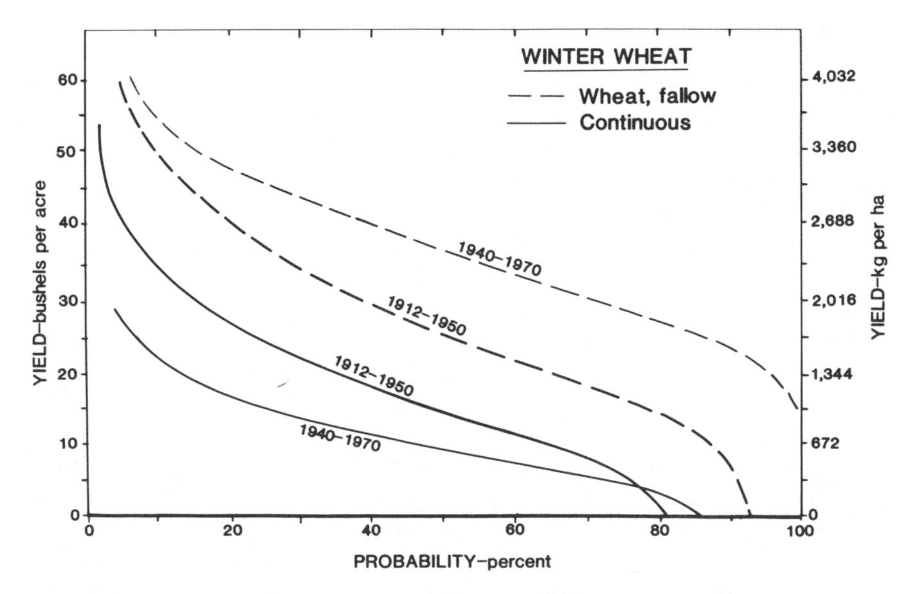

Source: U.S. Department of Agriculture, *Summer Fallow in the Western United States*, Conservation Research Report No. 17, 1974, p. 17.

Overall water-use efficiency of wheat-fallow rotation has gradually improved throughout the eastern Colorado region from an all-time low of 0.90 kg/ha/mm in the 1930s to about 3.18 kg/ha/mm since the mid-1960s (Table 11-7). This trend was maintained even through the record-low rainfall of the late 1950s and 1960s, since new technologies offset the adverse moisture conditions.

CONTROL OF SOIL MOISTURE–SOIL EROSION–CROP YIELD RELATIONSHIPS

Dryland use stimulates the processes of soil erosion and deterioration of production potential on the basis of diminishing soil profile qualities if nature's conservation and regeneration needs are neglected. In order to assure long-term agroecological stability, dryland wheat farming systems must reduce soil loss rates to a level that is considered to be nondetrimental from an ecological rather than an economic point of view. These tolerable soil loss rates are lower than those given in the wind and water erosion equations (Skidmore & Woodruff, 1968; Wischmeier & Smith, 1965), which are based on economic considerations.

Long-term destructive wind erosion has been avoided and productivity levels have been maintained on a county scale over the past 20 years by controlling this system of agroecological interrelationships, which are described in equation 11-2 (Späth, 1979, 1980a, 1980b):

$$F_{\text{opt}} = \frac{100 \times \text{Re}_{\text{min}}}{b \times d(100 - \text{Ti})} - \frac{a + b \times c}{b \times d} \qquad (11\text{-}2)$$

(F_{opt} = optimized moisture supply in mm; Re_{min} = minimum residue rate in kg/ha for effective erosion control required on the field surface for a given soil texture at planting time, if no erosion control method other than stubble mulching is in use; Ti = residue loss due to mechanical tillage between harvest and planting time in percentage of residue harvested; a = Y-intercept; b = slope of local grain/straw yield function; c = Y-intercept; and d = slope of local available moisture/grain yield function).

Equation 11-2 can serve as a conservation planning tool on a farm level as well as a guideline for defining the agroecological dry boundary of nonirrigated agriculture in a spatial perspective. Components of this agroecological dry boundary concept are defined for northeast Colorado by equation 11-1 and equation 11-3:

$$\text{Re} = -1{,}750 + 2.5Y \qquad (11\text{-}3)$$

(Re = straw in kg/ha; Y = grain yield in kg/ha; valid for a yield range of 1,000–2,500 kg/ha).

For Sedgwick County the average minimum residue rate for effective erosion control (Re_{min}) on silt loam is 1,400 kg/ha. Residue loss rate (Ti) ranges from 30 to 40% per disk plow operation, from 10 to 20% per sweep operation,

TABLE 11-7. Yield and water-use efficiency of winter wheat grown continuously and after fallow at Akron, Colorado.

	Average annual precipitation (mm)	Continuous wheat yield (kg/ha)	Fallow wheat yield (kg/ha)	Water-use efficiency[a] Continuous wheat (kg/ha/mm)	Water-use efficiency[a] Fallowed wheat (kg/ha/mm)
1911–1920	464.31	776.1	1,465.0	1.64	1.59
1921–1930	428.50	208.3	954.2	0.48	1.11
1931–1940	387.43	194.9	672.0	0.53	0.90
1941–1950	485.14	856.6	1,975.7	1.70	2.04
1951–1957	388.87	342.7	1,391.0	0.88	1.80
1958–1966	380.75	–	1,767.4	–	2.33
1967–1970	360.68	752.6	2,284.8	2.09	3.18
60-year average	419.10	497.3	1,424.6	1.67	1.70

[a]Water-use efficiency on a harvest-to-harvest precipitation basis.

Source: U.S. Department of Agriculture, *Summer Fallow in the Western United States*, Conservation Research Report No. 17, 1974, p. 17.

and from 5 to 10% per rodweeder operation. Using data for a minimum tillage or ecotillage systems with a 30% mechanical straw reduction and a requirement of 1,400 kg/ha standing stubble for effective wind erosion control as input for equation 11-2, the system calls for an optimized moisture supply (F_{opt} of 396 mm (15.59 in.) per crop. If this amount of moisture is made available for one crop, dry farming by means of minimum tillage produces enough straw to prevent destructive erosion and to maintain agroecological stability. If a given tillage practice cannot provide for this required optimized moisture level (F_{opt}), either alternative wheat varieties with another moisture/grain and/or grain/straw function must be introduced in order to balance equation 11-2 by changing factors *a, b, c*, or *d*, or additional erosion control methods (other than mulching, such as permanent grass wind barriers or reduction of field widths along the main wind direction) must help reduce locally required Re_{min} factors. If these options do not exist, the tillage system has to be changed in order to provide for a reduced Ti-value. If none of these management options balance equation 11-2, agroecological considerations call for a replacement of the wheat–fallow system by, for example, reestablishing grassland with an appropriate conservative approach to range management. These latter changes have become necessary only in areas where soils have been truncated because of high erosion rates.

PRODUCTION COST–RETURN COMPARISON

The economics of wheat–fallow systems change dramatically from year to year in response to real interest rates, commodity prices, and export demands; other variables that determine economic prosperity of individual farmers are yield levels, taxes, government land use and pricing programs; management practices; cost, size, and depreciation of farm implements; hectares farmed, planted, and harvested; and number of fallow operations performed. Most of these variables are beyond the farmers' immediate control. In 1981 and 1982, the gross value of farm (dryland wheat) production was about $21.50/ha (Table 11-8); Production costs per hectare, on the other hand, were about $2.00 for seed, fertilizer, and pesticides; $9.00 for power and equipment; $1.50 for buildings; $3.00 for labor; and $12.00 for other production-related costs. The overall loss was $6.25/ha. Farm returns and costs have been unbalanced since the late 1970s. Increasing numbers of foreclosures on family farms indicate that the dryland wheat system is in the third boom–bust cycle of this century. At present the only farms that survive are those that have succeeded in reducing loan payments or maintaining their creditworthiness.

The data on economic performance can be interpreted more objectively when output–input information is related exclusively to the production process. For conventional stubble mulching, the tillage costs alone reached $7.10–$8.10/ha for the 1980 and 1981 crops (see Table 11-5). At yield levels of 2,352 kg/ha (35 bu/acre) and 2,688 kg/ha (40 bu/acre) and wheat prices of $2.65/100 kg and $3.03/100 kg, performance ratios of 7 and 8 show that direct production costs are high in relation to wheat prices and that pricing for basic inputs requires reorientation in order to make returns worthwhile. So far land purchases, equipment investments, and harvesting contracts use up the small returns. Operational costs (see Table 11-8) need to be deflated and restructured, and land values stabilized, before farm budgets can be balanced and operational profit margins reestablished. Credit policies and government price support programs play a vital role in farmers' intricate endeavors to achieve a long-term budget balance. One such option that is within the farmers' control is a shift from semiclean tillage toward minimum tillage, which achieves a maximum performance level at a minimum inventory requirement and direct production input. Time savings can be invested in off-farm cash-raising activities, which could generate a daily income of $300.00 and $12,000–$15,000 per year in 1980 (Anderson, 1981).

Family living expenses are indicative of the economic well-being available to families of the wheat-farming system. Food purchases are the highest of all cash expenses (Table 11-9). Family living expenses show a decline of about 2% since 1978, but family size has also decreased each year, so that living expenses per person actually increased slightly. Comparing total living expenses with farm income–expense budgets suggests that the present economic structures leave little room for failure.

PRODUCTION ENERGY BUDGETS

Energy budgets disaggregate the energy inputs for a farm. They help to identify the overall energetic performance from which farm management changes can result. These budgets are also a means by which seasonal production energy

TABLE 11-8. Farm returns and costs per tillable hectare on dryland farms (in U.S. dollars).

Farm returns and costs per tillable hectare	1981	1982
Farm returns:		
Crop returns	$ 15.78	$ 15.37
Livestock return above feed	1.09	1.61
Custom work	0.72	0.99
Other farm receipts	4.37	3.29
Gross value of farm production	21.97	21.26
Farm costs:		
Soil fertility	0.77	0.64
Pesticides	0.34	0.62
Seed	0.75	0.87
Crop total	1.85	2.12
Utilities	0.30	0.42
Machinery repairs	2.52	1.81
Machine hire	0.96	1.14
Fuel and oil	2.59	2.12
Auto–farm share	0.12	0.13
Machinery depreciation	2.96	3.69
Power and equipment total	9.45	9.32
Drying and storage	0.30	1.40
Building repair	0.15	0.20
Building depreciation	0.42	0.58
Building total	0.87	2.18
Labor, unpaid	2.10	1.91
Labor, paid	1.16	0.92
Labor total	3.26	2.82
Livestock supplies and services	0.17	0.26
Insurance	0.33	0.35
Taxes	0.78	0.75
Miscellaneous	0.32	0.36
Interest charge, land 9%, other 9%	11.33	9.21
OTHER COSTS, TOTAL	12.93	10.93
Total nonfeed costs	28.37	27.37
Gain or loss on machinery and buildings sold	0.01	0.03
Management returns	– 6.39	– 6.08

TABLE 11-9. Family living expenses (in U.S. dollars), 1979–1982.

Item	Expenditure			
	1979	1980	1981	1982
Food	$ 3,372	$ 3,452	$ 3,629	$ 3,419
Savings and other investments	2,011	1,826	2,220	2,369
Health (doctor, drugs, medical insurance)	1,581	2,074	1,507	2,089
Housing, new and remodeling	1,554	1,318	1,236	1,593
Home furnishings and equipment	1,460	1,444	1,190	1,175
Insurance, life	1,387	1,157	1,132	1,006
Income tax (federal and state, 1978)	1,304	1,065	1,094	958
Miscellaneous (dues, cash, etc.)	1,294	1,113	1,067	904
Clothing	1,151	803	940	820
Contributions	778	851	849	784
Household operation	722	925	805	755
Personal items	593	734	617	665
Gifts	415	586	557	540
Recreation	323	417	518	473
Education	308	215	436	333
Total	$18,253	$17,980	$17,797	$17,883
Average size of family	3.5	3.33	3.11	3.17
Average expenditure per person	$5,215	$5,399	$5,723	$5,641

flows can be monitored and production assessed, unaffected by regional and temporal fluctuations of costs for services, commodities, real estate, or inventory. Energy budgets, therefore, provide an objective and time-independent means of evaluating and planning food production on both an economic and an ecological basis. Food production energy inputs are fuel, labor, machinery, seed, herbicides and pesticides, fertilizer, and tillage- and planting-related on-farm transportation. All other cultural energy inputs are not directly related to the production process.[1]

Field energy budgets of a 700-ha farm serve as an example (Tables 11-3, 11-4). This farm size was selected because farmers in Sedgwick County consider it the size necessary for a profitable operation. Field energy budgets are calculated as kcal/1000 ha and percentage of total input. The remaining straw residue level (in percentage) available at the field surface at planting time serves as a numeric expression of the tillage system practiced. The energy output/input ratio describes the energy reproduction efficiency level for the field under given tillage practices.

The results indicate that energy efficiency for unfertilized wheat can best be achieved if fuel consumption can be reduced. Fuel constitutes 60–75% of total input. Seeding rate performance tests might show potential savings of almost 25–40%. Farmers' estimates of potential savings in seed energy range from 10 to 30%. Fertilizer application comprises 50–60% of total cultural energy input; it improves grain yields by 5–15%. Energy efficiency levels, however, can be reduced by up to 50% (Späth, 1984).

Production energy budgets can be converted to production cost budgets by converting the individual energy inputs and the output into standard units (e.g., liter, hour, kg/ha) and by combining them with cost data applicable at the particular location and time of observation. A combination of both energy and fiscal budgets allows for an objective and comprehensive evaluation of production performance on a comparative basis for a multitude of planning and management tasks.

County energy efficiency levels depend on the tillage system implemented (see Table 11-6). The fewer mechanical operations involved, the higher the rates of remaining straw. High crop residue rates lead to high fallow efficiency and yield levels; in turn, high yields and low cultural energy inputs result in high energy output/input ratios. Table 11-6 indicates a positive trend toward improving performance with decreasing agrotechnical impact. At least two-thirds of all farms operate within the energy efficiency range from 12 to 20. As an expression of desertification, caloric energy output per unit of production energy input is lowest on fields that have been subject to wind erosion and on fields with lighter soils.

These results suggest the following. First, dryland winter wheat systems operate much more efficiently with respect to energy use and production than previous studies have shown (Heichel, 1967; Marchetti, 1980; Patrick, 1977; Pimentel & Pimentel, 1979). The energy output/input ratios of 1.99–3.25 resulting from these studies are based primarily on the inflated data published by Pimentel and colleagues (1973). Their depreciation models reflect equipment lifetimes as established by the Internal Revenue Service rather than actual lifetimes; they include unknown amounts of additional indirect (and not production-related) inputs; and they refer to smaller (and in some cases unspecified) farm sizes. These factors contribute to the drastically reduced energy ratios, which are inconsistent with those of Sedgwick County. Second, energy efficiency stands in sharp contrast to dollar efficiency for the same farming operation (see Tables 11-5, 11-6, 11-8). Finally, crop production energy ratios cannot be explained conclusively without an evaluation of the impact of surrounding geoecological forces. Soil characteristics; rainfall frequency, timing, and duration; and wheat varieties are the major variables and parameters that determine the production energy ratios of given tillage systems.

Conclusion

Eco- or chemical tillage, minimum tillage, stubble mulching, semiclean tillage, and clean tillage represent the diversity of the present-day dryland wheat farming system in Sedgwick County (see Table 11-6). Energy efficiency and socioeconomic welfare are highly variable within each tillage category. Available soil moisture and soil quality, followed by farm size and crop variety, are the prime internal variables determining economic and energetic performance levels within each category. Given the same agroecological setting, the various tillage systems achieve decisively different energy efficiency levels (see Figure 11-5). Clean and semiclean tillage (black fallowing and conventional tillage) consume the highest cultural energy inputs and produce the lowest food energy outputs. Output/input ratios of stubble mulching reflect substantially reduced consumption of production energy, with savings due to decreased fuel use and lower investments in implement inventories. Where feasible, minimum tillage and ecotillage achieve optimized soil moisture and nutrient availability and require minimal cultural energy inputs. On this basis they achieve the highest production performance and energy efficiency levels.

Adoption of the most recently developed tillage alternatives is slow. Changes from stubble mulching to ecotillage cannot be expected to diffuse throughout the region in the same fashion as the shift from moldboard to disk

FIG. 11-5. Energy efficiency levels and cultural energy used under various food production systems in Sedgwick County, Colorado (Späth).

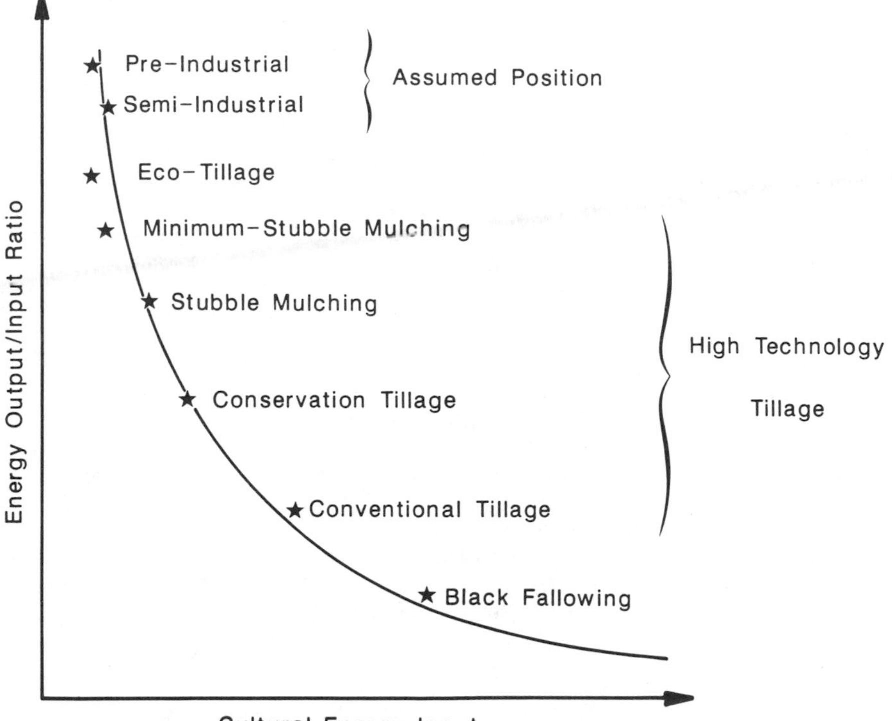

plow and later to large V-blades. Local soil characteristics, rainfall patterns, and specific weed problems prevent a uniform adoption of chemical and minimum mechanical practices or even of no-tillage fallow practices. Additional impediments to no-tillage are set up by extremely high costs for planting equipment that can drill seed through the standing wheat stubble. It may be expected, however, that continued constraints in production energy affordability and economic returns will support a continued general shift in soil and water management practices from traditional energy-intensive to more energy-conservative alternatives. These practices provide a key means for the dryland wheat system to mitigate the impacts of externally imposed political-economic constraints. The internal flexibility of this farming system will allow it to adapt to future external stresses caused by severe drought and persistent strong winds. Resulting spatial and temporal changes in land-use patterns will be small, however, compared to those changes exerted by external political and economic influences that are largely beyond the farmers' control. Future boom-and-bust cycles, therefore, will most likely be linked to such external forces rather than to internal deficiencies.

Note

1. The method for calculating production energy budgets on a field scale and the need for a more universal agreement on budget input components and on updated energy conversion factors are discussed in Späth (1984:37–38). The relationship between production energy output/input ratios, soil and climatic variability, and tillage systems, as well as temporal and spatial patterns of energy efficient distributions are discussed in Späth (1984, 1985b). This budgeting approach differs from the one proposed by Pimentel and colleagues (1973), which has been widely used in other energy analyses. Späth's approach includes only inputs that are directly related to the production process and that can be subject to change in an endeavor to improve output/input ratios. Energy expenses for crop harvest, drying, and transportation to the site of use are excluded because they occur after the production process in the field is completed.

Embodied energy in the Pimentel group's approach includes virtually all energy expenses necessary to produce and deliver raw metal and other materials to farm implement manufacturers, using the industrial production consumption rates in use before the first national energy crisis. Since then energy conservation has been implemented and improved in many industries, and energy requirements for mining ore, raw metal production, and manufacturing have declined sharply. Doering (1978) introduced the *value-added* concept, suggesting depreciation of only those energies invested in the manufacture of farm implements from raw materials, because metals can be recycled at the end of their useful life. Doering's conversion factors have been further reduced by 10% to account for the addtional progress in energy conservation in industrial processes. This reduced added energy (embodied) has been depreciated over the life spans given by farmers for various equipment. In dryland wheat farming hectarages are 10–100 times larger than the the 25 ha mentioned for corn by the Pimentel group (1973). Embodied energy per productive unit area in wheat farming, therefore, will be substantially lower than for corn. For all other inputs the lowest published conversion factors have been chosen to incorporate the progress that has been made since 1973 in energy conservation in industrial production processes.

References

Anderson, E.G. Economics of eco-fallow versus conventional tillage in wheat–summer fallow system. *Proceedings: Eco-Fallow Conferences*, Lamar, Colorado, 1981, 17–22.

Black, A. L., & Siddoway, F. H. Tall wheatgrass barriers for soil erosion control and water conservation. *Journal of Soil and Water Conservation*, 1971, 26, 107–111.

Bowden, M. J. Desertification of the Great Plains: Will it happen? *Economic Geography*, 1977, 53, 397–406.

Doering, O. C., III. *A general procedure for counting energy in farm machinery*. Department of Agricultural Economics, Purdue University, West Lafayette, Indiana, August 1978. (Mimeograph).

Elder, J. A. *Soils of Nebraska*. University of Nebraska Research Report 2, 1969.

Evans, M. Non-resident ownership: Evil or scapegoat? *Land Policy Review*, 1938, 1, 15–20.

Fischer, L. K. *Environment and farm size*. Paper presented at the Perry Foundation on Agricultural Policy and Marketing, University of Columbia, Missouri, December 10, 1978.

Greb, B. W. *Reducing drought effects on croplands in the west-central Great Plains*. Agriculture Information Bulletin, USDA, 1979, 420, 1–31.

Greb, B. W., & Black, A. L. Vegetative and artificial fences for managing snow in the central and northern Plains. *Symposium on Snow and Ice in Relation to Wildlife and Recreation*, Ames, Iowa, February 11-12, 1971, 96–111.

Greb, B. W., Smika, D. E., & Black, A. L. Water conservation with stubble mulch fallow. *Soil and Water Conservation*, 1970, 25, 58–62.

Greb, B. W., Whitney, R. S., & Tucker, R. H. *Commercial fertilizer experiments with non-irrigated crops in eastern Colorado in 1952*. Colorado Agricultural Experimental Station General Series Paper 526, 1953.

Greb, B. W., Whitney, R. S., & Tucker, R. H. *Commercial fertilizer experiments with non-irrigated winter wheat in eastern Colorado in 1953*. Colorado Agricultural Experimental Station General Series, Paper 586, 1954.

Heichel, G. H. Agricultural production and energy resources. *American Scientist*, 1967, 64, 61–72.

Hewes, L. *The suitcase farming frontier: A study in the historical geography of the central Great Plains*. Lincoln: University of Nebraska Press, 1973.

Hewes, L. Early suitcase farming in the central Great Plains. *Agriculture History*, 1977, 51, 23–27.

Hoover, L. M. *Kansas agriculture after 100 years*. Kansas Agricultural Experiment Bulletin 392, 1957.

Johnson, W. C. Some observations on the contribution of an inch of seeding-time soil moisture to wheat yield in the Great Plains. *Journal of Agronomy, 1964, 51, 29–*35.

Lee, L. K. Land tenure and adoption of conservation tillage. *Journal of Soil and Water Conservation*, 1983, 38, 166–168.

Lockeretz, W. The Dust Bowl: Its relevance to contemporary environmental problems. In M. P. Lawson & M. E. Baker (Eds.), *The Great Plains perspectives and prospects*. Lincoln: University of Nebraska Press, 1981.

Marchetti, C. Wieviel Öl kostet unser täglich Brot. *Bild der Wissenschaft*, 1980, 2.

Olsen, S. R., Cole, C. V. Watanabe, F. S., & Dean, L. A. *Estimation of available phosphorus in soils by extraction with sodium bicarbonate*. U.S. Department of Agricultural Circular 939, 1954.

Ottoson, H. W., Anderson, A. H., Birch, E. M., & Henderson, P. A. *Land and people in the northern plains transition area*. Lincoln: University of Nebraska Press, 1966.

Patrick, N. A. Energy use patterns for agricultural production in New Mexico. In W. Lockeretz (Ed.), *Agriculture and energy*. New York: Harcourt Brace Jovanovich, 1977.

Pimentel, D., Hurd, L. E., Bellotti, A. C., Forster, M. J., Oka, I. N., Sholes, O. D., & Whiteman, R. J. Food production and the energy crisis. *Science*, 1973, 182, 443–449.

Pimentel, D. M., & Pimentel, M. *Food, energy, and society*. New York: Wiley, 1979.

Skidmore, E. L., & Woodruff, N. P. *Wind erosion forces in the United States and their use in predicting soil loss*. USDA Agriculture Handbook 346, 1968.

Smika, D. E., Black, A. L., & Greb, B. W. Soil nitrate, soil water and grain yields in a wheat-fallow rotation in the central plains as influenced by straw mulch. *Journal of Agronomy*, 1969, 61, 785–787.

Smika, D. E., & Wicks, G. O. Soil water storage during fallow in the central Great Plains as influenced by tillage and herbicide treatments. *Soil Science Society of America*, 1968, 32, 591–595.

Späth, H.-J. Problems of optimal utilization of dry steppes with cold winters. *Natural Resources and Development* (Tübingen: Buchdruckerei Eugen Göbel), 1976, 3, 72–92.

Späth, H.-J. Ecological re-definition of the dry-boundary of non-irrigated agriculture and its applicability in developing countries. *Tecnologia Apropiada para Paises Sub-desarrollados*, Segundo Simposio Internacional de Ingenieria, Universidad Centroamericana Jose Simeon Canas, San Salvador, El Salvador, 1979.

Späth, H.-J. *The agro-ecological dry boundary of non-irrigated agriculture in the central Great Plains of North America.* Erdwissenschaftliche Forschung, Band XV. Wiesbaden: Franz Steiner Verlag GmbH, 1980a.

Späth, H.-J. Die agro-ökologische Trockengrenze. *Erdkunde*, 1980b, Band 34, 224–231.

Späth, H.-J. Energiewirksamkeit der Weizenproduktion unter semi-ariden Klimabedingungen. *Erdkunde*, 1984, Band 38, 35–44.

Späth, H.-J. The energy equation for dryland wheat, central Great Plains, North America. *Stuttgarter Geographische Studien*, Stuttgart, 1985a, Band 105, 141–154.

Späth, H.-J. Bilanzierung des Produktionsenergieflusses in der Trockenlandwirtschaft—Erster Schritt zur einzelbetrieblichen Energieumsatz-Optimierung und zur agroenergetischen Regionalanalyse (Budgeting of the agricultural production energy flow in dryland farming—First step to optimize the energy turnover on a farm basis and to an agroenergetic regional analysis). *Der Tropenlandwirt*, 1985b, 141–171.

U.S. Department of Agriculture. *Soil survey for Sedgwick County, Colorado*, 1969.

U.S. Department of Agriculture. Conservation Research Report No. 17, 1974, 17.

U.S. Department of Agriculture, Soil Conservation Service, Denver, Colorado, personal communications, 1984.

U.S. Department of Commerce, Bureau of Census, 1910–1980.

U.S. Department of Commerce, Bureau of Census, 1982 Preliminary Report, Sedgwick County, Colorado, February 1984.

Webb, W. P. *The Great Plains*. New York: Grosset & Dunlap, 1931.

Widstoe, J. A. *Dry-farming: A system of agriculture for countries under a low rainfall*. New York: Macmillan, 1911.

Wischmeier, W. H., & Smith, D. D. *Predicting rainfall erosion losses from cropland east of the Rocky Mountains*. Agriculture Handbook No. 282, USDA-ARS in cooperation with the Purdue Agricultural Experimental Station, 1965.

12

Growing against the Grain: Mechanized Rice Farming in the Sacramento Valley, California

MARY BETH PUDUP AND MICHAEL J. WATTS

> *Capitalist development in agriculture is neither unilinear nor convergent. . . . The need to discover these diverse forms [of agriculture] is implicit in Kautsky's classic statement of the "agrarian question": "In what ways is capital taking hold of agriculture?" This reflects an awareness . . . of capitalism's adaptability and the possibility of its appearance in divergent forms . . . [not] only in terms of hired labor.*
>
> —Patrick Mooney (1982:280)

One-quarter of the world's people spend much of their lives growing rice. According to the Food and Agricultural Organization (FAO), rice (*Oryza sativa*) is the staff of life for half the human race; it is unusually ubiquitous, cultivated from 36° north of the equator to 40° south. Moreover, rice is unique among cereals insofar as it is grown in standing water and is cooked and consumed as a whole grain (Kahn, 1985). In Asia, where 90% of the world's rice is grown, wet-rice production practices are especially labor intensive: 1,111 person-hours/ha in parts of Taiwan, and 1,729 hours/ha in some *sawah* systems in Bangladesh (Ruthenberg, 1980:207). In these regions, rice is frequently double-cropped and is capable of supporting enormously high population densities—almost 200/km^2 for Asia as a whole—but less than 2% of production enters the world market. Conversely, in the United States, which accounts for only 1.6% of global production, rice is principally an export commodity—in 1980 accounting for roughly 25% of the world rice trade—and its production is rendered distinctive by the dramatic displacement of farm labor and the almost complete mechanization of most farm operations. For example, total labor averaged only 4.4 persons-hours/ha in California rice production in the 1980s,

Mary Beth Pudup. Regional Research Institute, West Virginia University, Morgantown, West Virginia.

Michael J. Watts. Departments of Geography and Development Studies, University of California, Berkeley, California.

and 12.8 hours in the most labor-intensive areas of Texas (Mullins & Krenz, 1981:22; California Department of Food and Agriculture, 1984).

The fascinating story of the genesis of highly mechanized rice production in the United States has yet to be told. Introduced into the colonies in 1609, rice flourished in South Carolina and in the southeastern states as a slave-based, export-oriented plantation crop. On the eve of the American Revolution, 30 million kg of rice was exported from the American colonies, primarily to lucrative Caribbean markets. In 1879 almost half of U.S. production still originated in South Carolina. By the turn of the century, however, Louisiana rice grown on abandoned sugar plantations accounted for almost three-quarters of U.S. paddy production. Immigrant midwestern farmers, cultivating long- and medium-grain rice and familiar with mechanized wheat production, transformed large parts of the Mississippi floodplain and the Louisiana prairies into a "new Iowa." Indeed, it was the successful experimentations with rice culture in southwestern Louisiana that later spawned the growth of irrigated rice in Texas and Arkansas and, during the early 20th century, in California (Daniel, 1985). By the end of World War I, the latter three states alone grew well over half of the U.S. rice crop, having produced virtually no paddy 20 years earlier (Richards, 1969).

The contrasting historical legacies of rice development in the context of quite different local ecologies and growing conditions are pivotal in an understanding of the regional variation among contemporary rice production systems in the United States. Yields, capitalization, the intensity of mechanization, land values, scale of operation, subsidiary crops, forms of land tenure, share leasing arrangements, cultural practices, and rice varieties all vary markedly among five principal producing regions: Grand Prairie–northeast Arkansas, Gulf Coast of Texas, Mississippi (River Delta), southwest Louisiana, and the Sacramento and San Joaquin Valleys in California (Table 12-1). Although the largest harvested hectarage is currently to be found in northeastern Arkansas and the Mississippi Delta regions, the California rice culture, since its inception in 1912, has consistently yielded the highest crop output per hectare and the lowest costs of production, and has experienced the highest degree of state (both regional and national) intervention. Despite the fact that California rice growers are relative neophytes, they have been in the forefront of farm mechanization, technical innovation, and varietal improvement. By the 1930s aircraft were employed for aerial seeding, large investments had been made in seed experimentation and research, new forms of harvesting and drying had been rapidly introduced, and capital had been invested in land development and irrigation infrastructure on a huge scale.

In its genesis and development, California rice farming was originally connected to landed interests and land speculators, to large-scale finance capital, to eastern business and the railroad companies, and to systematic public-sector sponsorship of research and water development. In short, California rice production has always been capitalist through and through. At the same time,

TABLE 12-1. Comparative U.S. rice production systems, 1978–1979.

Variable	California	Mississippi	Texas	Arkansas	Louisiana
Area harvested (hectares)[a]	244,939	374,089	234,413	462,753	223,886
Yield (kg/ha)[a]	8,083.6	4,928.7	5,276.8	5,097.2	4,502.1
Production (1,000 t)[a]	1,978.34	391.3	991.9	2,411.5	960.05
Total cost/ton ($)[a]	208.12	274.12	301.40	260.48	251.24
Total ha/farm	292.7	534.	549.4	294.3	223.5
Value of land and buildings/ha ($)	3,304.86	2,260.05	1,852.50	2,190.89	2,437.89
Value of machinery/ha ($)	481.65	348.27	266.76	405.08	397.67
Percentage of land in rice per farm	58	31	56	35	44
Major secondary crops	Wheat/cotton	Soybeans	Soybeans	Soybeans	Soybeans
Percentage of rice acreage on farms 810+ ha.	36	53	40	24	18
Percentage of rice production on farms with sales $0.5 million +	39	34	21	12	9
Ownership (percentage of total)					
Individual/family	42.5	49.1	57.3	61.3	70.8
Family held corporation	28.5	22.2	11.2	15.9	6.2
Other than	5.2	2.8	1.2	1.5	5.2
Partnership	23.8	25.9	30.3	21.3	17.7
Tenure					
Percent of rice on owned land	45	50	13	36	14
Percent of rice on rented land:					
(i) Cash	6	20	28	11	3
(ii) Shares	49	30	59	53	83

(*continued*)

TABLE 12-1. Continued.

Variable	California	Mississippi	Texas	Arkansas	Louisiana
Shares of production received (percent of leases):					
Less than 22%	4	15	41	3	37
23–27%	27	25	8	65	5
27–43%	79	22	13	19	47
43% +	0	38	38	12	11

Sources: Flora & Flora (n.d.); Grant & Musick (n.d.); Mullins, Grant, & Krenz (1981); Rutger & Brandon (1980).
[a]Refers to 1981 data.

this extreme capitalization and mechanization of the rice-farming system has witnessed the near disappearance of wage labor in production and the persistence of sharecropping.

Between 1980 and 1984, California rice production averaged 1.694 million t per year, drawn from 201,214 ha (California Department of Food and Agriculture, 1980–1984). Over the same period, the value of the crop has declined by 58%, but rice is still the most valuable grain produced in the state ($216 million in 1984) and the 17th ranking agricultural crop by value (Security Pacific National Bank, 1985). In 1984, 1,300 California rice growers produced 23.5% of the total U.S. crop and accounted for 1.9% of the state's gross cash receipts. In the same year, rice yields reached a record level of 9,179 kg/ha, the highest in the world. Unlike the southeastern producer regions, only pearl (short-grain) and medium-grain rice is cultivated in California. Although some 17 California counties grow rice in the state (Table 12-2), almost 90% of production takes place in the Sacramento Valley, particularly Glenn, Colusa, Butte, and Sutter counties (Figure 12-1). This area is, and always has been, the heart of the California rice industry.

Ricelands and Aquatic Landscapes in Northern California

Rice is grown in California under flooded soil conditions. A single crop is sown in late April or early May and grown under continuous irrigation until the fields are drained in September. Rice is highly water-consumptive—the third most water-consuming crop in the state—accounting for more than 3.5 million acre-feet each growing season.[1] The use and function of water in the agroecology of rice production presupposes certain hydrological conditions, such as large volumes of low-cost water of quite specific quality, temperature, and chemical composition, coupled with the environmental prerequisites of an adequate growing season and appropriate soil structures. The result is a distinctive, indeed unmistakable, "rice landscape."

The heartland of California rice lies in the Sacramento Valley in a 150-km corridor between Davis and Chico. Flanked by the Klamath Mountains to the north, the Coastal Range to the west, and the southern Cascades to the east, the valley is bisected longitudinally by the Sacramento River and its levees. Trapped between the raised levees and foothills are low-lying and gently sloping flood basins. Beyond this central axis of heavy, impermeable clays lie parallel sequences of alluvial fans, older terraces, and—toward the perimeter of the valley—upland soils (Figure 12-2). The basin soils, which account for 25–30% of all land in the major rice counties, are ideal for rice cultivation in terms of topography and permeability.[2] Their textural and alkaline qualities also imply limited crop adaptability: In fact, the high clay and silt content, and the management problems associated with basin soils (tillage, drainage, organic matter, and soil structure), meant that the basins were rarely used for cultivation, even during the 19th century wheat boom, until the advent of rice. The basin soils are variegated, consisting of at least 19 soil series, but after generations of inundation and evaporation, they all contain high levels of alkaline (Richards, 1969; Storie & Weir, 1926). The most important and extensive rice soils are located in the Colusa, Sutter, and Butte basins. In any one year, however, perhaps only one-third of these soils will be cultivated in rice. At least one-third will lie fallow, and the remainder are in a variety of other grain and field crops.

Alluvial soils may also be used for rice production. Better drainage and resultant increased seepage, however, imply higher water costs. Loamy qualities and a mature soil profile also open the possibility of cropping alternatives such as tomatoes, vegetables, and orchard crops. The recent alluvial and floodplain

TABLE 12-2. Sacramento Valley Rice Production, 1982.

County	Harvested hectares	Gross production (t)	Yields (t/ha)	Value of production ($000s)	% of state production	Rank by value (statewide)
Butte	40,769	329,893	8.1	52,928	17.6	2
Colusa	53,381	406,824	7.62	62,721	23.0	1
Glenn	32,186	253,207	7.86	40,680	13.9	4
Sutter	42,657	323,095	7.57	50,066	18.4	3
Yolo	15,384	112,731	7.33	17,343	6.6	5
Yuba	13,777	110,550	8.0	17,008	5.9	6
Sacramento Valley totals	198,160	1,531,451	7.73	$240,745.7		
California totals	229,520	1,668,581	7.27	$278,947.7		

Source: U.S. Bureau of Census, *Census of Agriculture, 1982,* Part 5: *California.*

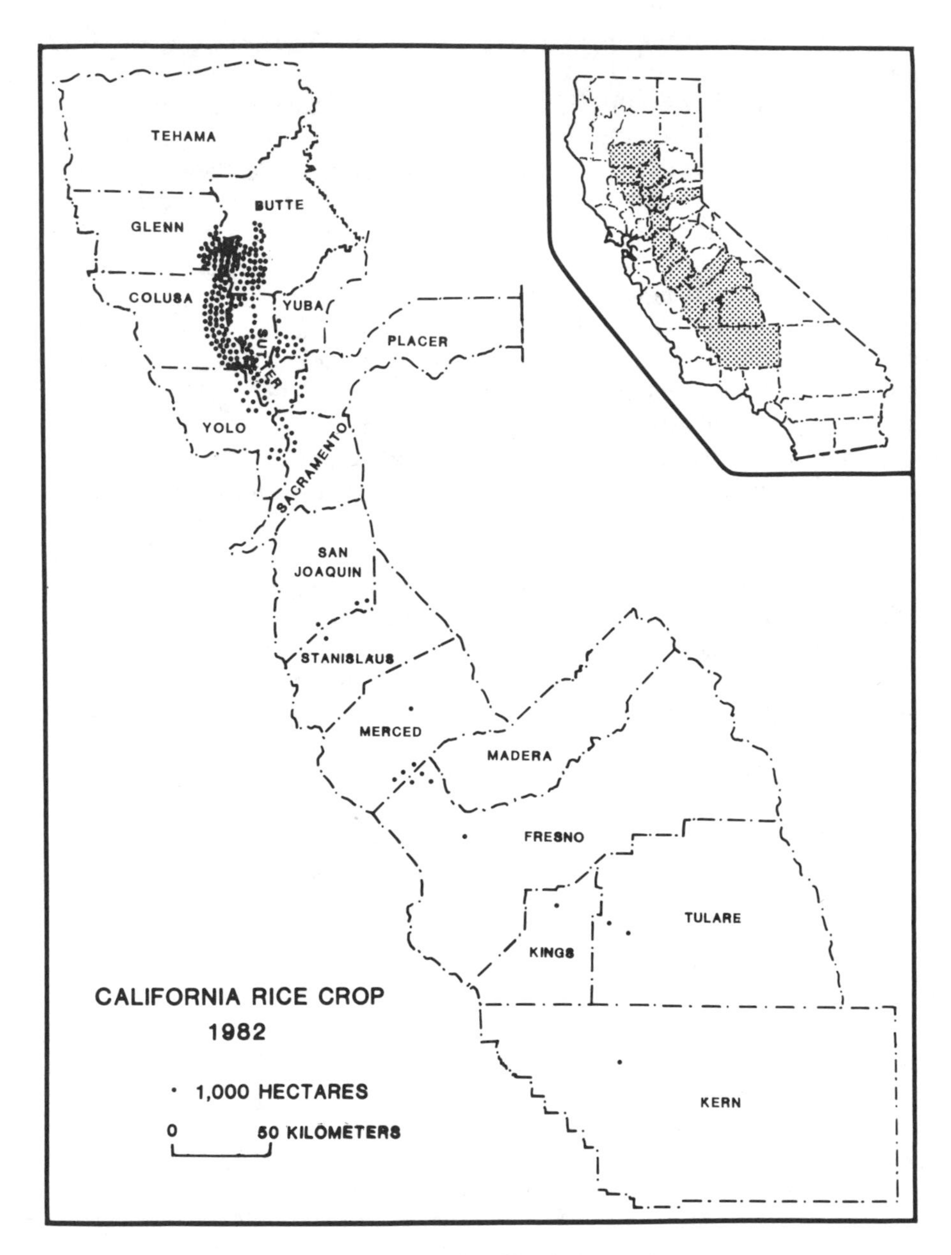

FIG. 12-1. The distribution of California rice production, 1982.

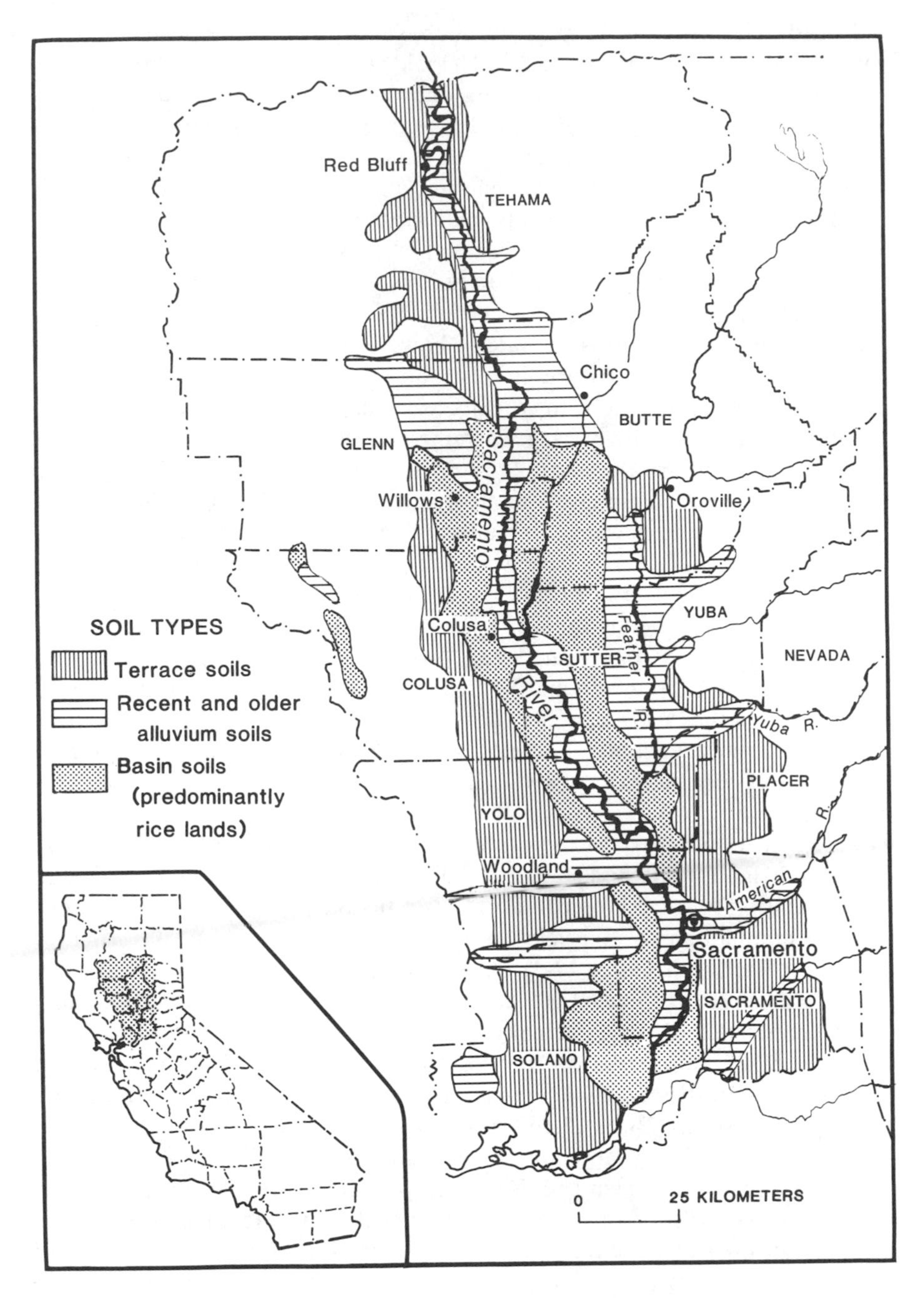

FIG. 12-2. Sacramento Valley rice soils.

soils tend to be in orchards, except for those soil series that overlie basin soils or have a hardpan near the surface. The old alluvial fans, conversely, often present cost problems associated with slope and land leveling (Hedges, 1974).

Wet rice has a higher yield under flooded conditions even if the non-flooded fields have adequate soil moisture. Inundation regulates air and soil temperature, reducing diurnal fluctuations, and also facilitates nutrient supply, conditions alkaline and saline soils, and suppresses weed growth. At a minimum, therefore, water supply must meet three fundamental requirements: (1) bring soil moisture in the root zone to holding capacity, (2) cover transpiration requirements, and (3) maintain a continual flow of floodwater through the checks (paddies) and out as tailwater. The volume of water required for these demands is enormous but varies in relation to rice variety (flooding depth and duration), percolation rates, microclimate, spillage, soil structure, evapotranspiration, and so on. On average, Sacramento rice, which relies completely on irrigation since the growing season occurs in the dry months between April and October, requires 8 acre-feet of water per growing season,[3] eight times greater than wheat and twice as water-demanding as tomatoes and most orchard crops. Shallow flooding (10 cm) on basin soils will generally only require 5.1 acre-feet; deep flooding (25 cm) on recent alluvium, conversely, may demand close to 9 acre-feet. The Sacramento Valley is fortunate in having access to large volumes of high-quality (i.e., oxygenated and low in dissolved salts) water derived mainly from the Sacramento and Feather River as snow melt. Two large dams, the Shasta and the Oroville, under the jurisdiction of the State Water Resources Board, store and regulate flow. Local irrigation districts (and occasionally mutual water companies and private firms) furnish the water to individual rice growers.

Real costs of water to the grower depends on water source (and hence subsidy), irrigation method (flood, furrow, border, sprinkler), and application rate in relation to local crop conditions. Yet the Sacramento Valley has ample water at nominal costs, especially compared with the southern California rice regions in Westlands (Table 12-3). A large irrigation district (ID), such as the Glenn–Colusa ID of some 67,000 ha, draws on average 1 million acre-feet each year; almost 20% of this is actually recaptured drain water. The remainder is currently (1985) supplied from the Sacramento River at $5.00/acre foot. In fact, water quantities available to many of the irrigation districts—Biggs–West Gridley (11,740 ha), Butte (5,668 ha), Colusa County (12,145 ha), Richvale (10,121 ha)—are clearly large in relation to the area served.

Although water is plentiful, it is often perilously cold for rice cultivation, since the snow melt is stored and drawn from the lower (colder) lake levels at the onset of the rice season. Extensive surface flow, warming baths, and stratified offtake systems (removing some surface waters first), however, raise the temperature to roughly 15.5°C by April (Chang, 1971:34). Rice will not germinate below 12.7°C, and some short-maturing varieties will fare badly if temperatures are 18.3°C or less. Rice also has quite restrictive climatic tolerances.

TABLE 12-3. Sacramento Valley and Westlands regions, California: Water sources, application methods, and irrigation costs for rice, 1975.

Area	Crop	Harvested acreage	Nonwater production costs $/acre	Irrigation method			% of irrigated crop acres percent	Water cost $/acre-foot	Application rate acre-feet	Application cost	Total irrigation cost $/acre	Total production cost
				Surface/ ground	Source/lift	Application method						
Sacramento Valley	Rice	440,214	539.84	Surface	Bureau[a]	Border	30	3.00	8.0	62.40	86.40	626.24
				Surface	State[b]	Border	35	2.00	8.0	62.40	78.40	618.24
				Surface	Local ID[c]	Border	10	3.00	8.0	62.40	86.40	626.24
				Ground	80 ft	Border	25	16.68	8.0	62.40	195.84	735.68
Westlands	Rice	14,250	539.84	Surface	State[b]	Flood	30	12.00	7.0	54.60	138.60	678.44
				Ground	200 ft	Flood	70	24.96	7.0	54.60	229.32	769.16

Source: *Agricultural water use costs in California*. Giannini Foundation, Bulletin #1896, 1980.

[a]Central Valley Project, Bureau of Reclamation.

[b]State water project.

[c]Local irrigation district.

Temperatures below 15.5°C retard seeding development, slow plant growth, and reduce grain yield; rain, wind, and excessive dryness can be debilitating at certain crucial moments in the life cycle (especially germination, tillering, and grain development).

The climate of the Sacramento Valley is Mediterranean (Köppen system)—hot and dry summers, winter rain—permitting one rice crop between April and October. The long growing season and equable temperature levels are perfectly suited for rice, as are the high August temperatures and warm summer nights necessary for pollen formation and floret fertilization. Winter rains have no affect on rice cultivation, but the onset of heavy storms in October can cause damage, lodging, and increased costs due to delayed harvest or excessive drying needs. A particularly dry spell after harvest, after the fields have been drained, may also be a problem, as occurred in October 1985 when desiccation of the grain caused excessive kernel shattering during milling.

These peculiar preconditions and the aquatic basis of rice production itself have produced an agricultural landscape of unmistakable character. In the late spring, huge swaths of the Sacramento Valley are awash with water; flooded paddies defined by sharp black levees and crisscrossed by water-bearing irrigation channels and ditches fill the rice basins to their brims. The large expanses of riceland, broken only by the Sutter Butte and interrupted occasionally by huge rice storage bins, affords the whole area a silent and strangely underpopulated quality. The leveled fields—usually rectangular or square and often in excess of 60.7 ha—are broken into checks or paddies, marked by the usual signs of rice culture: contour levees, levee boxes, inlets and drains, and irrigation canals. Service roads perhaps 1 m above the level of the field often mark the outside levees; inside the field, checks vary in size (from 1 to 10 ha) and shape depending on the slope of the land, and are delineated by smaller levees (40–50 cm high)—usually covered in unruly weeds during the growing season—which break the slope of the land and keep water within the paddies. For effective water control, all fields are graduated so that water enters through an inlet valve at the higher end and drains out into a ditch. Levee boxes—wooden boxes situated firmly within the levee—control the flow and depth of water. The entire artifice is, as Geertz (1966:31) once put it in describing an entirely different part of the world, "the fabrication of an aquarium."

By early summer, the pale green shoots unexpectedly emerge from their aquatic environment and within weeks have hidden the water beneath. The basins become seas of green. By September, they have been transformed to a golden brown and their tranquillity is shattered by the drone of enormous harvesting equipment and the first real evidence of intense human activity. In order to harvest before the first rains, fields are cleared and burned, and the levee boxes ripped from their dikes and left scattered as debris on the field until the following spring. Clouds of thick smoke from the huge quantities of rice straw mark the end of the aquatic cycle and of the diverse faunal life that it sustains. Carp, minnows, mosquito fish, and crayfish litter the canals and fields in September, lending the valley the olfactory atmosphere of an open-air fish market. In the off season the fields are usually left fallow and dry, unless they are briefly reflooded for waterfowl hunting.

The Agroecology of Rice Production in the Sacramento Valley

Rice growing has a telling effect on both the environment and the economy of major rice-growing counties in the Sacramento Valley. Colusa County, for example, whose alkaline, low-adaptability rice soils have been heavily exploited since the 1950s and which currently account for over 20% of state production, has at least one-third of its cultivated area in rice every year (U.S. Bureau of Census, 1982). As is the case throughout the state, rice growing is dominated by household- or family-based units of production. Two-thirds of production is controlled by what Buttel (1980) calls "larger than family farms" (i.e., units with annual sales of $200,000 or more with perhaps a full-time hired laborer), but many are household-based, if corporatized, forms of production—what we shall call petty commodity producers. As we seek to demonstrate in the following section, however, rice growers are really a heterogeneous and diverse community marked also by sharecropping on at least half of the ricelands, and by the presence—though certainly not the domination—of large-scale industrial-corporate (agribusiness) interests. Despite the family/individual focus of rice production, the rice agroecosystem has been fundamentally shaped by the scale and capitalization of its cycle of production. Even in the 1920s, one man alone could cultivate 100–120 ha of riceland up to the point of harvest (Federal Land Bank, 1924), but the average rice farm has increased in size dramatically, from perhaps 67 ha in 1919 to 130 ha in 1949, and from 230 ha in 1970 to over 325 ha today (Bleyhl, 1955; Cervinka & Chancellor, 1975; U.S. Bureau of Census, 1982).

The landholding question is, of course, connected to the rapid and relatively complete development of farm mechanization and to the ability of farmers, working in conjunction with the state, the state university system, and grower organizations, to overcome the technical and agronomic barriers of rice ecology (Figure 12-3). By 1936, for example, 60% of rice was sown by air, and by the 1950s harvest operations had fully adopted the integrated combine. Problems of pests, fertilization, land leveling, and water control have elicited intense research, experimentation, and technical change in creating new conditions for growth and increases in productivity. A 250-ha rice farm currently requires, for farm equipment alone, a financial outlay of close to $600,000, almost 50% of which is accounted for by harvesting machinery. Indeed, equipment costs (15.89%), mechanized harvesting (21.00%), and fertilizer–insecticide (15.23%), constitute over half of annual production costs (see also Table 12-4). The technification of the rice agroecosystem comprises a twofold movement: (1) the continuous growth of productivity (Table 12-5) and (2) a dramatic decline in hand labor in the production cycle. Yields increased from 2,806 kg/

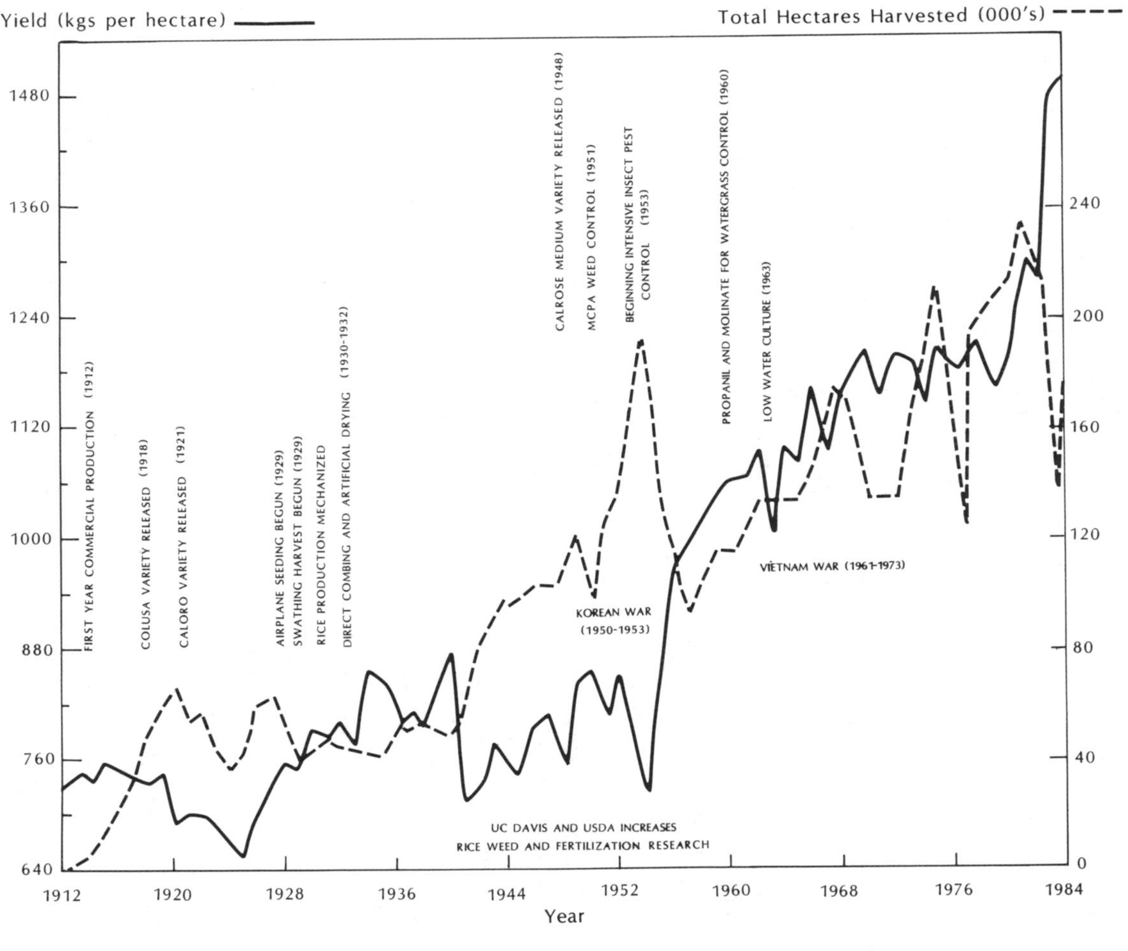

FIG. 12-3. Technical innovation and productivity changes in California rice production, 1912–1984.

TABLE 12-4. Rice production, Colusa County, 1983: Cash cost/hectare summary.

	Labor	Fuel	Repairs	Materials	Services	Overhead	Total	% Total
Land preparation	12.74	15.78	21.24	–	–	–	49.76	4.0
Fertilization	1.53	2.99	1.83	101.56	10.91	–	118.82	9.6
Weed and invertebrate control	–	–	–	95.68	27.66	–	123.34	10.0
Planting	–	–	–	40.95	16.82	–	57.77	4.7
Water management	14.91	–	–	62.96	–	–	77.87	6.3
Harvest and irrigation	24.60	19.11	70.48	–	47.79	–	161.99	13.1
Drying and storage	–	–	–	–	193.40	–	193.40	15.6
Post harvest operations	3.16	.44	.22	–	–	–	3.82	0.3
Cash overhead	–	–	–	–	–	450.82	450.82	36.4
Total	$56.94	$38.32	$93.78	$300.85	$296.58	$450.82	$1,237.59	
% total	4.6	3.1	7.6	24.3	24.0	36.4		100.0

Source: Scardaci & Olson (1983).

Note: Assumes a 243-ha farm, average yield 17,290 kg/ha.

TABLE 12-5. California rice farming acreage, production, yields, 1912–1982.

	Harvested hectares	Production (t)	Yield/hectare (t)
1912	405	118	0.30
1915	12,145	38,083	3.13
1920	65,587	169,169	2.58
1925	41,700	98,280	2.35
1930	44,534	148,876	3.34
1935	40,486	151,515	3.74
1940	47,773	193,284	4.04
1945	95,141	284,921	3.00
1950	96,356	376,285	3.90
1955	133,198	516,425	3.87
1960	115,599	625,716	5.36
1965	132,388	729,046	5.50
1970	134,008	858,448	6.40
1975	212,550	1,369,004	6.44
1980	227,125	1,582,087	6.96
1982	231,174	1,760,926	7.60

Source: California Department of Food and Agriculture, (1921–1982).

ha in 1912 to over 8,981 kg/ha of today; this compares with current wet-rice yield estimates of 6,287 kg in Taiwan and 4,554 kg in Java (Ruthenburg, 1980). Labor inputs in the Sacramento Valley have plummeted from close to 20 hours/ha in the 1920s, when 500 Chinese or Indian day laborers could be seen at harvest time on the large ranches, to 1.6 hours or less per ha in 1985 (Bainer, n.d.). Even in the late 1940s, labor costs constituted 30% of annual production costs, absorbed principally at harvest. By 1984 a Sutter or Colusa rice grower of 250 ha spent less than 5% of cash costs of labor (California Department of Food and Agriculture, 1947–1984). Although only 6% of all rice farmers now own their own combine harvesters, many of the smaller rice producers hire in (because they cannot afford the capitalization costs) or hire out (to make efficient use of farm machinery if their own hectarage is low) custom harvest operations.[4]

All agroecosystems involve structural and functional changes in natural communities. Agriculture generates high net productivity, but trophic chains, species and genetic diversity, mineral cycles, and habitat heterogeneity are all impoverished by the goal of maximum productivity (per unit area) and minimum interannual instability in production (Altieri, 1984). Flooded rice in the Sacramento Valley is a singular case in point, involving major transformations of local ecosystems. The rice agroecosystem is relatively specialized, monocultural, often permanently cultivated, and heavily subsidized by external inputs (such as fertilizers and herbicides) to compensate for nutrient depletion and the lack of self-regulatory mechanisms associated with floristically poor monocultures. At the level of the farm enterprise, the degree of crop–ecological specialization will vary considerably. Most large (½S400 ha) production units are somewhat polycultural and commonly engage in the cultivation of other cereals, orchard crops, and safflower. A part of the agroecological character of the farming system is framed, nonetheless, by soil structure and, by extension, by whether rotations are part of the rice cycle. In this sense, there are two discrete biotic environments: one associated with continuous rice cultivation (nonrotation) and the other with rotational cropping, a bifurcation that has direct implications for yield, farming operations, and costs of production.

Rotation and Cropping Systems

In the San Joaquin Valley most rice production is associated with diversified farm operations in which rice is cultivated in complex rotations with orchard crops, cotton, and vegetables. Conversely, in the Sacramento Valley, significant proportions of the rice crop are grown on basin clays and highly alkaline soils with very limited crop adaptability. In the early years of rice production, the basin soils were found to produce 25% less in the second year of cultivation (Federal Land Bank, 1924). Many of these soil series, for example the Willows series in Colusa County, are currently continuously cropped, however; in the off (winter) period, leguminous crops such as vetch (*Vicia atropurpurea* Desf.) and burclover (*Medicago hispidia* Gaertn.) may be planted in the checks, and are currently grown on roughly one-fifth of California riceland (Richards, 1969:n.p.). In 1955, with the reintroduction of the acreage allotment scheme, the California rice area dropped by 40%, and the excess land generated a renewed concern for rotation crops. Fallowing was found to be attractive because it eradicated paddy field grasses and increased yields. Green crop manures were also recommended; seeding a leguminous crop in the fall after a summer fallow, and plowing the cover crop under during the spring prior to rice seeding underwrote high yields without the addition of nitrogen fertilizer. On clay–loam soils with greater crop adaptability, other cereal crops are rotated with rice, especially barley, wheat, and other cultigens, whose work routines permit the complementary deployment of rice–farm machinery. Other crops included in such rotations include milo, beans, safflower, and maize. The most common rotations on soils with lower alkalinity are: rice, spring/early summer plowed fallow, fall wheat/oats/vetch, rice; or rice, spring-sown wheat/barley/safflower, fall wheat/oats/vetch, rice. For heavy rice soils of limited fertility, rotation crops in any case perform badly. Hence continuous croppings with a winter vetch, interrupted occasionally by a fallow, is commonplace, especially if weed problems are severe. Many of the basin soils in Colusa County have been continuously cropped since World War II. This biotic dichotomy between rotation and nonrotation has implications for understanding subtle distinctions in on-farm operations. First, continuously cropped, nutrient-depleted rice soils often require greater applications of fertilizer (140 kg/ha of aqueous nitrogen as opposed to 101 kg/ha on rotated soils). Second, seedbed preparation and tillage operations vary, since regularly inundated soils are "tight," demanding additional chisel plowing and often intensive herbicide control to regulate weedy grasses whose proliferation can be partially regulated through fallow and rotation. Third, rotation soils generally have higher percolation rates as a result of increased tillage; hence water application rates are generally greater (9 acre-feet as opposed to 7 acre-feet on the basin soils).

Polyvarietal Strategies and the Activity Profile

Until quite recently, the limited California growing season has prevented the cultivation of long-grain (indica) rice. The key to the rapid improvement in California rice productivity has been the proliferation and refinement of short- and medium-grain (japonica or temperate) rice varieties. The broad objectives of rice varietal development, sponsored by an active collaboration between the University of California, rice growers, and the state, have been hardy, high-yielding seeds, preferably with short-maturing growth cycles. In the past decade the critical development has been the release of semidwarf varieties, short-maturing, resistant to lodging (falling) and cold, and capable of thriving in shallow water conditions (Rutger & Brandon, 1982).

Commercial rice varieties are classified on the basis of growing period, grain size and shape, and chemical composition. The length of growing season is of direct significance for cultural practices, particularly water management. There are currently about 20 principal rice varieties cultivated in California—each of differing height, grain and hull type, straw strength, and heading period—that are conventionally distinguished as very early (100–115 days growing period), early (116–135 days), midseason (136–150 days), and late (over 150 days).[5] The most important rice varieties, accounting for over 75% of all California hectarage, are: (1) Caloro, a midseason short-grain; (2) Colusa, an early short-grain; (3) Calrose, a late medium-grain; and (4) M9/M101, an early-maturing medium-grain (Chang, 1971; Willson, 1979).

Early-maturing varieties, despite their lower yields than the full-season late rices, are attractive because they enable the grower to minimize the risks of inclement weather, specifically low temperatures at planting and moisture at harvest. In practice, virtually all rice systems are polyvarietal, at the very least, with one early variety and one late. Varietal choice will be shaped, of course, by several calculi, including local edaphic conditions and prevailing prices, but the principle of polyvarietal planting strategies is imperative, not only as a risk-averse strategy characteristic of many intercropping systems (e.g., Norman, 1974) but also because it extends or, more properly, spreads land preparation and harvest dates. Since 75% of all labor demand is accounted for by two 3-week operations—land preparation and harvest—it is desirable that the labor profile be smoothed out both to limit bottlenecks and to ensure efficient use of farm machinery. To plant huge areas of rice, all ripening simultaneously close to the termination of the growing season, would be hopelessly risky and impossibly "lumpy" for the optimal scheduling and employment of machinery.

The Farming Cycle and the Microecology of Rice

Rice production entails continuous inputs throughout the year, but in-field activities exhibit a pronounced bimodal pattern: land preparation/flooding, and harvest/bankout. Careful water management is required between these two moments in the farming cycle, but the labor profile from planting to harvest tends to be smooth. The broad contours of the farming calendar are nevertheless subject to constant modification and manipulation in relation to the empirical realities of the farming season (Table 12-6).

Most rice fields are large, usually in excess of 60 ha, and subdivided into contoured, and increasing rectangular, paddies or checks, 4–8 ha in extent. The leveling and bunding of the checks are crucial and have become an exact science with the proliferation of semidwarf shallow water (20–30 cm) varieties. Continuously cropped fields are generally leveled only every three years, but precision is demanded for each cropping cycle. A slight slope of 0.05 m per 30 m is needed to facilitate water movement, water depth control, and efficient drainage prior to harvest. Land grading has harnessed electronic innovations, in particular the use of laser leveling. The soil is tilled—perhaps two or three times, depending on the frequency of rotation—by large tractors pulling discs or chisel plows to break the clay clods, and then leveled carefully with a land plane or float. Earth levees 1 m high are "pulled" using a checker that scrapes the loosened soil from an area roughly 5 m wide and discards it at the rear of the dikes through a throat 1.5 m wide. The levees, which may absorb 10–12% of the cultivated area, are punctured by redwood levee or weir boxes that regulate the flow and depth of water. All the preparatory activities require dry conditions and hence can be complicated and, more crucially, delayed by the late termination of winter rains.

Nitrogen, the most common limiting plant nutrient, is applied before planting and worked into the soil with a spike harrow to a depth of 10 cm. Application rates are in large measure determined by cropping frequency and soil structure, but 134 kg/ha is common. On the basis of plant tissue analysis, nitrogen and other phosphate and zinc fertilizers may be applied by air in mid- or late growing season.

The fabrication of an aquarium is central to the productivity of wet-rice systems; inundation regulates air and water temperatures, facilitates nutrient availability, conditions alkaline soils, and suppresses weed growth. Flooding must, however, be conducted with great rapidity—a 40-ha field is flooded to a depth of 10 cm in 96 hours—in preparation for the immediate aerial seeding of soaked and pregerminated (and fungicide-treated) rice (Figure 12-4). Presoaking of seed both initiates germination and provides weight to ensure immediate

TABLE 12-6. Sample calendar of California rice production operations.

Months/days	Field days available	Rice growth stage	Typical field activity
January 1–31	4		Repair and rebuild equipment; office and management
February 1–28	4		Repair and rebuild equipment; office and management
March 1–10	2		Seedbed preparation – plow, disc
11–20	5		Landplane
21–31	7		Survey, mark, and make levees
April 1–10	7		Three-wheel land plane, fertilize
11–20	7		Preplant weed control; disc or harrow; close levees
21–30	8	Seeded	Flood; treat and soak seed; sow
May 1–10	8	Seedling emer-	Rice stand established
11–31	19	gence	Continue irrigation; tadpole shrimp control; postflood barnyardgrass control; algae control
June 1–10	9		Rice leafminer control
11–20	9	Tillering	
21–30	10		Continue water management
July 1–10	10	Internode elongation	Broadleaved weed control; nitrogen fertilization as needed
11–31	21	Boot	Raise irrigation water 2 inches higher than normal 3 weeks before heading: Prepare and check fallow fields; continue irrigation
August 1–31	31	Heading and flowering	Continue water management; prepare and check fallow fields
September 1–30	28	Grain formation	Drain fields and open checks; seed vetch cover crop
October 1–31	25	Maturity	Harvest, bank-out, and haul to dryer
November 1–30	19		Residue disposal, disc
December 1–31	10		Maintenance and repair of equipment; office and management

Source: Miller (1977).

sinkage contact with the flooded checks. Prompt germination is crucial because seedlings, to avoid weed competition, must emerge quickly (within 21 days) through the water before the seed's nutrient reserves are exhausted. Cool temperatures can have a devastating impact both on early seedling establishment and somewhat later prior to heading during panicle formation.

Water depth will vary in relation to rice variety, stage of plant development, weather, and pest problems. Ideally, shallow water (2–10 cm) favors stand establishment and tiller development. Deep water, conversely, may assist

Photo credits: Jack Kelly Clark and William Wildman.

FIG. 12-4. Aerial seeding in the Sacramento Valley.

in the dampening of weedy grasses, while water turbidity induced by wind is likely to be less problematic in deeply flooded paddies. Farmers may therefore carefully manipulate water depth throughout the growing season in relation to the cropping history of the field (i.e., weed infestation), and the specific conjunction of local climatic events. From the completion of tillering (60 days after planting) to 2 weeks before heading, water depth is of little practical consequence. During the 14-day period of panicle formation up to heading, however, warm water may be raised by 5–10 cm to protect the plant from low nighttime temperatures and to control "blanking." In the period of ripening, water depth is not important, provided the soil surface is covered to suppress weed development.

Chemical weed control is essential in the life history of the rice plant, increasingly so with the proliferation of shallow flooding. Broadleaf and grass herbicides are aerially administered from May to July to control infestations of barnyard grass (*Echinochloa crusgalli*) and sprangletop (*Leptchloa fasicularis*). While California rice is relatively free of plant diseases associated with humid climes, stem rot, water weevil, and tadpole shrimp, and occasionally leafhoppers and midges, may demand molinate and parathion.

Floodwater is drained roughly 20–30 days before harvest, when the grain on the lower stalk is still soft. The timing of draining is crucial in terms of both the moisture content of the kernels and the costs of harvesting in muddy conditions. Combines are employed when the moisture content of the grain drops to 18–26% to ensure first-class milling and limited shattering. Combines work continually, often through the night, to ensure a prompt harvest. Bankout wagons simultaneously shuttle to and from the dried paddies, ferrying grain to roadside trucks, which haul the rice to commercial dryers. Harvested grain is reduced to 14% moisture on off-farm facilities for storage and milling. Forty-five kilograms of rough paddy rice should produce 22 kg of head rice, 9.5 kg of brokens, 3.6 kg of bran, and 1.3 kg of polish (Willson, 1979).

The development of short-stature, robust rice varieties has produced enormous quantities of postharvest residues, roughly 7,859 kg of straw and 3,368 kg of roots per hectare! While burning is practiced subject to pollution controls, sometimes the residues must be plowed into the soil (for example, if early rains or a late harvest prevents burning) or hauled from the checks. The latter implies considerable cost, and the former can create enormous biotic problems as a result of the resiliency of the straw. This can take 7 years to decompose and can generate high-toxicity scums and algae and necessitate intraseason drainage in subsequent growing seasons. The end of the rice-growing cycle is signaled by the removal of the levee boxes; occasionally by light plowing; and, for the nonrotation fields, the seeding of vetch or a cover crop.

Orchestration, Timing, and Regulation

The farm plan constitutes only the broad canvas into which actual decision making and farm management must be worked. Indeed, the mechanization and technification of the rice-farming system should not obscure the careful scheduling and on-line adaptations prompted by the inevitable environmental risks of a short Sacramento Valley growing season. To assume that capitalization and large-scale farm machinery reduce risk or obviate the harvest variability is to misunderstand the contingent character of biological production systems and their fundamental dissimilarity from the industrial labor process.

The complexities of the rough and tumble of everyday rice farming can be ascertained through three basic aspects of the human ecology of rice production. The first, *orchestration*, refers to the sequential scheduling of operations across rice varieties to permit the loosening of bottlenecks and the temporal spreading of mechanized operations (especially land preparation and harvesting). Figure 12-5 depicts the sequencing of operations for one variety, 155-day Colusa. Superimposed on this decision regime will be others of differing periodicities and patterns, demanding district forms of water management, monitoring, fertilization, and harvest.

Second, the *timing* of operations in a short growing season is crucial. Since rice is especially vulnerable to temperatures below 15.5°C, the period between planting and rain-free harvest is short. A delay in tillage or seeding (due to cold or late rains) has deleterious consequences for yield and hence for profit; equally, the decision to drain—itself dependent on the conjuncture of local weather, soil structure, and plant growth—determines harvesting date, moisture content, and grain quality. In Colusa County, for example, planting generally occurs between April 15 and April 22; a 1- and 2-week delay, respectively, produces a 5% and 10% respective reduction in yield. Similarly, every week's delay in harvest after October 10 will cause mean yields to decrease by 5% (Cervinka & Chancellor, 1975: 33–35). Such lateness often implies higher moisture content at harvest and hence heavier drying costs (which even in normal years will absorb 15% of annual cash outlays per hectare). The salience of timing also extends to field flooding and seeding, which must be conducted rapidly to dampen weed competition. Sluggish and tardy operations again imply lower yields and higher herbicide costs.

Third, growing season cultural practices demand constant manipulation—*regulation*—and occasionally complete rupture through the rather drastic measure of field drainage. Ordinarily, field drainage should not be necessary, and indeed is problematic because soil drying runs the risk of nitrification, and field exposure stimulates weed growth. Water depth and movement are carefully monitored and regulated depending on local conditions; low temperatures and poor seedling performance may demand a sharp reduction in water depth to prevent "stretching" (rapid plant growth in deep conditions) or suffocation, while heading is facilitated by a 5- to 10-cm increase in water levels to ensure warm ambient temperatures to facilitate flowering. Although inflow and spillage rates vary with each field, a maintenance inflow rate of 0.030 cubic feet per second per acre (Chang, 1971:134) is ensured to prevent algal toxicity. Fields may, nevertheless, be totally drained for a variety of reasons, including control of algae or pests, limited stand establishment, control of water turbidity, overfertilization, and salinity management (Scheuring, 1983; Willson, 1979:114).

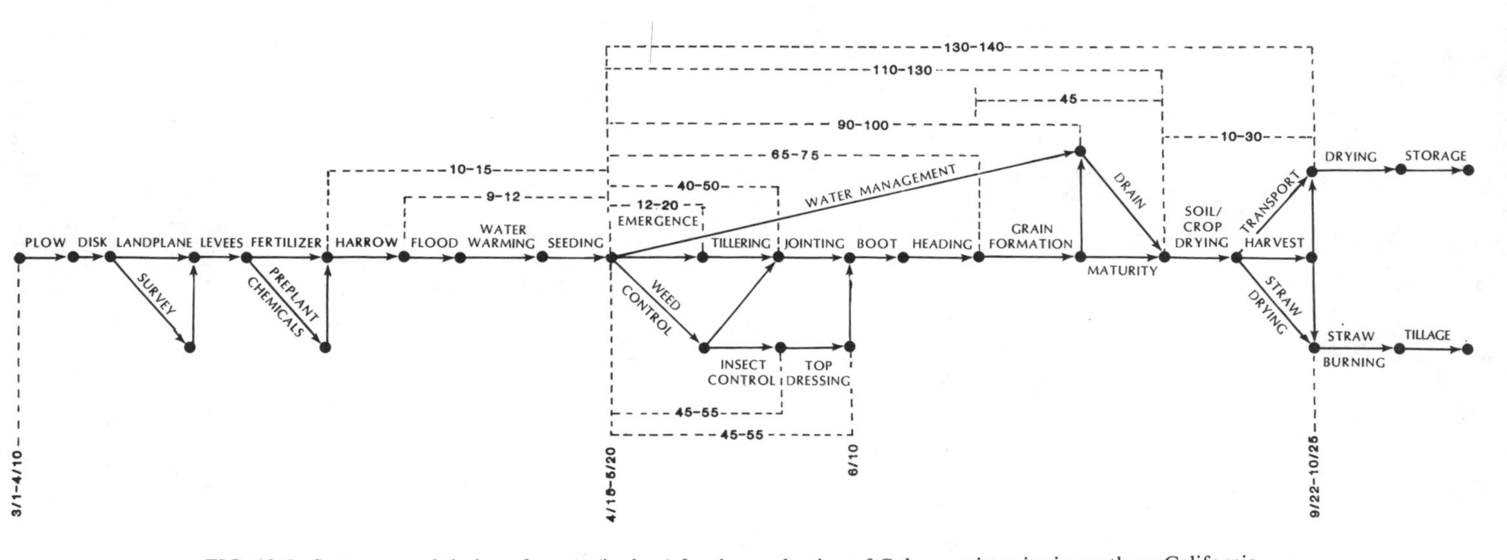

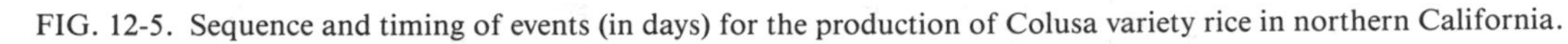
FIG. 12-5. Sequence and timing of events (in days) for the production of Colusa variety rice in northern California.

Drainage farming implies a modification of an ideal farm plan, which highlights the plasticity and internal adaptability of rice-farming systems even under conditions approaching a sort of agricultural fordism.

Rice Production and Energy Subsidies

The internal dynamics of rice production have, in a relatively short period of time, been shaped by mechanization and other technology (i.e., scientific water control, seed improvement, the use of biochemical inputs). Assisted by state intervention and large capital in the context of high seasonal labor demand, rice production has followed a "revolutionary road" with respect to the means and forces of production. In 1918, for example, intensive hand weeding using Filipino and Indian labor at a cost of up to $49.40/ha was widespread, and harvesting technologies entailed little more than the sickle and stationary thresher. By 1985, 20 kg of water grass control is applied by air, and combines reap and thresh 8 ha/day. The genesis of new farm machinery; the harnessing of aircraft for seed and fertilizer application; and the provision of water on a large scale through state projects, irrigation districts, and private companies, have collectively transformed the form and magnitude of farm inputs. During the past 70 years, the focus of technological improvements has shifted rice varieties in the 1920s, artificial drying in the late 1920s, aerial seed application and binder threshing in the 1930s, combine harvesting in the 1950s, and particularly chemical weed control and fertilization in conjunction with seed improvement in the postwar period (see Figure 12-3). As in other commodity sectors, the outcome has been a heavy reliance on fossil fuel inputs, high energy intensity, and a sharply negative net energy of production (Buttel, 1980; Steinhart & Steinhart, 1984).

The genesis of external energy subsidies is contained within the mechanization process and the character of water provision, but also in the rapid growth of biochemical inputs. In 1912 no fertilizers were employed in the Sacramento Valley ricelands; by 1962, 78.5 kg/ha of nitrogen was usually applied to continuously cultivated rice land, and in 1984 a 1-ha check in Sutter County could expect to receive 134 kg of nitrogen, 202 kg of ammonium sulphate, 134 kg of top dressing, and 50 kg of pest control chemicals. In 1978, 25% of the total energy budget was accounted for by fertilizer alone. Rice was the most energy demanding of all field crops (with the exception of cotton) in California, absorbing 33.5 million Btu/ha, almost equally divided between irrigation, equipment, and biochemical inputs (Bayliss-Smith, 1983; Cervinka, Chancellor, Coffelt, Curley, & Dobie, 1974). The process of energy intensification, moreover, has been self-propelling. The use of shallow flooding required more precise land leveling (by laser technologies), while the semidwarf, nutrient-demanding rice varieties appropriate to low flood levels provided less shade and a better growing environment for certain grasses. The result is greater dependence on biochemical weed control. As a consequence, energy input/

output ratios have deteriorated: According to our computations based on 1978 statistics, the figure for northern California was 1 : 2.49, and close to 1 : 2.0 in the Westlands (Cervinka et al., 1974; Chancellor Avlani, Thai, Cervinka, Rupp, & Yee, 1981). Bayliss-Smith's (1983:167) estimate for the Sacramento Valley is slightly less, a ratio of 1 : 1.75. Comparable ratios for other cereals in California are 1 : 6.0 for barley and 1 : 3.2 for corn.[6] Only the intensively cultivated vegetables and fruits, often subsumed within heavily mechanized and vertically integrated agribusinesses, fare worse, with ratios of 1 : 0.5 or less. In energetic terms, the rice agroecosystem is dependent on larger ecological and political-economic systems and, indeed, has been critically shaped by them throughout its brief California history.

Rice Production and the Specificity of the Farming Unit

In this section we seek to describe and theorize rice cultivation in California from the viewpoint of the social organization of production. The forms of production units and the organization among them shed light on the wider, and knottier, problem of the development of agriculture under capitalism. To this extent our discussion rests on two fundamental propositions. First, agricultural production can be distinguished from other economic, and particularly industrial, production processes. Perishability, the nonidentity of production time and labor time (that is, the critical gap between production activities due to the biological growth process), the uncertainties of plant biology and weather, and the relative immobility of enterprises contribute to—but do not in themselves explain—the great diversity of social organization and production relations in agriculture. There are, for example, important distinctions in the internal structure and operation of production units by commodity. Lettuce and tomatoes present different possibilities and constraints for farm mechanization and for upstream integration. Lettuce producers, growing two crops annually, must deliver a commodity of unusual delicacy to dispersed centers over 52 weeks each year, and are predominantly consumer-market grower–shippers who, as highly integrated agribusinesses, sell directly to supermarkets and retail outlets (Friedland, Barton, & Thomas, 1981). Conversely, tomato production, which is highly seasonal, concentrated, and heavily mechanized, is dominated by raw material producers that cultivate the crop under contract to firms who stand strategically between producers and the market (Thomas, 1981, 1985).

The second proposition concerns the nature of capitalist development itself: both its episodic and irregular quality, and its historically contingent character. Insofar as the long-term historical process of capitalist accumulation has served the general development of commodity production, it has not necessarily involved the universalization of the capital/wage relation in all branches and sectors—what Kitching (1985) calls the "unilinear proletarianization" view of capitalist development—nor has it eliminated all forms of noncapitalist production, even in the heartland of late capitalism. Such a historical proposition rejects the suggestion that there is one mature form of agrarian capitalism or, alternatively, that forms of agricultural production (such as household farms and sharecropping) seen to deviate from such an ideal type are simply residues from some precapitalist past. Rather, it is to assert the complex variability in the social organization of agriculture in capitalist societies and to recognize that agricultural labor processes and relations of production are much more differentiated than is commonly acknowledged. Indeed, the so-called anomalies of contemporary agriculture—the persistence of household forms of production, the resurgence of sharecropping—must be seen as constitutive of the current phase of capitalist development and of the broad pattern of modern class structure (Wells, 1984b).

The case of California rice farming must be situated in relation to these two propositions because rice growing is marked by a historic decline in on-farm labor use and the relative insignificance of wage labor in contemporary production, and by the widespread adoption of sharecropping arrangements and the dominance of petty production. Rice growers are in reality markedly differentiated, running the spectrum from small owner-operators for whom rice production is only one aspect of family income, to large-scale, heavily mechanized, capitalist (and sometimes corporate agribusiness) enterprises. A significant part of production, however, resides in the hands of a specific form of producer, what are referred to as *petty* or *simple* commodity producers (Friedmann, 1980; Kahn, 1980; Roldan, 1984; Smith, 1984; Whitehead, 1985). Petty commodity producers (PCP) are distinguished by the market determination of factors of production, by the control of their means of production, and by the dominance of nonwage (i.e., household) labor. These producers in general, as Friedmann (1980) shows in her study of family farms in the United States prior to 1945, employ irregular wage labor. Wheat farmers on the northern plains organized into household units of production proved competitive through capitalization (without hiring wage labor on a large scale) and by intensively harnessing household labor. Petty commodity production embraces a class of producers with quite different scales and capitalization in a variety of sectors and branches: It would subsume, for example, the fully commercialized French "peasant" farmer.[7]

Among California rice growers what is significant is not only the extent of petty commodity production but also that of sharecropping, the origins of which lie not only in the internal dynamics of household production, but also in the characteristics of the wider political-economic environment. We now turn to the specific form and character of production among California rice producers.

Rice farming in the Sacramento Valley is distinctive both as a form of commodity production and in its wider social organization. Unlike some of the labor-intensive crops in California agriculture—vegetables and the orchard crops—rice is marked by (1) a remarkably low labor input/ha and hence a low significance of labor costs as a percentage of product value, and (2) the employment of limited numbers of hired regular and casual workers by each rice production unit. In the case of the former, the person-hours/ha in the rice system have dropped from about 120 hours in 1930 to fewer than 10 hours in 1985 (Bainer, n.d.; California Department of Food and Agriculture, 1984), and from labor as 30% of production costs in 1947 to 5.6% in 1985 (California Department of Food and Agriculture, 1947–1984). On the latter score, over half of all rice farms in the Sacramento Valley hire only one regular and/or seasonal hired laborer each year (Hayes 1984:368).

The centrality of household labor and the limited use of wage labor in rice production is tightly bound to the scale and capitalization of farm operations. The average size of the rice farm in California (i.e., the acreage of those farm holdings on which rice is the most important crop) has increased from 107 ha in 1919 to 229 ha in 1970, and to 292 ha in 1978 (Bleyhl, 1955; Flora & Flora, n.d.); the statewide average in 1982 was in excess of 325 ha (U.S. Bureau of Census, 1982). The average value of capital assets among California rice growers in 1982 was $1.9 million for land and buildings and $181,000 for machinery, higher than for virtually all other crops except tobacco and sugar. Only 6% of all rice growers lacked combines, and only 21% relied on custom harvesting services (Cervinka & Chancellor, 1975). The new cost prices of machinery alone for a 250-ha rice farm in 1984 was close to $500,000 (California Department of Food and Agriculture, 1984). The labor and equipment costs have thus evolved differently (Table 12-7).

These purely quantitative measures do not, of course, constitute the form of production itself; rather, one must account for why specific dominant forms of production in the rice sector have come to assume particular sizes and mechanization levels.

Seventy-five percent of all California rice farms are individual or family enterprises (including family corporations), accounting for 70.9% of total harvested acreage in 1982 (U.S. Bureau of Census, 1982:38). This does not imply a homogeneity of either size or techniques of production, as Friedmann suggests. Indeed, 25% of the total number of rice farms (1,140 ha) account for only 4.2% of statewide harvested hectarage, and 50.8% of farms (1–100 ha) account for only 14% of total rice hectarage. Many of these smaller, less capitalized enterprises—some 300 farms of less than 40 ha—either have no expensive farm equipment and hire custom leveling and harvesting teams, or may hold some equipment cooperatively with other small producers. Conversely, 10.1% of farms (those over 400 ha) produced 41.2% of all rice. According to Flora and Flora (n.d.), 36% of California rice in 1978 was produced on farms of over 800 ha. The differentiation of rice producers—and variations in capitalization and levels of mechanization—does not alter the fact that the dominant form of production is the relatively specialized commodity producer, as thoroughly determined by the logic of the market. It is the nature of competition and state-supported mechanization that has permitted the individual or family commodity producer to expand the area of household production with

TABLE 12-7. Comparison of rice costs, Sutter County, 1966–1984.

Variable	Costs per hectare	
	1966	1984
Item		
Labor (hourly + salaried employees)	$ 42.36	$ 63.00
Fuel and repairs	50.51	182.31
Hauling and drying	49.27	199.62
Cultural practices		
Fertilization	34.08	121.30
Water and water management	40.26	98.80
Seed and seeding	32.60	63.53
Pest control	3.95	98.40
Weed control	20.62	1.23
Cash overhead		
Miscellaneous	17.16	41.99
Management	32.11	not included
Investment overhead		
Land (including rent)	59.28	302.57
Taxes	17.41	94.25
Buildings	3.21	14.69
Equipment	102.01	244.75
Drains and roads	–	not included
TOTAL	$504.83	$1,526.44
Yield – tons/hectare	5.8	7.8
Cost/wt	$87.04	$195.69

Source: California Department of Food and Agriculture (1966–1984).
Note: Assumes an approximately 284-ha farm with a 202-ha rice base.

the addition of limited quantities of hired labor.[8] As a purely theoretical figure on the basis of average labor requirements, a single farmer with minimal family support can cultivate over 200 ha each season.

The prevalence of simple commodity production should not obscure the fact that capitalist and industrial interests also participate in rice production. Although it is impossible to compute the overall area (see Liebman, 1983), partnerships and nonfamily corporations (22% of all farms) accounted for 25% of the cultivated rice hectarage in 1982 (U.S. Bureau of Census, 1982:38). Liebman (1983:139) lists some 16 corporate farms with holdings ranging from 1,000 to 15,000 ha, some of which are classically diversified and integrated businesses operating on an international scale, for which rice production may be a small, relatively insignificant part of total production. For example, McCarthy Farming, Inc., in southern California, with total cropland in excess of 25,000 ha, grows rice in Kern County (and also in Colusa County in the Sacramento Valley) but is also involved in the growing and processing of at least 15 cultivars in conjunction with 11 other corporate affiliates (including insurance, landholding, and construction). In northern California, M. and T., Inc., with landholdings in excess of 7,000 ha and annual turnover of $54 million in the late 1970s, is especially active in Butte County rice production. It is also active in real estate, oil and gas drilling, and plastic tray manufacturing for supermarkets (Villarejo, 1980). The Parrott Ranch Company, a San Francisco–based corporation controlling 15,000 ha in the Sacramento Valley, is a major rice producer that is also active in livestock production and processing. Some of these large agribusiness are also active in the Sacramento Water Contractors Association as private parties (not as irrigation districts) and have received considerable benefits as diverters (i.e., with riparian rights they have received enormous quantities of nonproject water after the construction of the Shasta Dam by the State Water Project).

Nonetheless, while the rice shows land concentration (though this has almost certainly decreased since 1914) the most significant production unit (by area and production) is the family or household system, relatively specialized in rice production, mechanised but hiring limited wage labor (one year-round employee plus harvest help), and not integrated either vertically or horizontally into other crop-related productive activities.

The scale of input costs, and hence the depth of market relations, in rice production is considerable (Table 12-7). Cash costs per hectare are currently in excess of $1,235. This excludes the relatively fixed machinery costs—roughly $500,000 for a 360-ha farm amortized over the 10-year life cycle of most heavy farm machinery. A major part of the fixed costs of production lies, of course, in land, since prime ricelands currently cost $7,400/ha. The extreme capitalization of production has increased the dependence of rice growers on credit and banking. Large-scale finance capital has become critical in three phases of production cycle (Hopkins, n.d.): (1) long-term credit for real estate acquisition and development (leveling, pump installation); (2) seasonal loans to facilitate what have become enormous cash operating expenses (close to $1,500/ha in 1985); and (3) intermediary loans associated with marketing and costly machinery purchase or leasing. It is, of course, precisely this debt financing of farm mechanization during the 1970s, often encouraged by the Federal Farm Bureau, that has underwritten the current fiscal crisis of many rice growers who face tumbling real estate prices, a thin rice market, and the termination of federal rice subsidies.

A distinguishing characteristic of petty commodity production systems in California rice production, however, is the prevalence of leasing as opposed to owner–operatorship. Indeed, over half of California rice is currently grown on rented land; the vast majority (97%) of land leasing operates on a variety of share–lease bases (irrespective of holding size and scale of operation). This so-called capitalist sharecropping is a system of separation of management from ownership based on crop–revenue division, paradoxically a system conventionally seen as a form of precapitalist rent associated with factor immobility (Wells, 1984a).

Leasing is, of course, widespread in California agriculture as a whole; in 1978 close to 50% of aggregate California cropland was leased (Villarejo, 1980:13). Many of the largest landowners in the state are also important farm operators (e.g., Tejon Ranch Company, J. G. Bothwell Company, South Lake Farms, Inc.), but they may also simultaneously rent on a huge scale. In addition, large landowners such as Southern Pacific, Standard Oil of California, Prudential Insurance, and major California banks (which collectively own close to 80,000 ha) lease the vast majority of their cropland, often to large-scale leaseholders. Southern Pacific Land Company, for example, leases 39,128 ha to 41 operators (Villarejo, 1980:13). Riceland leasing, especially common in the Sacramento Valley, has been widespread since the renting of irrigated properties by land companies at the onset of the rice boom in 1912. In 1950, 49% of all rice farmers operated solely as tenants; in 1969, 52% of riceland harvested was rented, and in 1979 this figure was 49% (Grant, Amarel, & Johnson, 1971; Mullins & Krenz, 1981; Richards, 1969). In 1982, 66% of all rice farms (accounting for 75% of total California rice area) rented some rice land; 19.3% of the rice hectarage was cultivated by pure tenants, and 55.7% by part owners, part tenants (U.S. Bureau of Census, 1982:30). Furthermore, leasing is significant in each landholding cohort. In each size category, for example, roughly 21–25% of all rice farms are pure tenants (U.S. Bureau of Census, 1982:30), while a detailed study conducted in 1969 in the Sacramento Valley showed that in the largest landholder category (over 600 ha), 52% of rice hectage was rented; the corresponding figures for 0–242 ha and 243–599 ha were 45.7% and 53.4%, respectively (Grant *et al.*, 1971:5).

The significance of land allocation, however, resides not only in its areal extent but also in its social form. Cash renting, in which the landlord assumes low risk, appears to be declining. In 1964 for example, 33% of all rice leases in Butte, Glenn, Colusa, and Sutter Counties were on a cash basis (at roughly $125.00/ha), whereas today 3% of all farms (6% of all rice hectage) is rented for cash. The vast majority of rice production is on a share lease or sharecropping basis. *Sharecropping* implies some form of contract whereby access to one or more factors of production is provided by one party (Byres, 1983). This usually involves land but also some part of the other means of production. As Pearce (1983:52–53) suggests: "[S]harecropping contracts constitute the provision of the means of production, or some of them, by one party in return for a share of the crop yield . . . [and] is one mechanism through which owners of the means of production acquire access to others' labour."

Sharecropping has been historically equated with precapitalist rent as a method of surplus appropriation, particularly landlord–peasant relations (Marx, 1962; Mill, 1915), and theoretically as a form of agrarian social structure incompatible with capitalism (Pearce, 1985). Share leasing, as it is called locally, however, has been a central scaffold in the architecture of California rice production from its outset in 1912; indeed, recent research (Wells, 1984a, 1984b) reveals how sharecropping has waxed and waned throughout California's agricultural history yet remains fundamentally constitutive of some labor-intensive branches (for example, strawberries). Rice sharecropping is unusual because landowners often provide land and water. Indeed, insofar as many lessors directly control water through riparian rights, they are also de facto waterlords.

The share contracts in rice growing are enormously varied and have changed over time. Prior to 1981, when federally regulated allotment systems influenced rice area, some landholders provided land and water plus a proportion of the allotment and inputs. In the late 1960s there were five principal varieties of lease providing land, or land allotment, in which the landlord received that proportion of the rice crop corresponding to the proportion of fertilizer–herbicide–drying costs he contributed (usually 15–40%). The most common lease arrangements (50.3% of the total in the Sacramento Valley) chanelled 25–33% of the crop to the landlord (Grant *et al.*, 1971:9). Prior to 1981 almost one-quarter of the leases involved only the lease of an allotment (i.e., to permit the expansion of rice hectarage under federal price conditions) in return for 15% of the crop. Currently, however, with the abolition of the allotment system, over 90% of all contracts specify a 23–37% share of the crop for the landlord, including subsidiary crops such as barley, milo, and safflower grown as a doublecrop or in rotation. Most of the share contracts provide for a sharing of specified operating costs in the same proportion as revenue sharing. In Sutter and Yuba Counties, for example, the two most widespread sharecropping arrangements are associated with rotation versus continuous or nonrotation, and hence with soil type/crop adaptability and production costs (see earlier).[9] They are as follows: (1) a 40/60 contract for rotation land in which the landlord provides land and water, plus 40% of the herbicide, insecticide, fertilizer, hauling, and drying costs for 40% of all crops (revenue), and (2) a 25/75 contract for nonrotation land, with the landlord providing 25% of the same production inputs and appropriating 25% of crop output. Leases are short-term, at least 50% being 1–3 years in duration (Grant *et al.*, 1971:5).

Sharecropping rests on specific relations of power and authority between both parties, and, correspondingly, share contracts contain differential elements of risk, costs, and decision making for lessor and tenant. In the most common rice contracts in the Sacramento Valley, the landlord assumes a significant part of the production risks by virtue of supplying 40% of some parts of the means of production. The continuing feature of the share lease is, however, the incentive for both the landlord and the tenant to maximise efficiency (Reid, 1976:574). Wells (1984a) shows how sharecropping in the labor-intensive California strawberry industry operates as a "quasi-wage" system whose reappearance has been framed by the increasing militancy and leverage of farm workers and the role of the state in formalizing farm workers' rights. In strawberry production, sharecroppers are employed by highly capitalized, rationalized, vertically integrated firms; they have limited decision-making autonomy and appear as informal-sector, disguised wage workers. This is quite unlike the rice system, which appears as a subcontracting system in which sharecroppers are heavily capitalized, relatively autonomous, and not organically part of vertically integrated firms. Furthermore, it is to be recalled that one-quarter of rice growers are "pure" sharecroppers, and just under half are of a "mixed form"–part owners and part sharecroppers.

Risk is the warp and weft of the complex fabric of sharecropping arrangements in California rice production because share leasing can operate as a means of dispersing risk under conditions of an uncertain market and a large fixed investment. The wider political economy, however, shapes the specific costs, benefits, and risks to landowners and producers. It is to the political and economic relations in California, specifically the intersection of the state, the world market, and certain social interests, that have shaped rice growing that we turn in the final section.

The Political Economy of Capitalist Sharecropping in Rice

California possesses a unique type of agriculture. For example, conditions governing land and labor in the state have been cited as exceptions to the dominant pattern of U.S. agricultural development. From the start, large landholdings were the rule, and labor was supplied from a shifting base of temporary and migratory workers to degrees not found elsewhere in the nation (Liebman, 1983). Perhaps more surprisingly, state and federal assistance has played an unparalleled role in the introduction and eventual success of many California crops.

This section examines those conditions, historical and contemporary, on which rice farming units have depended. Since an exhaustive treatment of the subject is impossible here, we wish to single out only three aspects for attention: the evolution of land ownership patterns in the Sacramento Valley; state intervention (in the forms of technical assistance and production control); and the structure of the world rice market. Rather than discussing these as purely *external* to the rice-farming unit, we attempt to demonstrate the extent to which such conditions are constitutive of the units themselves.

Historical Patterns of Land Ownership in the Sacramento Valley

Prior to the introduction of rice to the Sacramento Valley, as in most other parts of California's central valley during the late 19th century, bonanza wheat farms occupied most of the cropland. Such farms, often totaling more than several thousand hectares, were operated by resident managers while actual landowners lived elsewhere. The prevailing scale of land ownership originated first in the massive grants made by the Spanish Crown during colonial rule, and second in abuses of federal policies governing the disposition of western U.S. resources. In the area eventually subdivided into Glenn and Colusa Counties, for example, six Spanish land grants amounted to 62,263 ha, while in neighboring Butte County the comparative figure for eight such grants was 60,680 ha (USDA, 1909:10).

The collapse of the wheat market during the depression of the 1890s wrought fundamental changes in the state's agricultural economy, particularly in crop choice and land ownership structure (Liebman, 1983; Pisani, 1984). Within the Sacramento Valley, the introduction and expansion of rice cultivation constituted a microcosm of such changes. Farmers to the south devoted increasing acreage to a range of fruits and vegetables. But Sacramento Valley farmers were constrained in their choices. Both local soil and climate were poorly suited for many of the new specialty crops. Rice cultivation was simply one of the few land use alternatives to wheat or pasture.

It was in this general context that rice was first grown in the Valley, but its introduction was enmeshed in a set of concurrent local events surrounding land ownership. The first of these represented an attempt to attract a new, ostensibly healthier breed of farmer. Chambers of Commerce from towns throughout the Valley joined forces with landholding companies in elaborate colonization schemes targeting midwestern farming families. Both their goals and their methods were akin to those seen 20 years earlier in efforts to create a "new Iowa" in southern U.S. rice regions (Daniel, 1985). Visits to the area by prospective colonizers were subsidized, and farmers were given tours of demonstration farms, towns, newly planted orchards, and sites of ongoing rice experimentation. As Willson (1979:29) has written about the colonization work of the Richvale Land Company, its purpose in visiting the rice experiments was to indicate "that rice might develop as a quick cash crop that could be planted until their orchards and vineyards became productive."

Colonization schemes of the early 20th century have left an indelible mark on the Sacramento Valley rice landscape. The area within Butte County surrounding the town of Richvale, which reportedly received the largest number of midwestern smallholders, has long enjoyed a reputation for having the most well tended rice fields of any in the Valley. Today Butte County remains the bulwark of smallholder production, to the extent that nearly two-thirds of its rice farms measure less than 20 ha, while fully 90% are less than 200 ha (California Department of Food and Agriculture, 1982). Perhaps more critically, colonization schemes helped create a legitimate local political constituency with and for whom state and federal agricultural agencies could undertake rice improvement programs. Finally, as technological innovations, many of them achieved through state-sponsored research, lessened labor requirements on individual farms, these smallholders were one producer group that actively expanded its operations through sharecropping.

All this is not to suggest that the introduction of rice ushered in a halcyon era of small-scale family farming. Although some large land parcels were broken up during the early 20th century, others remained essentially intact. In 1924 the Federal Land Bank reported a high degree of land ownership concentration remaining in the Valley after the first rice boom and bust. For example, in the newly formed Glenn–Colusa Irrigation District (embracing 47,206 ha), 47 of the total 403 landowners held 34,138 ha, or 72% of the land. By contrast, 84% of the landowners averaged only 36 ha per holding (Federal Land Bank, 1924:87).

Rice was a panacea for landowners wishing to keep their large holdings intact. Some of them became rice growers in a fashion familiar to them from the wheat era, as absentee owners delegating local management to farm operators. Willson (1979) notes that some of the earliest private-sector rice trials were conducted on the Valley's largest farms. Rice also invigorated what had become a sluggish local land market. Along with the buying and selling of land during the World War I riceland scramble, leasing was established as a form of easy entry into rice farming.

State Intervention: Expansion or Control?

Government involvement in rice farming commenced at an early date, indeed prior to cultivation of the grain, when in 1862 the California legislature began offering bounties for successful rice harvests. Financial incentives failed to stimulate production, however, since the concurrent wheat boom deflected attention from rice. Yet during the latter decades of the 19th century, rice experimentation was conducted in various areas of the central valley. These attempts met with repeated failure. It is no exaggeration to claim that during these years California farmers simply did not know how to grow rice. That they learned at all must be attributed in part to the collapse of the wheat boom during the 1890s, when a large number of landowners and farm operators in the Sacramento Valley were bereft of a readily available alternative crop. This group formed the nucleus of a constituency that began petitioning the federal government for assistance in introducing rice to the Sacramento Valley. During the 1890s and 1900s this constituency widened to embrace newly arrived smallholders from the Midwest and merchants and financiers from throughout northern California. Their efforts, including political agitation in Washington and practical rice trials locally, led to the founding of the Rice Experiment Station at Biggs (Butte County) in 1912.

The experiment station combined federal technical personnel with local capital. During the same year, the Sacramento Valley Grain Association was organized for the express purpose of raising capital for the Biggs Station. The association contributed working funds, while the Sutter Butte Canal Company provided 22 ha of land at a nominal annual rent and "topped off its overwhelming support of the Station by supplying free water for the original acreage in perpetuity. . . ." (Willson, 1979:43). The experiment station is still located in Biggs but is linked with the federal government through the USDA Extension Service at the University of California, Davis.

The work of the experiment station has centered on increasing the productivity of rice farmers. During its formative years, of course, its most important tasks involved identifying and distributing japonica varieties adapted to local environmental conditions; concomitantly, federal personnel imparted knowledge about proper cultivation practices, especially irrigation. Varietal experiments at the station (see Figure 12-3) have sought two principal goals: first, higher yields per ha, and second, and more recently, development of locally adapted long-grain varieties. The station also undertakes typical extension work like solving weed and pest control problems and advising on fertilizer application.

Entirely apart from the state intervention tied to the Rice Experiment Station are the federal programs covering rice as one of the nation's basic farm commodities (Kincannon, 1956). These programs have their origin in New Deal legislation, in particular the Agricultural Adjustment Act of 1938 and its subsequent amendments. Like most other federal farm programs, the rice program has sought to stabilize supply through a battery of methods including surplus storage, acreage allotments, and marketing quotas. The parameters of each have been adjusted over time in response to changes in world market conditions. During the boom market of the Vietnam War era, for example, marketing quotas were suspended and farmers were free to cultivate acreage in excess of their allotments (USDA, 1985:25–26). Most recently, in the Food and Agriculture Act of 1981, the rice allotment and quota systems were eliminated entirely in favor of deficiency payments (equivalent to the parity price) based on current plantings. These payments, based on output, benefit large-scale rice producers, since less than 10% of rice farmers (holding 400 ha or more) receive over one-third of the payments (USDA, 1985:29). Zumwalt Farms in Colusa County, for example, with 6,500 ha in rice, received deficiency payments of $1.5 million in 1984 (Sinclair, 1985:34)!

The single case of rice makes clear some of the contradictions embodied in state intervention in U.S. agriculture, which derive from the nature of the USDA itself. Through its technical and promotional work, the USDA has helped to raise productivity on rice farms, allowing farmers to counteract the intended effects of supply control programs devised by the department's economists. With the USDA working *internally* at such cross-purposes, small wonder that rice farmers now face an unprecedented crisis of overproduction (Schact, 1985:27).

Rice Market Structure

Of course the USDA does not operate in a political-economic vacuum. The crisis facing rice farmers, state interventions notwithstanding, is compounded by a world market for rice that has been called "thin" and "residual" (Siamwalla & Haykin, 1983). Little of the world's rice enters into international trade, since it is typically produced and consumed within national borders. To the extent that the largest producing nations, located principally in Asia, do participate in a world market, either on the demand or the supply side, they do so only sporadically in response to adventitious circumstances.

U.S. rice producers have been struggling for more than a century to secure a place in this notoriously thin market. Relative to Asian nations, the domestic market absorbs only a tiny fraction of domestic production. The absence of a national or world market has meant that halcyon days for rice farmers arrive only during wartime (cf. Figure 12-3). U.S. international food aid has absorbed some surplus rice stocks, serving as a surrogate market, though one subject to political vissicitudes. The achievements of Green Revolution high-yielding varieties—emblematic in India's emergence as a net rice exporter—make the world market prognosis for rice even less sanguine.

California rice farmers suffer from these structural problems plus a few peculiar to themselves. The market for California's rice has always been smaller than those for other U.S. rice regions because of the state's limited varietal adaptability. Although during the immediate postwar era Japan provided one steady source of demand for California's japonica varieties, Japan has not only become self-sufficient in rice, but it now actually competes with California in the same limited market. For these reasons the Rice Experiment Station, along with privately funded researchers, has been eagerly searching for a locally adapted long-grain variety (Anonymous, *Rice Journal*, 1985).

Perhaps because they were relatively late entrants in what was correctly understood as a limited market, California rice growers eagerly and early seized upon cooperative marketing associations as one source of stability. The first, the Pacific Rice Growers' Association (PRG), was founded in 1915 with control of 70% of state rice hectarage. Political differences within the PRG led to the establishment in 1921 of the Rice Growers' Association (RGA) of California, which over the years has become the dominant rice cooperative in the state. Its mission has extended far beyond maintaining quality control in milling and securing markets. During the bust immediately following World War I, for example, RGA successfully bargained for production financing on behalf of its members (Willson, 1979). RGA also finances varietal experiments, promotes domestic consumption, and has become a powerful political voice for California rice farmers at the federal level.[10]

Conclusion

In this essay we have endeavored to show how forms of rice production have arisen in the heartland of what French geographer Dorel (1980) calls "l'entreprise capitalist et la grande agriculture." Rice production in the Sacramento Valley is dominated by large, mechanized, and heavily capitalized production units, on large, family-based —and often corporatized—farms fully integrated through the market, many of which are petty commodity producers. Moreover, the particular evolution of these systems has witnessed a radical decline in the use of hired labor—for some, a measure of agrarian capitalism—and the persistence of dynamic forms of sharecropping, themselves often posed as precapitalist forms of agrarian social relations. We have sought to ground these characteristics not simply in the history of California and its landholding patterns, but also in the character of the world rice market, the role of the state, and the cost associated with highly capitalized aquatic crops such as rice.

Although the ecology of rice production and the environmental conditions of the northern Central Valley presented various opportunities for large-scale, mechanized rice production based on household units, the form and internal dynamics of rice-farming systems have been fundamentally shaped by the wider political economy. California agriculture is distinctive because it has been shaped de novo by capital; in the same way, large banking interests, holding companies, and industrialists operating in conjunction with the state were present in California rice farming from the beginning. These interests plus the growers themselves succeeded in establishing the political conditions of growth rather than simply exploiting natural conditions. Indeed, the growers and the growers' associations (for example, the California Rice Growers' Association, which markets rice) have successfully gained state support to protect their interests, sought out world markets, and creatively overcome barriers to growth.

This elaboration of the social dynamics of rice farming leads to two important conclusions. First, one should be aware of universal models of agrarian change that posit unilinear, predetermined changes of whatever hue. In rice production in California, for example, landholding and agrarian social relations patterns have been especially complex. Some enormous holdings were broken up and sold to smallholders or leased and sharecropped, but then saw a period of concentration and continued leasing in which cash renting underwent a rapid demise after 1960. Any grand theory of agricultural intensification must be handled with extreme care, not least because when it is broadened to accommodate its obvious weaknesses, it becomes a wider—and, we would argue, more contingent—theory of resource mobilization—in short, a theory of political economy.

Second, this study reveals the diversity and coexistence of a variety of agricultural production forms under capitalism (Pearce, 1985). These forms, such as sharecropping, are not relics or survivals but, rather, are constitutive of present political economy. Lenin pointed out that our views of agrarian change under capitalism are much too stereotyped and that its transformations are varied and heterogeneous (Lenin, 1964). Which is perhaps an apposite place to conclude, precisely because the California rice sector may be in the midst of a major crisis. Faced with limited world demand and domestic surpluses (Schacht, 1985), many heavily indebted producers who overcapitalized under state direction in the 1970s now face the legacy of high interest rates and foreclosure. The possibility of major withdrawals of rice subsidies in the 1985 Food Security Bill would, of course, result in a fundamental restructuring of the rice sector, euphemistically referred to as a "shakeout." Whatever the long-term consequences, rice production systems will bear the traces of this contemporary array of political and economic forces.

Acknowledgments

The authors acknowledge the assistance and kindness of Vern Ericksen and his family of Woodland in the course of conducting field research in Colusa County. In addition, this chapter has been fundamentally shaped and influenced by the important work of Miriam Wells, Harriet Friedman, and Peter Gibbon and Michael Neocosmos.

Notes

1. This nonmetric water measure will be employed throughout the chapter since there is no conventionally designated metric measure.

2. In Colusa and Butte Counties, the two largest producers, the extent of basin and older alluvial soils is, respectively, 100,000 and 64,000 ha (Hedges, 1974:8).

3. One acre-foot of water, the conventional measure, is equivalent to 325,851 gallons.

4. Two combines and one bankout vehicle are assumed to be necessary for the efficient operation of a 200-ha rice farm (California Department of Food and Agriculture, 1984).

5. Currently, 65% of all rice grown in California is of the medium variety, the remainder being short or pearl varieties.

6. The coefficients used in this calculation are taken from Pimentel (1973) and from the California Food and Agriculture Energy Study. Comparable input/output energy ratios are as follows: Taiwan, 1 : 5.0; Thailand, 1 : 8.0, (semi-industrial, that is, partly mechanized); and Bangladesh, 1 : 15.0 (preindustrial). These figures compare with those of Bayliss-Smith (1983:166–167) as follows: Yunnan, China, 1 : 15.0 (preindustrial); Karnataka, 1 : 4.78 (semi-industrial). The Green Revolution apparently halved the input/output ratios of wet-rice production, and industrial agriculture halved it again.

7. Three important insights emerge from this generic view of PCP. The first, following Gibbon and Neocosmos (1985), is that "places" or "niches" for simple commodity producers can be generated in the course of capitalist development. Although these places, particularly in agriculture, can be a haven for the preservation or expansion of PCP, it is nonetheless important to try to understand why and where such places are generated in specific historical circumstances. Second, one must distinguish between the "fate of petty commodity producers—which is to systematically divide into capitalists and wage laborers" (Gibbon & Neocosmos 1985:178). This is especially apposite among California rice growers, who are markedly differentiated between smallholders (less than 40 ha) who may sell their labor (and hence, strictly speaking, are semi-proletarians and not commodity producers), and enormous capitalist enterprises (whether corporatized family or industrial agribusinesses). Third, to describe and account for a specific form of PCP, "one must consider the organization of production at the level of actual practice . . . [because] different forms emerge as different elements in the production process become commodities, as labor processes are organised in different ways, as labor recruitment changes" (Smith, 1984:202).

8. With such differentiation under conditions of market competition, the central question becomes class differentiation, namely, internal transformation, dispossession, and exit from the rice sector. The current rice crisis—excess world supply, the overcapitalization of production with high interest rates, the threat of farm subsidy removal—is simply the present manifestation of such processes of differentiation.

9. This reflects the different costs and returns between single-crop rice with perhaps only a winter cover crop, and effective double-cropping or rotational systems with other cereals/field crops that demands other landform inputs.

10. An example of the power and influence of the rice growers can currently be seen in what is locally referred to as the "rice wars" between the Rice Growers' Association (RGA) and the Farmers' Rice Cooperative (FRC) and the tactics pursued in the 1970s concerning rice sales to Asia, specifically the so-called Koreagate scandal of 1977. In the course of the bitter rivalries over the control of California rice marketing, a complex story is unfolding of massive corruption and backroom politics involving the Reagan administration, the Department of Agriculture, the former mayor of San Francisco (himself an executive director of RGA), and two South Korean presidents (Conner, 1985).

References

Agricultural water use costs in California. Gianinni Foundation, Bulletin #1896, 1980.

Altieri, M. The requirements of sustainable agroecosystems. In G. Douglass (Ed.), *Agricultural sustainability in a changing world order*. Boulder: Westview Press, 1984.

Anonymous. N. F. Davis and his efforts with new varieties. *Rice Journal*, February 12, 1985, 12–14.

Bainer, R. *The evolution of rice mechanization in the United States*. Paper presented at the FAO Rice Year Conference, Rome, n.d.

Bayliss-Smith, T. Energy flows and agrarian change in Karnataka: The Green Revolution at micro-scale. In T. Bayliss-Smith (Ed.), *Understanding green revolutions*. Cambridge: Cambridge University Press, 1983.

Blehyl, N. *A history of production and marketing of rice in California*. PhD dissertion, University of Minnesota, 1955.

Buttel, F. Agriculture, environment and social change. In F. Buttel & H. Newby (Eds.), *The rural sociology of advanced industrial societies*. Totowa, NJ: Allanheld, 1980.

Byres, T. (Ed.). *Sharecropping and sharecroppers*. London: Cass, 1983.

California Department of Food and Agriculture. *Crop and livestock reports*. Sacramento: various dates.

Cervinka, V. & W. Chancellor. *Machinery management for timely planting and harvest of rice in northern California*. Publication #3005, University of California Division of Agricultural Sciences, Davis, 1975.

Cervinka, V., Chancellor, W., Coffelt, R., Curley, R., & Dobie, J. *Energy requirements for agriculture in California*. Joint Study, California Department of Food and Agriculture and University of California, Davis, 1974.

Chancellor, W., Avlani, P., Thai, N., Cervinka, V., Rupp, N., & Yee, E. *Energy requirements for agriculture in California*. Joint Study, California Department of Food and Agriculture and University of California, Davis, 1981.

Chang, S. *The role of water resources in the wet rice agricultural systems of northern California*. PhD dissertation, University of California, Los Angeles, 1971.

Conner, P. Rice wars. San Francisco *Chronicle*, February 3, 1985.

Daniel, P. *Breaking the land: The transformation of cotton, tobacco and rice cultures since 1880*. Urbana: University of Illinois Press, 1985.

Dorel, G. *La grande agriculture aux États Unis*. Thèse de doctorat d'état, University of Paris, 1980.

Federal Land Bank. *Sacramento Valley rice production*. Berkeley, CA: Division of Rural Institutions, 1924.

Flora, J. & Flora, C. *The state and rice production, marketing and research in the U.S.* Unpublished paper, Kansas State University, n.d.

Friedland, W., Barton, A., and Thomas, R. *Manufacturing green gold: Capital, labor and technology in the lettuce industry*. Cambridge: Cambridge University Press, 1981.

Friedmann, H. Household production and the national economy: concepts for analysis of agrarian formations. *Journal of Peasant Studies*, 1980, 7, 158–184.

Geertz, C. *Agricultural involution*. Berkeley: University of California Press, 1966.

Gibbon, P., & Neocosmos, M. Some problems in the political economy of African socialism. In H. Bernstein & B. Campbell (Eds.), *Contradictions of accumulation in Africa*. Beverly Hills, CA: Sage, 1985.

Grant, W., Amarel, R., & Johnson, S. *Leasing on California farms*. Davis: University of California Extension Service, 1971.

Grant, W., & Musick, J. *Regional adjustments and regional competitive advantages in the U.S. rice industry*. Paper presented to the Rice Technical Working Group, Hot Springs, AK, n.d.

Hayes, S. The California Agricultural Relations Act and national agricultural labor relations legislation. In R. Emerson (Ed.), *Seasonal agricultural labor markets in the United States*. Ames: Iowa State University Press, 1984.

Hedges, T. *Water supplies and cost in relation to farm resource use decisions and profits on Sacramento Valley farms*. Gianini Foundation Research Report #320, University of California, Berkeley, 1974.

Hopkins, J. *Financing the California rice industry*. Paper presented to the 1964 California Agricultural Extension Meetings, n.d.

Kahn, E. J., Jr. *The staffs of life*. Boston: Little, Brown, 1985.

Kahn, J. *Minang Kabau social formations*. Cambridge: Cambridge University Press, 1980.

Kincannon, J. *Legislation affecting the rice industry, 1933–1956*. Bulletin #839, Texas Agricultural Experiment Station, College Station, 1956.

Kitching, G. Politics, method and evidence in the "Kenya debate" In H. Bernstein & B. Campbell (Eds.), *Contradictions of accumulation in Africa*. Beverly Hills, CA: Sage, 1985.

Lenin, V. *The development of capitalism in Russia*. Moscow: International Publishers, 1964. (originally published 1956)

Liebman, E. *California farmland: A history of large agricultural holdings*. Boulder: Westview Press, 1983.

Marx, K. *Capital* (Vols. 1 & 3). New York: International Publishers, 1962. (originally published 1867 and 1894)

Mill, J. S. *Principles of political economy*. London: Longmans, 1915. (originally published 1848)

Miller, M. D. *Rice production in California*. University of California Division of Agricultural Sciences Publication #75-LE/2236 Davis, 1977.

Mooney, P. Labor time, production time and capitalist development in agriculture. *Sociologica Ruralis*, 1982, 22, 279–282.

Mullins, T., Grant, W., and Krenz, R. *Rice production practices and costs in major U.S. rice areas, 1979*. Bulletin #851, Agricultural Experiment Station, Fayetteville, AK, 1981.

Norman, D. Rationalizing mixed cropping under indigenous conditions. *Journal of Development Studies*, 1974, 11, 3–21.

Pearce, R. Sharecropping: Towards a marxist view. In T. Byres (Ed.), *Sharecropping and sharecroppers*. London: Cass, 1983.

Pearce, R. The agrarian question. In Z. Baranski & J. Short (Eds.), *Developing contemporary Marxism*. London: MacMillan, 1985.

Pimentel, D. Food production and the energy crisis. *Science*, 1973, 182, 443–449.

Pisani, D. *From the family farm to agribusiness: The irrigation crusade in California and the west, 1850–1931*. Berkeley: University of California Press, 1984.

Reid, J. Sharecropping and agricultural uncertainty. *Economic Development and Cultural Change*, 1976, 24, 578–595.

Richards, S. *Geographical aspects of rice cultivation in California*. MA thesis, University of California, Berkeley, 1969.

Roldan, M. Industrial outworking, struggles for the reproduction of working-class families and gender subordination. In N. Redclift & E. Mingione (Eds.), *Beyond employment*. Oxford: Basil Blackwell, 1985.

Rutger, J., & D. Brandon. California rice culture. *Scientific American*, 1982, 129, 44–51.

Ruthenberg, H. *Farming systems in the tropics*. London: Clarendon, 1980.
Scardaci, S., & K. Olson. *Rice production costs*. Colusa, CA: University of California Cooperative Extension, 1983.
Schacht, H. Bumper rice crop causes marketing worries. San Francisco *Chronicle*, November 23, 1985, 27.
Scheuring, A. (Eds.). *California agriculture*. Berkeley: University of California Press, 1983.
Security Pacific National Bank. *California's agricultural trends and issues*. Los Angeles: Privately printed, 1985.
Siamwalla, A., and S. Haykin. *The world rice market: Structure, conduct and performance*. IFPRI Research Report #39, Washington, DC, 1983.
Sinclair, W. *A farming lesson well learned*. Washington *Post* Weekly, November 11, 1985, 34.
Smith, C. Forms of production in practice: Fresh approaches to simple commodity production. *Journal of Peasant Studies*, 1984, 11, 201–222.
Steinhart, J., & C. Steinhart. Energy use in the U.S. food system. *Science*, 1984, 184, 307–315.
Storie, R. E., & Weir, W. *Generalized soil map of California*. Storie Index Bulletin #4, Series 1926.
Thomas, R. The social organization of industrial agriculture. *Insurgent Sociologist*, 1981, 10, 5–22.
Thomas, R. *Citizenship, gender and work*. Berkeley: University of California Press, 1985.
U.S. Bureau of Census. *Census of Agriculture*, Vol. 1, Part 5: *California*. Washington, DC: U.S. Government Printing Office, 1982.
U.S. Department of Agriculture. *Rice farming in the Sacramento Valley*. Washington DC: U.S. Government Printing Office, 1909.
U.S. Department of Agriculture. *Rice: Background for 1985 farm legislation*. Bulletin #470, Washington DC, 1985.
Villarejo, D. *Getting bigger: Large scale farming in California*. Davis, CA: Institute for Rural Studies, 1980.
Wells, M. The resurgence of sharecropping: Historical anomaly or political strategy? *American Journal of Sociology*, 1984a, 90, 1–29.
Wells, M. What is a worker? The role of sharecroppers in contemporary class structure. *Politics and Society*, 1984b, 13, 295–320.
Whitehead, A. *Gender and class in petty commodity production*. Unpublished manuscript, Sussex University, 1985.
Willson, J. (Ed.). *Rice production in California*. Richvale, CA: Butte County Rice Growers Association, 1979.

13

A Hungarian Cooperative Farm: Kondoros Village

IMRÉNÉ KARÁCSONY, ANTAL SZATHMÁRY, AND ISTVÁN SZÜCS

Agriculture plays an important role in the Hungarian economy. Seventy percent of the country's territory is devoted to agriculture, and in 1984 this industry's share in the net national production was 20.3%. Moreover, some 1.5 million families are engaged in some kind of agriculture. The large-scale socialist sector of agriculture is composed primarily of cooperative and state farms, cooperatives being the larger segment.

The socialist transformation of Hungary's agriculture was completed between 1959 and 1961. Agricultural output did not decrease during the transformation, which involved, among other things, the establishment of an organizational framework and proprietorship suitable for large-scale, modernized production. In the early stages of the transformation, cooperatives were small, about 1,000 ha, and clustered as one or two per village. In total, some 4,195 cooperatives existed, making up 61% of the country's agricultural land. By the mid-1970s, however, these cooperatives had been enlarged and centralized and had embarked on horizontal integration. The funds to do this were created by the cooperatives themselves and by the state.

Concomitant with spatial and economic expansion was the growth of other parts of the cooperative system. Cooperatives expanded to include as many as five or six communities, increasing the need for community services and infrastructure. Farm modernization and complexity increased, promoting the need for education among the cooperative members. As mechanization increased, labor decreased and migrated to the industrial and service sectors of the economy. By the 1980s, however, farm labor had stabilized.

Another essential feature of the transformation was to accelerate the development of the nonagricultural operations of cooperatives. Initially this was done primarily through on-farm construction operations. Later, vertical integration of production was promoted—for example, food processing. This integration was facilitated by the development of regional activities, including the manufacture of spare parts, small equipment and machinery, and textiles.

Imréné Karácsony, Antal Szathmáry, and István Szücs. Department of Planning and Economic Analysis, Mezögazdasági És Elemezésügyi Minisztérium Statisztikai És Gazdaságelemzö Központ, Föosztályvezetö, Hungary.

Cooperative agriculture also incorporates private, family farming. This type of small-scale farming helps families to meet their own demands and increases the supply of produce for the market. It concentrates on crops and produce that require high labor inputs, such as horticulture, pigs, and poultry. These operations are fully integrated within the socialist sector. The cooperative is responsible for distribution of the produce of family-managed farms.

This chapter discusses the workings of the United Cooperative of Kondoros (UCK), a large farm located in southeastern Hungary. The farm is a highly modernized and diversified operation, and the farming system extends beyond immediate production to include the social-welfare fabric of the cooperative members and the village as a whole.

Environmental Setting

The farm is located in the southeastern corner of Hungary; all of its land belongs to the village community of Kondoros, Békés County (Figure 13-1). Békés Country has 81 cooperative farms, operating on a total of 375,000 ha. About 83% of the land is considered arable, compared to the national average of 67%. Ecologically, Békés County can be divided into two microregions: the so-called Békés lowlands and the region of Körös rivers. The cooperative farm described here lies in the western part of the Békés lowlands.

This microregion covers about half of Békés County. Elevation above sea level is in the western part, around 100 m, whereas the eastern part is situated at 150 m. The dominating soil type is of meadow origin, with good conditions for water management. This type of soil covers some 80% of the area. Minor areas in the north and northwest contain the prairie soil of the Big Hungarian Lowland. In general, the soils boast one of the thickest humus layers (A horizon) in Hungary, at some places 150 cm thick.

Variations in groundwater level are considerable and not easy to forecast. Variations range between the surface and 3 m, but at certain points depths of 4 m or more may occur.

Climactic conditions over the region are rather uniform. The annual solar energy received amounts of 107 kcal/cm^2 practically everywhere the region; the number of hours with full sunshine is about 2,000 per year. Annual mean temperature at the southeast border of the region is 11°C, which changes to 10.5°C in areas lying west and north. The number of winter days is 25–30; days with frost range from 95 to 100, increasing toward the southwest corner of the region. First frost typically begins after October 31, and the last frost rarely comes after April 5.

Average annual precipitation ranges between 550 and 600 mm, of which 225–250 mm reach the soil during the fall and winter. Normal precipitation from March to June is about 225 mm, while 325–350 mm falls throughout the rest of the year. The annual number of days with full snow cover is 30, and the first snowfall usually occurs after November 20. The last snowfalls may be

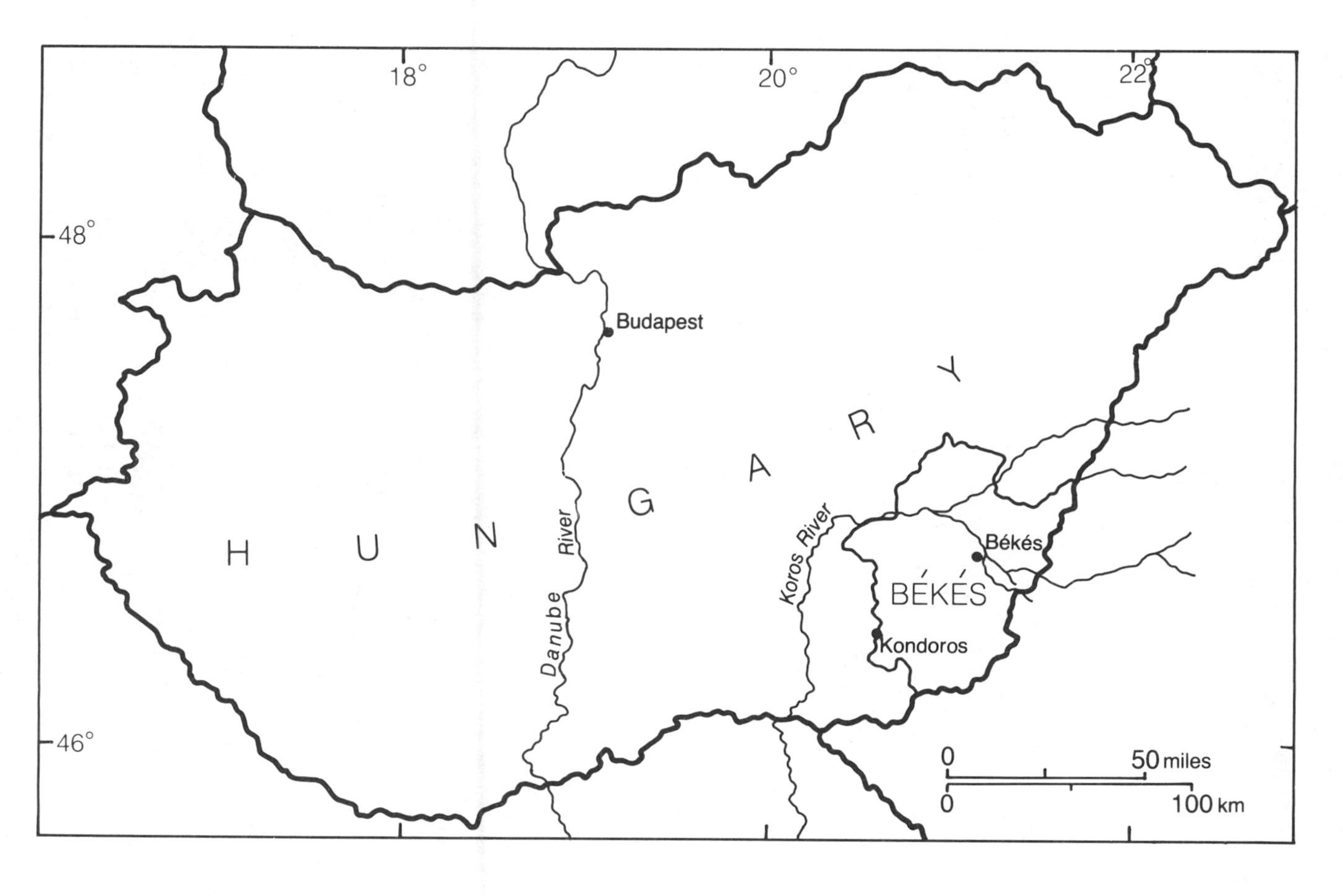

FIG. 13-1. Map of Békés County, Hungary.

expected between March 15 and 20. Air humidity during the vegetation period ranges between 66 and 68%.

The Farm

The Development of the Cooperative

During the socialist transformation of agriculture (1959–1961), three fully independent cooperative farms were established in the village of Kondoros. By the early 1970s, technological progress and mechanization of Hungary's agriculture had accelerated by rapid development, and large-scale production techniques called for the concentration of resources, both intellectual and material. This was the basic factor that led to the 1975 merger of the three cooperatives to form the United (Egyesült) Cooperative of Kondoros (UCK).

The origins of this cooperative farm go back to 1959, when 13 members (individuals who collectively own the assets of the cooperative) put together 49 ha of land. In 2 years the cooperative had grown to 114 working members and 321 ha. More smallholders joined, so that by 1965 the farm was 4,000 ha in size, with 951 members and employees. By 1975 the UCK had 2,410 members (1,514 working members and the remainder retired) and 8,555 ha.

At first the cooperative had 3 pairs of horses, 10 cows, and 20 sows. Major tillage work, threshing of cereals, and the bulk of transportation were done by the local machine station on a contract basis. Machine stations were state-owned contractors for mechanized agricultural work. They were effective, ensuring maximum utilization of equipment at both national and regional levels. With relatively few machines they performed a larger portion of basic chores for several nearby cooperatives and even for smallholders.

Originally the cooperatives did not have the facilities for operating and maintaining agricultural machinery; they did not possess either the capital to buy, or the skilled staff to operate, them. But as the cooperatives grew, their purchasing power increased. There was a gradual change to more intensive methods of production, which required more mechanization and called for establishing machine stocks within the cooperatives. Eventually machine stations could not meet the demand raised by the cooperatives at peak working periods. Therefore, some of the machine stations continued as machine repair companies, which contracted for repairing and maintaining agricultural equipment owned by the cooperatives, and others were purchased by the cooperatives or other state companies. In retrospect, mechanization primarily replaced horses and oxen and manual work, and thus played only a secondary role.

Economic and technical conditions for establishing their own machine stock became reality for the UCK in the mid-1960s. The level of mechanization matched the scale of operations by the mid-1970s. At present the cooperative owns 104 tractors with a total combined power of 5,800 kw, 48 trucks, and 22 combine harvesters. The number of units of self-propelled large equipment (e.g., sugar beet harvesters, hay mowing and handling machines, green pea viners) is over 100. Some of the machinery originated in Western countries, and some is made in Hungary or in other socialist countries.

Land Use, Cultivars, and Rotations

The cooperative's total area now amounts to 8,715 ha, of which 1,178 ha are allotted to members for their private use. All of the latter is arable land worked with the cooperative's machinery. Furthermore, these 1,178 ha are integrated with the cooperative's own rotation of crops and are usually cultivated to maize (*Zea mays*). About 20% of the members do not want private plots and receive 2.4 t of feed grains free, or the equivalent in cash. This cash corresponds to the difference in current selling price and direct production costs (i.e., from the income that could be realized on the respective private plot with the crop in question).

Those members of the cooperative who insist on receiving feed grain use it to feed livestock kept around their houses. A certain portion of the land allotted to members is used for growing medicinal herbs and spices that require a large amount of manual labor and special care. These special cultivars are sold directly to companies engaged in processing herbs and spices. The income realized in this specialized activity is then used for paying the cooperative for the mechanized work, work transportation, and seeds or other propagation materials.

Of the cooperative's total arable land, 215 ha are pastures, 18 ha are orchards, 87 ha are occupied by forests, and 316 ha are covered by farmsteads and roads.

This cooperative is a typical grain-producing entity of the Great Hungarian Plain, part of the wheat–corn–alfalfa belt. This fact is fundamental to the structure of field production and to the process of specialization. Wheat (*Triticum aestivum* [*vulgare*]) occupies 33% of the arable land. Maize is grown on 20% (30% if private plots are considered). An additional 2% of the arable land is used for seed corn, 4.6% for sugar beet (*Beta vulgaris* var. saccharifend), and 4.7% for green peas (*Pisum sativum*). Paprika pods (*Capsicum annuum*) are grown on 0.7% of the arable land, while tomatoes (*Solanum lycopersicum*) occupy 0.4%. Other vegetables and spices occupy something less than 4% of the arable land.

The cooperative has a relatively good supply of proteins for its livestock, as 11.4% of the arable land is allotted to soybeans (*Glycine soja*). To meet the roughage requirement of the cooperative's own livestock and of the livestock kept by members as part of their private farming, the land share of alfalfa (*Medicago sativa*) is 7.6% and a further 10% is allotted for silage corn and other green roughage crops. In summary, about one-third of the land is used for bread cereals (i.e., wheat); one-third for maize; and one-third for sugar beet, soybean, roughages, and vegetables.

Crop rotations are complex. Wheat precedes itself on about 40% of the land devoted to wheat cultivation. On the other 60%, soybeans, alfalfa, green peas, or fast-maturing maize precedes wheat. Land harvested in wheat is usually planted to sugar beets, followed by maize and roughages. Maize land is typically cultivated for 3 years consecutively in order to make full use of herbicides. Land used for vegetables is irrigated and involves a separate rotation from the cereal lands. Some wheat is grown on irrigated lands to permit the application of livestock manure.

The areas devoted to wheat and maize are fairly constant. Larger variations occur for the remaining crops, depending on market demand and the size of livestock. Agrotechnological requirements are also taken into consideration in determining crop rotations.

Production is intensive. All land is cultivated every year. No land is left idle, and no fallow is used. Wheat is sown in October and harvested in the first half of July (Figure 13-2). Sufficient equipment is available to complete the wheat harvest during an optimum 11- to 12-day period. Maize and soybeans are planted in the second half of April and are harvested, depending on crop variety, from the end of September to the end of October. Depending on weather conditions, sugar beet is planted in March and harvested during October.

FIG. 13-2. Cultivation periods for principal cultivars.

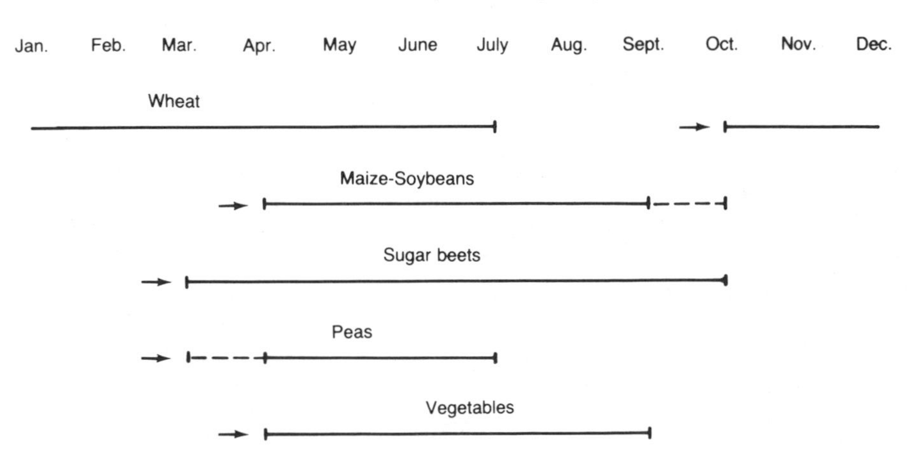

NOTE: Crop rotations are discussed in text.

Among roughages, alfalfa is left in production for 4 years; a new crop is sown in the fifth year. Depending on climatic factors, alfalfa is harvested four times per year. Green roughages are sown at different times to provide for a uniform supply, and they are harvested as feed demand arises.

Green peas are sown in March or April, according to the genetic heat demand of the variety in question. At present, a new system of sowing green peas is being tested in which the exact date of sowing is determined by the canning factory's computer program. In order to ensure a uniform supply of produce, the cooperative will follow the dates of sowing as set by the program. Harvesting takes place from late May to the end of June.

A major vegetable crop of the cooperative is green beans, covering a total of 170 ha. Green beans are planted similarly to green peas (i.e., in steps between May 10 and June 10. A self-propelled harvester is used, beginning in July and continuing to mid-September. Seedlings of other vegetable crops (tomato, paprika, cabbage) are transplanted during April and May. They are harvested when ripe, usually from July to the end of September, or the beginning of October. For the canning factory, the cooperative grows specific varieties of vegetables, which are usually harvested during the first days of August. Some 25% of green paprika is planted in place in order to save greenhouse costs.

The cooperative also operates a nursery of fruit trees, occupying 20 ha. Saplings are sold through a distributing company; the bulk of this produce is exported.

Mechanized Technology

The cooperative's stock of machinery includes 50- to 150-kw tractors for tillage works, grain harvesters with a throughput of 12 kg/sec, and trucks with payloads ranging from 5 to 10 t.

Sowing and transporting are fully mechanized. Selective weedkillers are in use, and the use of pesticides is common. Mechanized weed control is used once for corn and horticultural crops, and as required for sugar beets. The implement used is a tractor-mounted row crop cultivator.

Harvesting is fully mechanized (Figure 13-3). High-performance grain harvesters are used for wheat, maize, and soybeans. After cleaning and, if necessary, drying the grain, a part is sold directly to the company specializing in storing and distributing grains, and a portion is stored in the cooperative's own silos until selling. Wheat straw is baled and used as litter both in cooperative and in family-managed livestock facilities. Only a portion of cornstalks will be collected and used as roughage feed; the rest will be shredded (as with soybean straw) and left on the soil for underplowing, to raise the organic matter content of the soil.

A green pea viner is used for collecting this crop. The extracted pea is transported directly to the canning factory. To prevent any deterioration of the

FIG. 13-3. Harvesting of green beans.

produce, green pea harvesting and transporting operations are timed to ensure processing within 6 hours after entering the canning factory. Other vegetable crops and fruits are mainly harvested.

Sugar beet harvesting includes transportation to a railway transfer point where beets are deposited in a heap and then transported to a sugar refinery. Some 40% of alfalfa is stacked in the wilted stage for drying with cold air fans. Ten percent of the alfalfa crop is used for green conservation (ensilage). The rest is left on the field for natural drying, baling, and stacking.

A portion of green roughage is fed daily to livestock, mainly to cattle, during the summer. The green roughage is brought in daily by self-propelled forage harvesters. The resulting forage is hauled directly into the stables. Only young heifers are pastured. The rest of the green roughage will be ensilaged.

Cooperative Livestock Operations

Only a portion of the feed crops produced by the cooperative are sold directly for cash. The rest are used as feed for livestock. All hay, beet, and green roughage feed, as well as the litter straw required by the livestock operations, is produced by the cooperative itself. But the feed grain produced by the cooperative is first sold to a feed mixing and processing company formed by four neighboring cooperatives, and then repurchased after processing and enrichment with proteins, minerals, and vitamins. (The United Cooperative of Kondoros has a 65% share in this food mixing and processing company.)

The cooperative's livestock operations involve cattle, hogs, and poultry. The number of cattle is 3,100; the stock of cows is 1,110. The major line of production is dairying. The offspring exceeding the requirement for replacing animals culled from the breeding stock are fattened and sold on foot to the meat industry. This stock of cattle is kept in three housing facilities, one each for cows, young cattle, and beef cattle. Feeding and manure removal in these facilities are mechanized. A separate milking house with an Alfa-Laval herringbone milking parlor has been established (Figure 13-4). After milking, the milk runs over pipelines into cooled storage tanks from which it is pumped into sealed tank wagons for transportation to the milk processing company.

The number of hogs averages around 17,300, and the number of sows is 1,200. The cooperative has two hog production facilities, each with mechanized feed distribution. Manure is removed by flushing. After a two-phase separation, the slurry is stored in high-capacity ponds, to be pumped from time to time onto free lands. The average number of sows rearing is 2.1–2.2. Piglets

FIG. 13-4. Milking cow house.

are weaned at 8 weeks and are sent to raising and fattening houses. At 6 months, with weights ranging from 96 to 100 kg, they are sold to the meat industry. Processed feed for the hogs is purchased from the feed mixing and processing company mentioned previously.

Breeding stock is raised only to replace culled sows. To replace culled breeding boars, the cooperative turns to specialized farms engaged in raising and selling hybrid breeding hogs.

In addition to cattle and hogs, the cooperative's livestock sector includes broiler (chicken) operations. The 780,000 broilers sold annually are raised in two broiler housing facilities. Day-old, meat-type hybrid chicks are purchased from hatcheries. Protein-rich processed feeds, mixed in the cooperative's own mill, are fed to the broilers. Feeds are tuned to the biological requirements of the birds. Broiler feeding is fully mechanized. One batch of birds is kept for 53–55 days. Every 55th day involves hauling birds with a live weight of 1 kg to the broiler slaughterhouse, and placing new chicks into the stables. Thus six batches of broilers are raised per year, for a total of 700,000–780,000, requiring a 130,000-bird-capacity facility.

Private Farming Activities

Large-scale state- or cooperative-owned farming and family-managed private farming in Hungary cannot be considered separately, as the two types of management are complementary, even indispensable, to each other. The private farming activities of cooperative members should not be seen as remnants of yesteryear's smallholder farming, as in fact they constitute a new sector within the national economy, an addition to large-scale state- and cooperative-owned farming.

There has been substantial change in private farming since the foundation of the first cooperative. At first, it mimicked the cultivars grown on the cooperative, though by manual work. Nowadays, the crops grown on private farms are, in most cases, those that do not lend themselves to mechanization or to effective large-scale farming techniques, but are still profitable using small-scale, family-managed methods. For the Kondoros area, medicinal herbs are such crops.[1]

Small-scale private farming has another advantage: Disabled or retired individuals and parents caring for small children who cannot participate in the normal working process of large-scale farming, but who still have some spare time, may profitably use their time and skill. They work as much as they like and when they like, as a source of extra income.

The cooperative's large-scale operations and the members' small-scale private farming operations are linked mainly in the fields of livestock production. The cooperative provides the feed required for starting, raising, and finishing the privately owned animals. In addition, the cooperative organizes the selling of privately raised produce when the crops are harvested and the livestock is finished (e.g., milk, fattening hogs and beef, medicinal herbs).

The yearly average number of livestock kept by the members of the cooperative as their property amounts to 80 cows, 50 beef cattle, 750 sows, 300 sheep, and 30,000 poultry. The cooperative purchases the produce raised privately by members (typical annual figures are 16,000 liters of milk, 25 tons of beef, 1,500 tons of finished hogs, 1 million head of broilers and other poultry) and sells it, along with its own produce, to distributing companies.

Insufficient mechanization of small-scale farm work is a drawback to the private activity of cooperative members. Small milking equipment is needed, and feed distribution and manure removal work should be mechanized. Most of the private houses have no pipeline water. Finally, many kinds of equipment and investments are lacking that could facilitate the tiring work of private farming.

Industrial Activities

In addition to farming, the cooperative is engaged in industrial activities that supplement its farming sector. The cooperative's industrial sector comprises a machine repair shop and a building contractor unit. Both units perform work for other cooperatives or state-owned companies, too. This sector employs 400 people, or 27% of the working staff. Small industry's share in the cooperative's total income is 30–35%. The cooperative decided to set up these trades because they required minimal fixed investment.

Production

Large-scale farming requires increased manufactured materials. At present 60% of all material input for the farm originates in the manufacturing industries, the most important being fertilizers. This subsidy is indispensable for maintaining soil fertility and productivity, and the Kondoros cooperative purchases 2,000 t on a yearly average.

For every hectare of arable land, 301 kg of fertilizer is applied (1982–1984 average). In the preceding years more fertilizers were used, some 350–400 kg/ha/yr. Recent economic problems have necessitated security and restraint in application. The nutrient concentration in the soil was high, however, and with increased amounts of manure, the cooperative decreased the amount of fertilizers used.

Fertilizers are spread in a solid state by mobile spreaders or by airplane; liquid fertilizers are not used. This work, including spraying herbicides and pesticides, is performed on a contractual basis by several agrochemical centers that serve neighboring cooperatives and state farms (a total of some 30,000–35,000 ha.).

Fertilization involves scientific planning. Soil tests are made every second year to establish the soil's nutrient content. Soil nutrient data (nutrient composition and amount) and expected crop yield are used in determining the rate of fertilizer application for the coming year. The yield potential of the crop is also taken into consideration, of course. The cooperative applies manure from its large-scale livestock facilities and from some of the private livestock operations to maintain soil organic matter. On the average, 6.3 t/ha of manure are applied yearly. Considering that a given field receives manure at 6-year intervals or more, the rate of manure application should be increased further. This is offset, however, by including maize stalks.

Tractors (150 kw) are used for all basic tillage works, while 50- to 60-kw tractors are used for crop cultivation and plant protection works, for a portion of transportation works, and for pulling implements. Considering the cooperative's total hectarage (including the land allotted for private use), 69 kw of tractor power and 0.6 kw of truck power are available for every 100 ha. One combine harvester serves 203 ha of wheat, corn, and soybeans. Considering the cooperative's total acreage (including the land allotted for private use), the tractor stock's performance is 18.6 normal ha.[2] Effective truck hours/ha/yr amount to 9.4. These tractor and truck performance data cover also the industrial activities of the cooperative.

Horticultural production employs irrigation. Some 1.1–1.2 million m^3 of water is applied yearly on average.

The cooperative's agricultural and industrial operations consumed 2,961,000 working hours in 1984, of which 1,822,000 hours involved agricultural activities – 880,000 hours in crop production, 815,000 hours in livestock production. About 106 hours/ha of arable land, pasture, and orchard are used.

Variations in crop yields show the effects of favorable or unfavorable weather conditions. Wheat yields in recent years have ranged from 6.02 to 6.5 t/ha. The last two years (1983–1984) were classified as droughty, with consequent sharp drops in maize yields. Recent maize yields averaged 7.02 t/ha, while the 1981 and 1982 yields were 9.10 and 9.29 t/ha, respectively. In the drought year of 1984, yields fell to only 5.58 t/ha. The sugar beet yield was 36.9 t/ha, 7.8 t/ha less than the 1982 yield. Drought caused soybean yields to fall by more than 50%. The 1982 yield was 3.2 t/ha, while the 1984 yield amounted to only 1.2 t/ha, bringing the last 3 years' average down to 2.0 t/ha. Alfalfa yields ranged between 4.5 and 7.3 t/ha (see Table 13-1).

Crop yield variations are much smaller in the horticultural sector. Green pea yields range from 3.5 to 4.3 t/ha – the record yield is 5.9 t/ha. Tomatoes produced 58–66 t/ha, the record being something over 70 t/ha. Because of the drought, however, the 1984 yield was as low as 24.3 t/ha, while green paprika yielded 26–30 t/ha. Seed corn yield averages between 3.3 and 3.9 t/ha, but yields were as low as 1.7 and 1.6 t/ha, respectively, in 1983 and 1984.

In 1984 the cattle sector produced 4,835 liters of milk per cow and a total of 957 head of beef cattle (517 t live weight). Calfing was efficient, amounting to 1.05–1.07 calves/cow/year.

In the hog production sector, 1.01 t of finished hogs and sows per year was obtained, bringing the total output to 2,500 t. Farrowing figures were 1,773 piglets/100 sows and an average litter of 8 piglets. About 4.1 kg of feed were

TABLE 13-1. Crop yields of the United Cooperative of Kondoros in 1984.

	Tons	
Crops	Total	Per hectare
Wheat	13,822	6.0
Maize	7,689	5.6
Sugar beets	11,583	36.9
Soybeans	917	1.2
Alfalfa	2,371	4.5
Green peas	1,905	5.9
Tomatoes	730	24.3
Green paprika pods	666	1.3
Seed corn	196	1.6

required to produce a 1-kg increase in live weight of fattened pigs. The cooperative sold 768,000 head of poultry to the slaughterhouse, with a total weight of 1,206 t. About 2.42 kg of feed were used to obtain 1 kg of live weight gain.

Livestock feeding is based on strict scientific methods. Production processes involving all livestock are controlled by computers. Microcomputers are used for storing and recording animal data—daily weight gains, milk produced daily, milk quality parameters. These data are used to determine daily feed rations and ingredients for individual cows and flocks.

Fixed and Working Capital

The gross value of the cooperative's fixed capital is 478 million forints (in 1985, about U.S.$10 million at 1 forint = $0.0209): Buildings are 302 million ($6.3 million) and equipment is 166 million ($3.5 million). The level of depreciation of fixed assets is 65%. Pending loans repaid for investments amount to 56 million forints, 12% of the value of fixed assets. Circulating assets (working capital) are valued at 242 million forints ($5.05 million). Fixed capital of 57,000 forints is charged against 1 combined hectare of arable land, pasture, orchard, and forest. The same figure for working capital is 29,000 forints.

The effectiveness of the cooperative's economic activities is measured by changes in gross income. (Gross income equals total income less salaries and other expenditures without taxes.) Yearly gross income of the cooperative ranges between 100 and 120 million forints ($2.09–$2.5 million), equaling 14,000–17,000 forints ($293–$355) per combined hectare of arable land, pasture, orchard, and forest, or 68,000–82,000 forints ($1,421–$1,714) per full-time working person.

From the gross income (without tax on land), 10–18 million forints are paid for taxes; 12–20 million forints are used for fixed and working capital; and

7–8 million forints are put aside for welfare, cultural, and reserve expenditures. Staff salaries (i.e., salary income of staff paid by the cooperative) amount to 70–74 million forints, or 48,000–53,000 forints ($1,000–$1,100) per full-time working person. In droughty years, staff income is supplemented by amounts of reserve or by loans received from banks.

It must be noted that the conversion of forint figures by the offical rate of forint–dollar exchange does not provide a realistic comparison, as relationships between prices and income are different in Hungary than in developed Western countries. Perhaps a better indicator of the value of the incomes given here is the fact that the income for a full-time working family of the cooperative (father—60,000 forints/yr; mother—48,000 forints/yr) is sufficient to maintain a four-person family at a moderate standard of living. This family income is usually increased by 20–25% from income realized by private small-scale farming.

Cooperative Management

The members of the cooperative jointly and directly own the assets of the cooperative. It follows that the members manage their own affairs (i.e., the cooperative's direction may be termed *self-management* or *self-government*). To the outside world the cooperative is an independent legal entity managed by the principles of *membership control*. Each member has a right to vote, and all fundamental questions of management must be decided by a plenary session of members. At the same time, principles of one-person control are also in effect. Executives are entrusted by the members for a set number of years. They are responsible to the body of members. One peculiarity of membership control is that decisions made by a collective body must be executed under the control of a single responsible person. Thus the life and economic activities of the cooperative are controlled by fundamental decisions made by a collective body, and by day-to-day decisions—based on these fundamental decisions—made by one person responsible for running the affairs of the cooperative.

The managing of the cooperative's day-to-day operations is entrusted to its president, who must honor the decisions of the members' plenary session and of the council of members of the directorate. For the outside world (e.g., other economic entities and authorities) the president is the only lawful representative of the cooperative.

To enact cooperative democracy and to ensure wide participation of members in managing the cooperative's affairs, several elected commissions have been established. The members can control the entire activities of the cooperative and management through the *Supervisory Commission*. The commission reports to the plenary session of members but must also present its findings regularly to management.

The *Disciplinary Commission* plays an important role in maintaining working discipline and order within the cooperative. Disciplinary resolutions are passed by the management, and by the plenary session of members. The *Welfare Commission* monitors the welfare and cultural requirements voiced by members and the situation of the needy. The commission prepares recommendations for using funds set aside for welfare and cultural expenditure.

The *Arbitration Commission* decides disputes between the cooperative and individual members concerning employment and compensation. The *Commission on Family Managed Private Farming* prepares recommendations to management concerning the feed requirements of privately raised livestock, and conducts day-to-day affairs connected with private farming activities of cooperative members.

Modes and means of directing agricultural production entities are basically regulated by centrally passed directives concerning the national economy as a whole. Before 1968, production tasks and the means of purchasing equipment and investment for fulfilling these tasks were determined by breaking down the central plan of the national economy and the plans of the ministries that headed the given sector of economy. This was known as direct control of economic entities. Since 1968 these economic entities have been controlled largely by indirect means and methods called regulatory directives. To some extent, direct control remains in force and exerts its regulatory function in different ways (Figure 13-2).

The cooperative's highest decision-making organ is the members' plenary session, which is also the highest forum for putting cooperative democracy into action. Only the plenary session is allowed to act on the following: amendments of bylaws; election and dismissal of management officers and executives; fixing of salaries for top-level executives; approval of plans and balance sheet statements; directives for distributing the income realized by the cooperative's economic activities, and for the system of distribution of personal income; involvement with regional or professional organizations; and decisions on mergers, separations, and liquidations.

Between plenary sessions, problems are dealt with by the Management Council. This council meets fortnightly or at shorter intervals and enforces the decisions passed during the plenary session. The council must be active in all aspects of the cooperative's activities, but it concentrates on production. The cooperative's president is responsible for organizing and preparing council meetings. As the cooperative's life has many diverse segments, a certain division of responsibilities and tasks among members of the council is regular practice.

The highest-level executive of the cooperative is the president, who is personally responsible and is its legal representative. He or she is chosen for a 5-year term by the plenary session from among the members of the cooperative by secret ballot. His situation, however, differs from that of the directors of state-owned companies. As noted, all fundamental decisions of the cooperative are made by the plenary session or by the Management Council, and the president is only responsible for executing these decisions. Professional ability, good personality, and diligence are all attributes the president should have to promote the success of cooperative activities. He must respect the peculiarities resulting from cooperative democracy and should consult his partners in management before making any important decisions.

Presidential tasks include: directing and administering the cooperative's entire production activity and maintaining financial records. The president deals daily with problems of production (e.g., questions concerning loans, contracts, development plans) and with the preparation of agendas for plenary sessions. He also represents the cooperative in meetings with state and other authorities. The president exerts control and direction of production only through the managers responsible for the sector in question.

Production operations are directed by the head agronomist, who is the most important individual next to the president. The head accountant is in charge of the cooperative's administration and bookkeeping and is responsible for financial matters and bank loans. Directly supervised by the president, he is not entitled to make financial and banking decisions, but all such decisions of the cooperative presuppose the consent of the head accountant.

Managers of the different production sectors play an important role in directing the cooperative's operations, which are organized in sectors. The UCK cooperative has six major sectors: field crop production, horticultural production, livestock production, family-managed private farming, machine operations, and building construction. The operations of these six sectors are coordinated by the head agronomist. Each production sector, however, is headed by managers who have a university or high school degree in the relevant professional field.[3]

Top-level managers of the cooperative (the president, head agronomist, and head accountant) are concerned with forthcoming developments; they also deal with the substance of such business. The practical execution of work is entrusted to sector managers (managers of operation), who are in charge of materials, equipment, and work force as required for the given work task, and for the efficient use of such resources. Thus the essentials of control and execution are provided, and the spheres for independent decision making are ensured.[4]

Decision making is assisted by a variety of measures, including at least two yearly surveys of financial and production data, detailed working schedules for the various working peaks, extensive interaction between sector managers and top-level management, evaluations every 3 months of fulfillment of the sector's annual plan, evaluations of the effects of external and internal information and other factors, and the use of a bonus system to raise the efficiency of both workers and executives.

The cooperative's economic activity is analyzed twice a year. The resulting data are compared with planned figures on income, expenditures, profits, investment assets and liabilities, investments, and yields. Such surveys are helpful in deciding on corrective action if required to solve any financial or professional problems that may arise.

The working schedules for peak working periods (campaigns) deal with operational decisions extracted from the annual plan for the relevant period. This plan details the methods for executing said decisions and lists resources available for solving the jobs. The cooperative's annual operation includes three peak periods when substantially more decisions must be made.

Sector managers make regular reports to the council of managers, when the progress and problems of their sectors are fully discussed and evaluated. A weekly top-level meeting lasting about 2 hours includes the following officials: president, head agronomist, sector managers, legal adviser, deputy president, head veterinarian, surveyor of internal affairs, head accountant, head of economic department, secretary of the cooperative's party organization, and president of the supervisory commission. Within each sector there is a weekly meeting for informing and instructing the professional staff about that sector.

At 3-month intervals the activities of each sector are discussed using reports prepared by sector and subsector managers. These reports must cover actual and anticipated figures of costs, income, profits, and yields in order to show the deviations, if any, from planned figures. The economic department prepares these figures in advance for distribution to participants in the meeting.

For many years the bulk of the data concerning the economic activities of the cooperative has been processed by computer, in an effort to use up-to-date methods of analyzing and planning. There is a computerized subsystem for the management of stocks.

Stability of the System

The activities and success of an economic entity are influenced by all environmental factors that may be related to the entity in question. In shaping and conducting the economic activities of the cooperative, the most diversified effects and impulses come in the form of requirements and conditions created by the national economy as a whole. What the national economy expects from economic entities is conveyed to them by so-called regulatory measures. These regulatory measures or economic regulations are based on the country's economic policy as expressed in the national economic plan. The most important elements of economic regulations are price, market, subsidy, loan, and taxation, in combination with controlling profit and company and personal income. A change in any of these elements has immediate effects on the economic entity. Prices also have restricting effects, as those for manufactured products have risen considerably in recent years. Sectors using large amounts of industrial products are now at great disadvantage. Low prices for livestock produce also have unfavorable effects, as the cooperative is forced to prepare as many feeds as possible by itself. Although many channels exist for selling the products of the cooperative, substantial amounts can usually be sold only to the state-owned distributor and processing companies.

Stability of the system depends on the use of resources of production, and on honoring the costs of production either in form of produce prices or in the form of subsidies. Credits and loans play an increasing role, as does vertical and horizontal integration among the cooperatives and state farms. The Kondoros cooperative has placed capital investment into several joint ventures aimed at better utilization of assets and equipment and at raising additional income. An important factor in stability is the close link between private and large-scale farming, which adds more incentive for work.

Since the mid-1960s relations between state economic control and the cooperative have become increasingly indirect. State subsidy has been gradually decreased, and the market mechanism controls both production and sales. There are no more state-dictated production figures; therefore, the stability of the system is increasingly dependent on the soundness of independent management, on finding the optimum combination of factors of production, and on deciding weekly the rate of growth within the cooperative itself.

System of Distribution

The cooperative's products are sold by contract agreements, which are indispensable because the large amounts cannot be sold on the free market with certainty, neither to private retailers nor to members of the cooperative. Contract agreements usually contain the name of the contracting parties, the subject of the sale, a description of the quality of the produce as determined by relevant standard specifications, the terms of delivery, the price and the conditions of payment, and sanctions.

On the basis of available information, the cooperative is free to decide what and how much to grow or raise, with whom to make contract agreements, and what channel to choose as outlet for its products. The major distributors, processors, and retailers are located at an average distance of 25 km from the cooperative.[5]

In 1984, some 93% of all field products were sold by the cooperative. A mere 8% remained in stock as feed or seed, and a negligible portion had to be written off as loss. From the total sales, some 96% went to state-owned distributors and processors and to other agricultural producers, while 4% went to members or other private entities.

On a yearly average, some 89% of all horticultural products are sold, and the remaining 11% are kept to provide winter work for members. During the winter months onions and carrots are cleaned and then sold to the canning factory or the cold storage company. From the sale of horticultural products, 97% went to state-owned distributors and processors, 2% to other large-scale farming entities, and 1% to members or the retail trade.

The livestock sector of the cooperative produces some 4,272 t of beef, finished hogs, horses to be slaughtered, and broilers. Of these, some 96% have been sold to state-owned entities, while 4% went to members or have been processed by the cooperative. Of the 52,220 hectoliters of milk, 96% was sold to the Joint Milk Processing Company "Sárrét," 23% was sold to members in the cooperative's own shop, and the rest was fed to livestock.

The cooperative has no direct links with foreign companies. Imported stock and equipment are purchased via foreign trade companies. The same applies to products to be exported.

Social Contributions

The cooperative plays a decisive role within the village community of Kondoros, where slightly more than two-thirds of the village's population are members of the cooperative. The cooperative and the Village Community Council concluded an Agreement of Cooperation that lists community development projects (telephone exchange, gas pipeline, paved lanes for bicyles, school construction) that are financially supported in part by the cooperative. The cooperative's support also includes equipment works, provision of materials, and manual labor. Such support is organized according to actual requirements.

In addition to investment projects of the Community Council, which serve the entire village population, the cooperative donates some 500,000 forints (U.S.$10,450) per year to the Community Council. Coordination funds to support selected institutions and organizations of the community (e.g., the local branch of the Red Cross, the sports club, the cultural center, the local youth organization). The local fire-fighting unit is also financed by this fund.

To improve commodity supplies for members of the cooperative and other local customers, the cooperative runs a retail shop for meat and one for milk. A food canteen serves warm food to subscribing members for 10 months a year. Meals from this canteen are also transported to production sites for consumption. The price of meals is fixed to approximately equal the costs incurred in preparing the meals. Both active and retired members may subscribe for meals, and at present 500–600 persons take their lunch from the canteen.

Within the cooperative there is a subunit of the Red Cross, which is materially supported by the cooperative. Its tasks include family and youth care, protection of the natural environment, organization of blood donations, care of the elderly, and other welfare and emergency work.

The Club of Retired Members and Citizens, fully operated by the cooperative, supports the elderly. The elderly and retired men and women living in Kondoros are well cared for. They are kept informed about the everyday life of the cooperative, and occasionally inspect the fields and livestock facilities. The club has a television, radios, playing cards, and other game facilities. Films are shown at regular intervals, and there are publications on health matters and on popular science. Trips to health spas are organized for the elderly, along with sightseeing trips to other parts of the country. For elderly members with a low retirement income, the cooperative assumes some or all of the costs of machine works performed on the elderly member's allotment of privately usable land.

The cooperative supports the sporting activities of members by a donation of 20,000–30,000 forints (U.S.$418–$622). In addition, athletes may use the cooperative's bus for trips to engagements free of charge.

The cooperative has formed a Circle for Supporting the Local Museum, which collects objects illustrating the historical way of life of Hungarian peasants. These objects are on display in a well-preserved, traditionally styled house. The cooperative runs another circle for handicrafts where participating women preserve the traditional folk motifs of embroidery. The cooperative helps members of these circles visit regional and ethnographical museums by providing its bus free of charge and providing other material facilities.

The cultural life of the cooperative's members is the responsibility of the Cultural Commission, which has an annual allotment of some 1.6 million forints (U.S.$33,440). The Welfare Commission monitors the welfare of individual members. An annual welfare fund is established to pay for such items as assistance in case of death, marriage, childbirth, drafting into the army, and emergencies.

Summary

The United Cooperative of Kondoros is an example of the type of farming system that dominates contemporary Hungarian agriculture. Cooperatives have become highly neotechnic, managerially complex, and market-oriented. Both plants and animals are intensively produced in several combinations of cooperative and private operations. Moreover, cooperatives have emerged as socioeconomic units, the services of which transcend agriculture per se and deal with the welfare of the cooperative's members.

Notes

Editors' note: For a recent review of the region, see: T. Bernát, (Ed.), *An Economic Geography of Hungary* (Budapest: Akadémiai Kiadó, 1985).

1. Even if most of the land allotted to private use by cooperative members is occupied by maize, this cultivar could not be considered as privately managed because all maize production is fully mechanized. Maize received for livestock kept as property around houses, on the other hand, is genuine private farming.

2. *Normal hectare* is the term for measuring equipment performance. One unit of normal hectare equals the energy required for ploughing 1 ha of medium heavy soil at a depth of 18 cm.

3. The cooperative has 41 members who have a university or high school degree and 170 members who have completed secondary professional schools. A total of 570 skilled workers constitutes the bulk of the work force.

4. For example, one production sector has full availability of specialized tractors and implements, trucks and auxiliary equipment, and sufficient work force. Top-level management will set the amount of net income to be realized by the sector during the year's activities, but the activities must be planned by the sector itself.

5. They are: the Canning Factory of Békéscsaba, the Cold Storage Company of Békéscsaba, the Poultry Processing Company of Békéscsaba, the Meat Packing Combinate of Gyula, the Joint Milk Processing Company "Sárrét" of Szeghalom, the Orosháza Seed Growing and Distributing Company, the State Farm of Hidashát, the Feed Mixing and Processing Joint Company of Kondoros, the Seed Growing Association of Mezökovácsháza, the Sugar Refinery of Sarkad, the Agro-Industrial Association of Békéscsaba, and the Cereals Research Institute of Kiszombor.

14

High-Tech Farming Systems in Champagne, France: Change in Response to Agribusiness and International Controls

GÉRARD DOREL

The chalky plains of Champagne, also called dry Champagne, correspond to a wide crescent of "white lands" situated east of the Paris Basin. This plain is 200 km long but never more than about 50 km wide (Figure 14-1). It forms a step between Lorraine and the plateau of the Paris region (the Brie). These plains have experienced dramatic changes, which have turned the region—once so poor as to be nicknamed *pouilleuse* (literally, mean and wretched)—into one of the leading French and European farming regions.

After World War II this plain was a long, forested belt sprinkled with numerous cultivated clearings. The valleys, slightly embanked and covered with alluvium, were cultivated by smallholders practicing mixed farming. The plateau was largely covered with black pines, the forest having reclaimed thousands of farm plots and pastures previously situated on the outskirts of the villages. The owners began to plant them with pine trees, renting them as hunting grounds for lack of a better use (Dion, 1961). About 200,000 ha of woods existed at the beginning of the 1950s, while thousands of hectares lay in fallow, numberless *savarts* (land units) without any economic interest.

The post-World War II period was marked by a spectacular wave of clearings and land reclamation, an unprecedented event in France since the great clearings of the end of the Middle Ages. The first great reclamation began in 1947 in North Champagne and gradually spread over the whole plain by the 1970s (Figure 14-2). Within 15 years, from 1955 to 1970, the farming area in the chalky plains of Champagne increased by 135,000 ha, an increase of more than 20%. In some villages, 30–50% increases were registered.

The reclaiming operations were launched, at least at first, by farmers coming from other areas, as the local farmers lacked not only the necessary funds but also the inclination to get involved in what they considered a foolish venture. Initial successes convinced them otherwise, however, and they moved into the reclamation in great numbers (Dorel & Dumesnil, 1984).

Gérard Dorel. Département de Géographie, Université de Paris XII-Val de la Marne, Créteil, France.

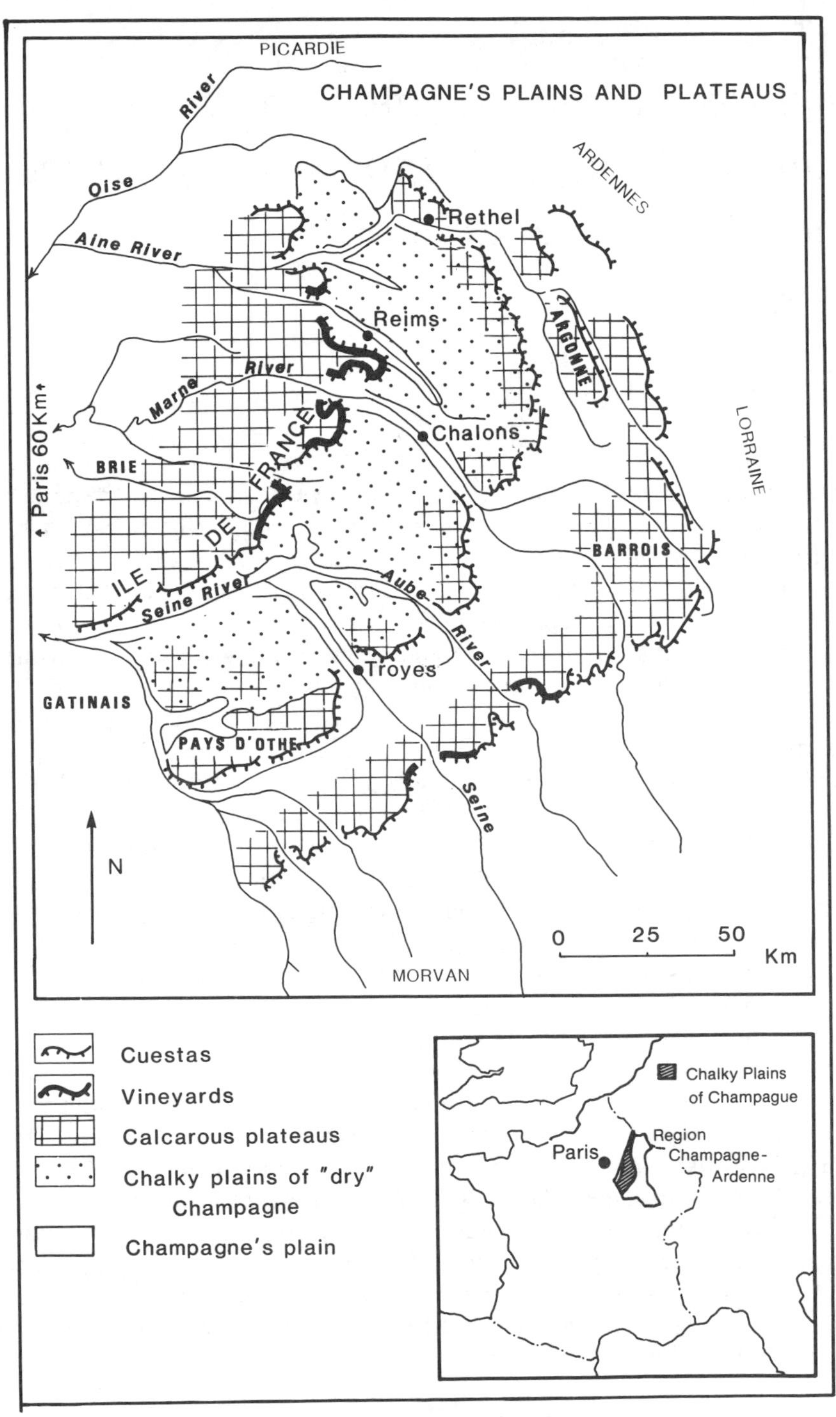

FIG. 14-1. Champagne's plains and plateaus.

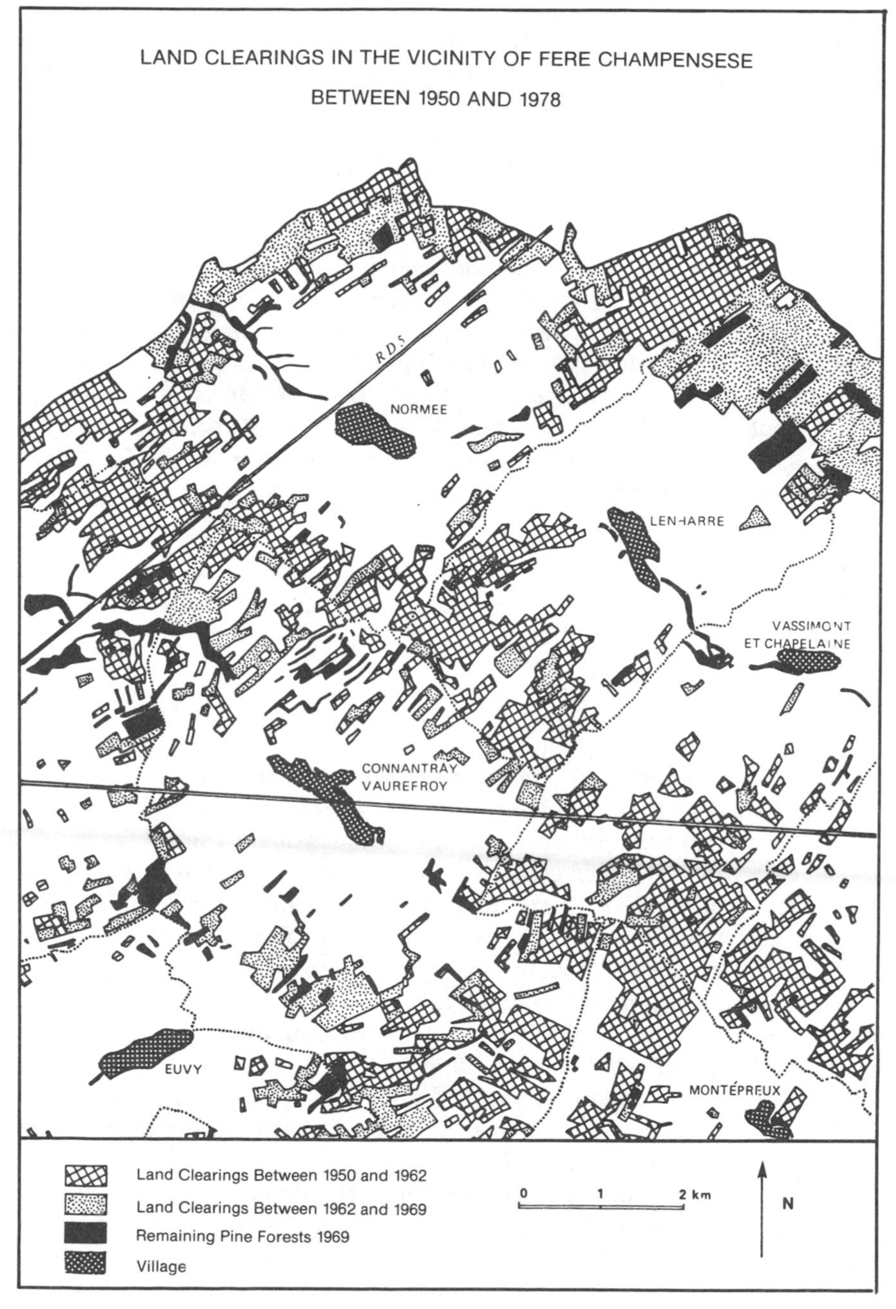

Source: Radet (1977). Reprinted by permission.

FIG. 14-2. Land clearing in the vicinity of Fere, Champagne, between 1950 and 1976.

The transformation was rapid—only 20 years—explaining the uniform, regional farm patterns reminiscent of the Great Plains of the United States (the general flatness, the small number of trees, the large cultivated fields, the scattered homesteads). Large-scale farms prevail in the Champagne area. A little over 2,000 farms, or 20% of the total number of agricultural units, are more than 100 ha in size and make up more than half of the 680,000 ha of the chalky area.

Champagne exemplifies agricultural intensification involving the integration of advanced technology and considerable contributions of farm capital. Yet in 30 years these farms have shifted from an essentially quasi-extensive cereal system to an intensive, mixed farming system in which sugar beets (*Beta vulgaris*) and alfalfa (*Medicago sativa*) were first introduced as wheat "go-byers" (supplements), but now compete with the wheat (*Triticum* spp.) for the best Champagne lands. Numerous processing plants raise the value of these farm commodities to a maximum.

An Intensive Rainfed System

The Farming Landscape and Climate

The reclamation of Champagne has created an original landscape in which the monotony of the expanses that are fully bare and apparently deserted by humans does not diminish its loveliness. A few country lanes mark the boundaries of long plots, which sometimes cover several tens of hectares, unusual in Europe. They form beautiful open fields, which take on the general feature of a mosaic, each patch of which consists of striped plots from 100 to 200 m wide and sometimes more than 1 km long (Figure 14-3). Buildings are rare. Traditionally, villages were hidden at the bottom of the small valleys, either boxed in at the bottom of a *noue* (marshy meadow) or stretched along a road. The valleys shelter modern farm buildings with big corrugated-roofed sheds, elevators, and repair shops. The farmers live nearby, usually in comfortable new houses similar to the middle-class bungalows that surround French towns. There also exist isolated farm homes in the middle of fields. It is not uncommon to find a cattle feed lot or even an alfalfa dehydrating plant, detectable by its heavy wreath of vapor among the fields.

This region is one of the driest in France, receiving no more than 600 mm of rain a year (Table 14-1). This situation is deceiving, however, because precipitation is spread over 140–160 days each year, with only slight annual variability. Also, the chalky soils have a good water retention capacity. Therefore, Champagne agriculture does not lack for water except for some crops like potatoes or field vegetables, which demand complementary watering in the summer. Indeed, one of the most severe droughts registered in Champagne, that of 1976,

FIG. 14-3. Southern Champagne landscape (near Arcis sur Aube). Notice the remnants of once vast (1950s) pine forests in the right foreground (G. Dorel).

TABLE 14-1. Average monthly precipitation and temperature in Champagne (Reims), 1931–1960.

Month	Precipitation (mm)	Temperature (°C)	Number of freeze days
January	49	2.0	16
February	44	2.9	14
March	36	6.5	12
April	41	9.6	4
May	52	13.3	1
June	51	16.4	0
July	57	18.3	0
August	63	18.0	0
September	58	15.3	trace
October	47	10.5	3
November	48	6.2	7
December	52	3.1	13
Total or average	598	10.2	70

did not appreciably affect the production of deep-rooted crops because of the moisture stored in the soils (Dumesnil, 1978).

The cropping season depends on spring temperatures, which are relatively unstable. It can start as early as the beginning of April but can also be delayed as late as mid-May. In that case cultivation is largely handicapped, especially if the rainfall has been insufficient. The cropping season ends in October.

Farm Patterns

Of the 680,000 ha of Champagne's chalky plains, 662,000 ha (97.4%) are cultivated. These lands are dominated by large farms. The average size of each farm is more than twice the national average (60 ha versus 25 ha), a figure that is lowered because nearly 4,000 "farms" in Champagne have less than 20 ha. Most of these are retirement farms, but some include the cultivation of slopes situated on the outskirts of the area of the famous "champagne" vineyards. In any case, of the 11,000 farms registered in the 1979–1980 agricultural census, 2,000 control more than half of the cultivated lands (Table 14-2).

This pattern of landholdings, to which the great wave of clearings in the 1950s and 1960s gave birth, has changed little since the 1970s. Some large farms have been divided for fiscal or inheritance reasons. In practice, however, fathers and sons work together, sometimes controlling several big production units. These large holdings are cultivated primarily in the highly profitable crops of wheat, sugar beets, potatoes (*Solanum tuberosum*), alfalfa, and some vegetables.

The Production System

Farmlands correspond to the distribution of rendzina (mollisol), a thin calcareous soil that is low in oligo elements (magnesia and boron). Nevertheless, the soils respond well to fertilizers, especially liquid nitrogen, and are very productive. A serious problem is the deficiency of organic matter, which must be compensated for by manures and the plowing in of wheat and barley straw. Even where the soils appear dry, the limestone subsoil retains moisture, acting like a sponge into which the roots of wheat and alfalfa plunge deeply.

Despite their good quality for cultivation, soils are fragile and require protective care, notably the addition of fertilizers for the reconstitution of their productive capacity. Inputs reflect this; the budget for "fertilizers" alone absorbs 18% of the total expense of a typical farm. The Département de la Marne (administrative subdivision), in the middle of the plains of Champagne, leads all French departments in average consumption of fertilizers per hectare: 317 kg NPK (nitrogen, phosphorus, potassium) for Champagne versus 194 kg on average. Finally, it should be noted that the soils were well rested until the recent outburst of cultivation. Some questions exist concerning the long-term productivity of the soils or their degradation under current cropping practices.

TABLE 14-2. Sizes of farms in Champagne, 1979–1980.

Farm size	Number	Percentage of total	Land cultivated (ha)	Percentage of total
<20 ha	5,745	50.7	13,500	2.0
20–50 ha	260	2.3	73,700	11.1
50–100 ha	3,218	28.5	232,100	35.0
100–200 ha	1,647	14.6	217,100	32.8
200–300 ha	299	2.7	71,000	10.7
>300 ha	132	1.2	54,700	8.4
Total	11,301	100.0	662,100	100.0

As in the case of large-scale farming in the Paris Basin, crop rotations are comparatively simple and are dominated by a single issue—the choice of the first crop in the rotation (Table 14-3). Wheat remains the principal cultivar in Champagne, but it could not profitably and steadily hold this position without a proper "heading crop" (first cultivar in the rotation cycle). The selection of the heading crop depends on many endogeneous and exogenous factors (Table 14-4).

Of the different heading crops in the region, sugar beets, alfalfa, potatoes, peas (*Pisum* spp.) and rape (*Brassica oleracea*), the former two are the most favored, but their production depends on a tightly structured agribusiness with numerous sugar plants, dehydrating mills, and, for sugar beets, a strictly enforced production quota. Cereals dominate the farming system, but "with a subtle skillfulness in the use of rotation crops whose result is that it is impossible to say which is the starter one: it is really a genuine system the element of which supports and pluralistically enrich each other" (Brunet, 1981). This is true especially on the large farms, where the significance of heading crops has increased for about 20 years (Table 14-5). In addition to this change in the cultivation system, dramatic advances had been registered in production (Table 14-6).

The data indicate that the agricultural system of Champagne has moved from a semiextensive to an intensive one, including (1) less emphasis on grains in general (59% in 1980 versus 64% in 1960) but an increase in wheat as a replacement for minor grains, and (2) an increase in heading crops that not only are highly profitable but also allow farmers to prepare their fields properly for the following wheat crop. It must also be noted that the amount of land on large farms devoted to beets has doubled, from 13 to 23%, a percentage that is much higher than the one registered for the whole of Champagne (15.3%). Alfalfa has kept its position (about 10%) among the heading crops, but peas have increased theirs. Fallowing has virtually disappeared, as evidence of the intense frequency of cultivation.

This change in the agricultural system testifies to the aggressive capitalist

TABLE 14-3. Regular field rotations and schedules in Champagne.

Year	Soil preparation	Seeding	Fertilizing and other	Harvest
First (wheat)	August–September	Mid-October	April and May: nitrogen	Late July–early August
Second (barley)	September	Early October	March: nitrogen: weeding April–May: fungicides	Late July
Third (sugar beet)	November (plowing) and April	Mid-April	February, March, and April: nitrogen April, May, and June: weeding	Early November
Fourth (wheat)	November	November	April: weeding May and June: fungicide	Late July
Fifth (peas)	November (plowing) and March	March	March and May: weeding May and June: fungicides June (2 ×): insecticides	Late July
Sixth (wheat)	August or September	Mid-October	February and April: nitrogen March: weeding May and June: fungicides July: insecticides	

TABLE 14-4. Main crops in Champagne's plains.

Crop	Hectares	Percentage of total cultivated
Grains (wheat, barley)	402,423	60.7
Potatoes	7,570	1.1
Sugar beets	101,614	15.3
Alfalfa	66,545	9.7
Rape	9,036	1.4
Vegetables (dry and fresh)	14,305	2.2
Others	61,007	9.6
Total	662,500	100.0

spirit of the big farmers who took the risks of venturing into beet production. These growers planted beets for some years when prices were not particularly high. The same spirit of enterprise can be found in those—often the same people—who ventured into building alfalfa dehydrating plants. These farmers were able to foresee that demand and a European open-door policy would eventually make these ventures profitable.

The figures for an average farm in 1982 in the Fère Champenoise District, in the southeastern part of the Département de la Marne, demonstrate the amount of these inputs (Table 14-7). This farm has 280 ha cultivated, no

TABLE 14-5. Trends in crop selection by large-scale farms in the plains of Champagne, 1960–1980.

	1960	1980
	(Percentage of all crops)	
Grains	64.0	59.0
Wheat	29.0	40.0
Barley	24.0	15.0
Oats	7.0	1.0
Rye	4.0	–
Maize (corn)	–	3.0
First rotation crops	32.5	41.0
Sugar beets	12.0	23.0
Alfalfa	10.0	10.0
Other forrages	5.5	1.0
Peas	–	3.0
Others	5.0	4.0
(Fallow land)	3.5	–

TABLE 14-6. Increase in yields for some crops in Champagne, 1961–1980.

Production	Average, 1961–1965[a]	Average, 1976–1980	Mean annual growth
Wheat	38.7 qx/ha	59.0 qx/ha	+3%
Winter barley	33.5 qx/ha	57.0 qx/ha	+3.8%
Sugar beets	38.2 t/ha	52.0 t/ha	+2%

[a]qx/ha = quintals/ha (one quintal = 100 kg); t/ha = metric tons/ha

livestock, and two hired workers. The main commercial crops are grains (58% of the land harvested), sugar beets (23%), alfalfa (7%), starch potatoes (4%), and several other crops such as rape, peas, or potatoes for food.

The data indicate the following:

1. The system relies on industrial inputs, especially those of petroleum products, fertilizers, pesticides, fungicides, and gasoline. These ex-

TABLE 14-7. Farm expenses for large-scale wheat and sugar beets in the southern plain of Champagne, 1982.

Expenses	F.F./ha	Percentage of total farm expenses
Fertilizers	1,150[a]	17.9
Agricultural chemicals	760	11.8
Seeds	550	8.6
Gas, diesel oil, greases	270	4.2
Miscellaneous	150	2.3
Subtotal—industrial purchases	2,880	44.8
Hired farm labor	730	11.4
Farmer's family Social Security	200	3.1
Subtotal—manpower	930	14.5
Rent	500	7.8
Machine repair	330	5.1
Machine hire and custom work	150	2.3
Insurances	170	2.6
Loan interest[b]	410	6.4
Depreciation	600	9.4
All Others	447	7.1
Subtotal—farm-related expenses	2,607	40.7
Total farm production expenses	6,417	100

[a]Figures in French francs (F.F.) per hectare (U.S.$1.00 = 8 F.F. in 1982).

[b]Cost of interest for machine and equipment loans. In 1982, the farm debt capital could be estimated at 2,800 F.F./ha.

penses have increased since the oil crisis of the 1970s. The price of fertilizers alone has increased fourfold.

2. Labor is cheap and has decreased in total inputs with the disappearance of seasonal labor, which was formerly indispensable for the thinning and hoeing of the beets. Some 15 years ago, the cost of labor meant expenses as high as those imposed by the purchase of fertilizers. Now, large-scale family farms employ about one man working full time for each 75 ha cultivated, including time necessary for management.
3. Special seeds, supplied by local dealers, represent a new selected input. These seeds are important to the high yields that have been registered for 10–15 years (7.6 t/ha for wheat or 112 bu/acre; 58.7 t/ha for sugar beets; 12 t/ha for alfalfa in four cuts).

The assets of the farms in Champagne are large, an estimated average of some 37,000 French francs (F.F.) per hectare or U.S.$4,600, in the spring of 1983 (Table 14-8). A large farm tilling 200 ha, of which only about half the land is owned by the farmer (an average for the region), has a total value of close to 5,000,000 F.F. or U.S.$625,000.

A considerable part of the working capital is commonly borrowed from the Crédit Agricole, the powerful mutualist bank of French farmers, which also finances the purchase of land and machinery (Table 14-9). The regional branch of the Crédit Agricole of Reims (capital of the region) has become increasingly involved with the short-term loans of the farmers in the northern part of Champagne. Hence Champagne agriculture depends more and more on outside financing sources for its current business.

Adaptation to the European Economic Community (EEC)

This farming system of the plains of Champagne, born in the spirit of enterprise, is no longer a speculative venture. As in the other grain belts of the Paris Basin, the farmers choose, as soon as they can, financially secure crops with a guaranteed price. Grains and sugar beets are examples. Indeed, farming in the plains of Champagne developed in size and economy in conjunction with the organization of the Agricultural Common Market in the early 1960s and the creation of the European Fund for Agricultural Orientation and Guaranteed Prices (Fonds Européen d'Orientation et de Garantie Agricole), and with the setting of European regulations concerning cereals (grains) and sugar (Dorel, Gauthier, & Reynaud, 1980). The opening of a continent-sized market and the high level of guaranteed prices for grains and beets promoted the boom for this farming system and region. Farmers belong to unions or leagues whose representatives lobby the French government and the EEC commissions for price and quota controls.Thus Champagne agriculture is the daughter of the agricultural common market.

TABLE 14-8. Typical farm assets in Champagne, 1982 (150- to 200-ha farms).

Asset	F.F./ha[a]	U.S.$/ha
Machinery and equipment (estimated market value)	5,300	662
Livestock (cattle, sheep, and lambs)	700	88
Cash and deposits	5,000	625
Investments in cooperative	1.032	129
Farmland	25,000	3,125
Total	37,032	4,629

[a]U.S.$1.00 = 8 F.F.

The Champagne region plays a remarkable part in France's agricultural production. From less than 4% of the harvested lands in France, these chalky plains produce more than 20% of the nation's beet production, about 30% of the starch potatoes, 10% of the feed peas, 60% of the dehydrated alfalfa (French production representing 75% of that of the ten members of the EEC) and 8–10% of the grain (wheat and barley). These performances are largely due to the efficiency of this capital- and technologically intensive agriculture, which is well managed and supported, and has been able to adapt itself better than other systems to the productive mold imposed by the EEC's farm prices system.

The average gross farm prices, based on harvest values, range from 8,000 to 10,000 F.F./ha (U.S.$1,000–1,250/ha). The higher figures belong to the farms with high sugar beet quotas. The average yields and profits (in 1982) are critical to cropping strategies (Table 14-10). The most profitable crop is sugar

TABLE 14-9. Credit availability, regional agricultural bank (Crédit Agricole de Champagne).

	Percentage of total credit granted		
	1979	1980	1981
Crop loans	16.2	18.8	25.9
Equipment loans	40.6	42.2	42.3
Real estate loans	34.2	32.5	27.4
Emergency loans	8.4	6.0	3.7
Miscellaneous	0.6	0.5	0.7

TABLE 14-10. Yields and market value of the main crops in Champagne, 1982.

	Yields[b]		Market Value[a]	
			F.F./ha	U.S.$/ha
First rotating crops				
Sugar beets	58.4	t/ha	11,878	1,485
Faeculia potatoes	33.8	t/ha	11,931	1,491
Peas	47.2	qx/ha	8,590	1,074
Alfalfa	11.4	t/ha		
Grains				
Wheat	72.3	qx/ha	7,620	953
Barley	70.6	qx/ha	6,608	826
Maize (corn)	70.0	qx/ha	7,266	908

[a]U.S.$1.00 = 8 F.F.

[b]qx/ha = quintals/ha (1 quintal = 100 kg); t/ha = metric tons per hectare.

beets, inasmuch as it has a quota A, paid at the highest price. Grains are the second most profitable crop. Potatoes for starch are grown by contract with local manufacturers. Alfalfa is dehydrated in cooperative shops, and peas are sold to dealers or cooperatives for use as cattle feed.

The average net income per hectare for large farms (280 ha) registered in 1982 was a little higher than 2,000 F.F. (U.S.$250), a figure that conceals remarkable differences between crops and strongly underlines the prosperity of farmers in the Champagne region (Table 14-11). This prosperity is largely

TABLE 14-11. Production expenses and net income for the main crops in Champagne, 1982.

	Wheat		Sugar beet	
	F.F./ha	U.S.$/ha	F.F./ha	U.S.$/ha
Direct production expenses	1,663	208	3,383	423
Seeds	382	48	598	75
Fertilizers	818	102	1,467	183
Agricultural chemicals	463	58	1,318	165
Other expenses and depreciation	4,000	500	4,000	500
Market value	7,620	953	11,878	1,485
Net income before taxes	1,957	245	4,495	562

[a]U.S.$1.00 = 8 F.F.

artificial because it is based entirely on European price levels that are much higher than that of the world market (Table 14-12).

Given that grain and sugar beets represent 50% and 21%, respectively, of total farm gross income, the prosperity of this agricultural system is largely based on a quasi "pension" provided by the EEC. This support, which is dramatic and decisive for grains and sugar, is also considerable for other crops. Alfalfa benefits from direct subsidies granted to its production and to the other commodities. The direct aid for the production of dehydrated forages represents 15% of the price received by the local alfalfa producers.

The goal of the EEC has been to create a common agricultural policy that would allow a double demand to be met: to feed the populations of the members of the Community and thereby reduce importations, and to secure a decent standard of living for the farm population, which was large at the turn of the 1960s. A common policy emerged between 1958 and 1966, largely as a result of the effectiveness of the negotiators the French sent to the Community. They managed to impose rules and prices consistent with the views of the powerful and influential grain and sugar beet growers.

The EEC system for agricultural products is based on three principles: (1) a unique market that supposes a free circulation of farm products, the prices of which are established by a common consent between the members; (2) the general preference given to EEC products, which means that an EEC nation must buy products from an EEC producer who can offer them; and (3) a financial solidarity that compels each member to support the cost of that agricultural policy.

The adoption of these principles induced the Community to establish rules

TABLE 14-12. World prices and European prices.

Agricultural commodities	European prices[a]	World prices	Disparity in F.F.	Disparity in percentage
Wheat (mt)	1,005 F.F. U.S.$126	799F.F. U.S.$100	206 U.S.$26	20.5
Sugar (mt)				
Quota A[b]	3,401 F.F. U.S.$425		1,666 U.S.$208	52.0
Quota B	2,360 F.F. U.S.$295	1,735 F.F. U.S.$217	625 U.S.$78	26.5
Quota C	1,735 F.F. U.S.$217		0	0

[a]U.S.$1.00 = 8 F.F.

[b]In France, quota A is equal to 63% of total production, quota B to 17%, and quota C to 20% (average figures).

and a common price for most of the main European agricultural products. The system is not the same for all products. A total guarantee is granted to grain growers, whatever the quantities brought onto the market. Public authorities are obliged to buy from the producers at a fixed price. Large grain farmers in Champagne can produce as much as they like and be ensured a price that is profitable, because the price is set so as to allow smallholders to survive.

The Community has an excess of wheat and must finance exports representing about 17% of total European agricultural production. The same is true for barley, especially the varieties used for livestock feed. Sugar beets are set on a strict quota system. The Community fixes an annual volume (9.5 million t in 1982) sufficient to satisfy Europe's needs. This is the basic quota A, which is highly profitable. Quota B is added to A, but at a slightly lower price. The remaining sugar beet production does not benefit from guaranteed price.

The quota is shared between the European countries; each country distributes its share among the sugar manufacturing plants, which in turn distribute their shares among the growers. This system obviously favors those farmers who are able to bind themselves by contract to provide the sugar mills regularly with the quantities of beets they need, using quota C if the world market justifies it.

The Sugar Beet System

Beet growing is a tradition in the plains of Champagne, as it is everywhere on the plateau and plains of the northern and eastern part of the Paris Basin. Sugar beets were introduced into France during the First Napoleonic Empire, at a time when France was cut off from its West Indies colonies. The best lands in Champagne were reserved for beets, especially the heavy and deep soils of the Rheims Basin and Rethel Basin in the northern part of the province. It was there, in the 19th century, that the first sugar mills of the region were built (in Attigny and St. Germainmont). Just after World War II, two new plants were settled in the vicinity of the city of Rheims (Bazancourt and Sillery). But the real boom of beet growing goes back to the pioneering years of the chalky plains in the 1960s, when those who cleared the land did not hesitate to use beets as the heading crop, despite the absence of profitable quotas.

The development of beet growing seems to have been an agronomic necessity, caused by the need for a heading crop that would ensure good wheat yields the following year (Montanbaux, 1971). Beyond this, the growers were well aware that the development of the national market, and soon the European one, would open outlets that would impose a better remuneration for this crop. As beet growing has spread, the number of sugar plants has multiplied. Since 1961 three new mills have been added in the Champagne area, bringing the total to four (see Radet, 1977).

The farmers gambled that the mills would ultimately be granted sugar quotas, creating prices that were more profitable than those offered by the world market. The gamble was rewarded. Since 1983, 80% of the beets delivered to the mills have been paid at the highest European price. This link between the farming system and industry has resulted in rapid growth of beet farming. In 1950, 28,000 ha were devoted to beets; in 1966, 67,000 ha; in 1975, 130,000 ha; and in 1981, a record 152,740 ha. Since 1981, the market was dropped because world production was in excess of 100 million tons, while consumption was only around 90 million tons. The result has been a decrease (16%) in hectares in beets in the Champagne area.

The cultivation of sugar beets follows a rather standard mode in Champagne. In April the land is plowed, followed by sowing in mid-May. Insecticides are applied at the time of sowing. June witnesses weeding and more insecticides. Fungicides are applied at the end of August. Harvesting is from October to the beginning of December.

The sugar beet is no longer a pioneering and speculative crop in Champagne but, rather, a no-risk production. But production expenses are high, particularly monogerm seeds, treatments, fertilizers, and harvesting machinery. And since the market for sugar beets is currently saturated, local farmers are looking for other heading crops until world market prices justify a return to the beet.[1]

One alternative is peas for feed mills, a development in Champagne for the past 2-years. This crop has the enormous advantage of requiring the same sowing and harvesting machinery as wheat. The market is presently good, especially for cattle feeding but also for human consumption (e.g., mashed peas, a common diet in Southeast Asia). Others grow poppy (*Meconopsis cambrica*) for the pharmaceutic industry, an alternative that is severely regulated and supervised. Another possibility is open-field vegetable cultivation, such as carrots or onions, but the region does not yet have any dehydrating plant to transform these products, which must be trucked to neighboring regions (e.g, southern Picardy).

If the demand for beets does not return to higher levels than currently exist, then the agricultural system could change in one of the following ways: the production, year after year, of grain, which is now possible because of the intensive use of fungicides; or a return to alfalfa, the production of which is supported by the European Community and the markets for which are good at the moment.

The Alfalfa System

Europe produces nearly a million tons of dehydrated alfalfa. France produces 768,000 t, or 76% of the European figure, and Champagne provides 81% of French production. Alfalfa is the main source of vegetal proteins after rape, a plant that is also cultivated extensively in these plains. Rape provides 236,400 t of protein, and dehydrated alfalfa accounts for 119,700 t in France.[2]

Alfalfa is a Flemish introduction, perfectly suitable for the physical conditions found in Champagne. It makes a remarkable heading crop, the yields of wheat after 2–3 years of alfalfa are 8–9 quintals/ha higher than when the heading crop is another grain.[3]

The number of alfalfa fields has increased enormously in Champagne, especially in the south of the Département de la Marne and in the recently cleared areas. Alfalfa production has risen from 45,000 t in 1967 to 65,000 in 1960, and reached 150,000 t in 1964, 320,000 t in 1967, and 530,000 t in 1975.

This dramatic increase is related to the farmer's development of dehydrating plants capable of transforming this forage production into dry produce that can be easily stocked and sold. The first plant was built near Chalons sur Marne in 1962. Producers organized rapidly, and within 10 years about 50 plants were opened in the region. An agroindustrial chain was created, including two commercial cooperatives that deal with sales.

Alfalfa fields are cut four times a year with special choppers belonging to dehydrating plants. Upon delivery to the shop, the lucerne enters a dehydrating process, a method that rapidly reduces the water held in the forage from 80% to 10%. The highly valuable finished product takes the form of small pellets rich in protein (+18%) and carotene (+125 mg/kg of dry matter).

The energy crisis of 1960s and 1970s seriously threatened alfalfa production because energy represents the largest part of the final cost of producing alfalfa pellets. From 1975 to 1982 the number of dehydration plants decreased from 50 to 39. The remainder were able to withstand the increased energy costs and to take over the processing of lost plants. Indeed, the region's plants are powerful and well organized and have better resisted the general decline in alfalfa production registered elsewhere in France and throughout Europe.

The powerful mills of Champagne are nearly an oligopoly, especially for exports. One of these mills, located at Pauvres in the northern part of the plain of Champagne (Département des Adrennes), produces nearly 10% of the total French production (Table 14-13). The main source of energy used in these alfalfa mills is coal, which has replaced fuel oil. New cultivation devices and harvesting methods have allowed important energy savings, with 33%–50% less energy consumed than in 1975.

Farmers in Champagne also benefit from considerable energy subsidies. These subsidies were granted by the French government (Agence Française pour la Maîtrise de l'Énergie). Alfalfa producers have also obtained the EEC's guaranteed high and profitable prices for dehydrated alfalfa. This production, not guaranteed until 1978, has thus entered the exclusive club of farm production under the protection of the Common Market.

Europe is badly in need of cattle feeds. European livestock feeders import huge amounts of cattle cakes and soybeans from the United States and Brazil, which are paid for in dollars, a currency whose exchange rate has been erratic since the late 1970s. The EEC decided in 1978 to grant alfalfa growers direct aid by fixing fairly high target prices and by granting subsidies for each ton produced, a lump sum deficiency payment that may be followed by a second

TABLE 14-13. Dehydrated alfalfa production in Champagne compared with total French production, 1975–1982.

	1975	1970	1982
Number of dehydration plants	50	44	39
Percentage in France	32	35	43
Dehydrated alfalfa production (t)	530,100	636,100	606,100
Percentage in France	64	70	81

payment if market prices are too far from the target prices fixed at the beginning of the year. This assistance, however, has not rescued the marginal growers of alfalfa, particularly those who could not or would not invest in cheaper energy operations.

The resourceful Champagne farmers have managed to secure markets for the nearly 1.2 million tons of alfalfa they produce annually. The growers in Champagne are the foremost in Europe. Champagne supplies 300,000 t out of 670,000 t of pellets imported by France's partners in the Community. The alfalfa is shipped by waterways and is competitive on every European market. Champagne and France provide Belgium with nearly all its needs (96,000 t), Germany with more than a quarter of its needs, and Italy and Netherlands with more than two-thirds of their needs.

The growers in Champagne benefit fully from assistance kept at a high level, accounting for nearly 15% of growers' gross product and 25%–50% of the net income for alfalfa. The direct aid from the EEC to Champagne amounts to about 20 million E.C.U. (European currency units, 1 unit = U.S.$0.85). Again, as with sugar beets and grains, the Champagne farmers have remarkably managed to perpetuate a crop system that was at first risky.

Comments

The high-tech agribusiness and agricultural system that has developed in Champagne is a product of the most notable mutual achievement of the European Community since 1958, its common agricultural policy. The rise of the Champagne system has been sustained by a subsidy program that is of no small cost to the European nations. Moreover, energy costs are crucial to the system, with fuel- and oil-related products accounting for nearly one-third of farming expenses in Champagne. The recent oil crises have increased the costs of chemicals twenty-fourfold, of fuel eightfold, and of fertilizers fourfold.

The decreased profit margins brought on by the change in oil prices or the projected changes in subsidies undoubtedly affect this high-tech, high-capital system. Although possible changes in it are speculative, the intensity of production would probably fall and marginal farms would be consumed by more profitable farmers. Intensive systems would need to turn more to grain production for livestock feed.

Finally, it must be remembered that the productivity of the Champagne soils is somewhat related to the relatively short time (35 years) in which they have been cultivated. Although the mollisols are good agricultural soils, they and the system imposed on them require intensive use of fertilizers. But the humic reserves are being exhausted, and agronomists have noted the long-term decline in productivity if sufficient animal manures are not applied. Also, the use of fungicides that remove the need for heading crops and allow continuous grain production raises the problem of increased parasites. The Champagne system might best be viewed as one based on "mining the soil."

Notes

1. The professional union of beet growers is now lobbying to obtain national subsidies to help the transformation of the sugar beets into alcohol for gas substitute. But so far its claims have not met with much success. The price of beets is too high to justify their use for fuel, even if the French government officially authorizes the sale of "carburol" in France.

2. Champagne's plain is also the number one region in France and in Europe for the production of dehydrated beet pulp (385,000 t out of 1,247,000 t in Europe).

3. One quintal = 100 kg.

References

Brunet, R. *Champagne, Pays de Meuse, Basse Bourgogne*. Atlas et Géographie de la France. Paris: Flammarion, 1981.

Dion, R. Le "bon" et "beau" pays nommé Champagne pouilleuse. *L'Information Géographique*, 1961, 25, 209–214.

Dorel, G., & Dumesnil, C. Trente ans de mutations de l'espace rural champenois. *Travaux de l'Institut de Géographie de Reims*, 1984.

Dorel, G., Gauthier, A., & Reynaud, A. *Genèse et Économie de la Communauté Européenne*. Montreuil: Breal, 1980.

Dumesnil, C. *L'environnement en Champagne Ardennes*. [Special issue] *Travaux de l'Institut de Géographie de Reims*, 1978, 35–36.

Montanbaux, G. La grande agriculture champenoise au sud ouest de Chalons sur Marne, *Travaux de l'Institut de Géographie de Reims*, 1971, 7, 17–35.

Radet, M. E. *L'évolution et le rôle de la grande exploitation en Champagne crayeuse depuis l'après guerre*. Master's Thesis, Reims University, 1977.

Index